INTERFERENCE POTENTIAL OF ULTRAWIDEBAND SIGNALS

INTERFERENCE POTENTIAL OF ULTRAWIDEBAND SIGNALS

MICHAEL COTTON
ROBERT ACHATZ
JEFFERY WEPMAN
BRENT BEDFORD
PAUL RUNKLE
ROGER DALKE

Nova Science Publishers, Inc.
New York

NOTICE TO THE READER

Library of Congress Cataloging-in-Publication Data
Interference potential of ultrawideband signals / Michael Cotton ... [et al.].
p. cm. -- (NTIA report ; TR-05-419)
Includes bibliographical references and index.
ISBN 13 978-1-60021-359-5
ISBN 10 1-60021-359-6
1. Broadband communication systems--Testing. 2. Radio--Interference. 3. Electromagnetic interference. I. Cotton, Michael.
TK5103.4.I556 2006
621.384'11--dc22 2006022229

Published by Nova Science Publishers, Inc. ✣New York

CONTENTS

PREFACE

The purpose of this book is to provide data needed to predict the interference potential of various UWB signals. The results will be useful for regulatory agencies that are currently charged with defining UWB emission limits and corresponding compliance measurement procedures. They will also be useful to the ultrawideband communications industry, which will only thrive if compliance measurement procedures fairly evaluate interference potential of all UWB signals. The approach is to inject carefully characterized interference into an operating narrowband receiver and measure susceptibility of the receiver with precisely defined signal quality metrics. Data is then analyzed to determine if there are common signal characteristics that predict interference potential. The victim receiver chosen for this interference susceptibility test is C-band satellite digital television (DTV). This receiver demodulates signals transmitted in the 3.7 to 4.2 GHz frequency range, which lies within the band allocated for UWB devices. The authors chose to generate the satellite signal in the laboratory rather than use an unpredictable live signal, to generate interference signals with a vector signal generator rather than rely solely on prototype devices, and to perform signal and system characterization measurements primarily with the vector signal analyzer to provide comprehensive data capable of being post-processed in many ways.

In: Interference Potential of Ultrawideband Signals
Editors: M. Cotton, R. Achatz, pp. 1-68
ISBN: 1-60021-359-6

PART 1: PROCEDURES TO CHARACTERIZE ULTRAWIDEBAND EMISSIONS AND MEASURE INTERFERENCE SUSCEPTIBILITY OF C-BAND SATELLITE DIGITAL TELEVISION RECEIVERS*

ABSTRACT

In this study, we hypothesize that ultrawideband (UWB) interference potential can be quantified in terms of UWB signal characteristics. To test this hypothesis, a test system was designed and built to inject UWB signals with known characteristics into a C-band satellite digital television receiver and quantitatively measure interference susceptibility via signal quality metrics (e.g., segment error rate, pre-Viterbi bit error rate, and modulation error ratio) taken from various points in the receiver signal processing chain. UWB signals are characterized by the amplitude probability distribution and power spectral density. Characterization measurements done with a vector signal analyzer provide amplitude and phase information to enable extensive post-measurement capability. This chapter describes the test setup and procedures in detail. Subsequent chapters will provide assessment of interference potential for gated Gaussian noise bursts (Part 2) and modern UWB systems (Part 3).

Key words: digital television; interference; satellite communications; ultrawideband.

1. INTRODUCTION

In April 2002, the Federal Communications Commission (FCC) released *FCC 02-48* [1] legalizing intentional, low-power UWB emissions between 3.1 GHz and 10.6 GHz for communications devices operated indoors. UWB emissions were limited to -41 dBm average power in 1-MHz bandwidth and 0 dBm peak power in 50-MHz bandwidth, where average power is measured over a 1-millisecond integration time. Related compliance measurements require continuous transmissions during the measurement period. The rules define a UWB device as one that emits signals with 10-dB bandwidth greater than 500 MHz or greater than

---

* Excerpted from Report TR-05-419, dated February 2005.

20% of the center frequency. The FCC rules do not specify how the bandwidth requirement is achieved, consequently allowing industry considerable breadth in choosing a modulation.

This breadth is exemplified by the development of Direct-Sequence (DS-UWB) and Multi-Band Orthogonal Frequency-Domain Multiplexing (MB-OFDM) ultrawideband technologies. Proponents of DS-UWB and MB-OFDM both seek standardization from IEEE (Institute of Electrical and Electronics Engineers) 802.15 working group 3a on high-rate (greater than 20 million bits per second) Wireless Personal Area Networks (WPAN). As the name implies, WPAN is intended for short-distance (nominally less than 10 meters) wireless networking of devices such as PCs, personal digital assistants, digital cameras, and mobile phones.

Both DS-UWB and MB-OFDM transmitters are based on state-of-the-art integrated circuitry. DS-UWB modulation controls pulse polarity and hence supports phase shifting modulations. It achieves its ultra-wide bandwidth by transmitting sufficiently narrow pulses. MB-OFDM simultaneously modulates 122 carriers spaced 4.125 MHz apart to achieve its ultra-wide bandwidth and frequency hops the modulated carriers between non-overlapping bands.

Since previous work performed at the Institute for Telecommunication Sciences (ITS) [2 - 5] did not specifically look at susceptibility of receivers to DS-UWB or MB-OFDM emissions, and since there is little published information regarding their interference potential, ITS entered into a Cooperative Research and Development Agreement (CRADA) with the Freescale subsidiary of Motorola, Inc. to study how interference potential of these emissions could be quantified in terms of signal characteristics that distinguish UWB signals.

1.1. Purpose and Approach of this Study

The purpose of this study is to provide data needed to predict the interference potential of various UWB signals. Results will be useful for regulatory agencies that are currently charged with defining UWB emission limits and corresponding compliance measurement procedures. Results will also be useful to the ultrawideband communications industry, which will only thrive if compliance measurement procedures fairly evaluate interference potential of all UWB signals.

Our approach is to inject carefully characterized interference into an operating narrowband receiver and measure susceptibility of the receiver with precisely defined signal quality metrics. Data will then be analyzed to determine if there are common signal characteristics that predict interference potential.

The victim receiver chosen for this interference susceptibility test is C-band satellite digital television (DTV). This receiver demodulates signals transmitted in the 3.7 to 4.2 GHz frequency range, which lies within the band allocated for UWB devices. We chose to generate the satellite signal in the laboratory rather than use an unpredictable live signal, to generate interference signals with a vector signal generator rather than rely solely on prototype devices, and to perform signal and system characterization measurements primarily with the vector signal analyzer to provide comprehensive data capable of being post-processed in many ways.

This study has been broken down into three parts whose results will be published in three separate chapters. The first chapter provides procedures to generate and characterize various interference signals and measure the susceptibility of a satellite DTV victim receiver to these signals. It provides sufficient detail to allow for thorough review and replication of the test. The third chapter will present results for various UWB signals, which will include MB-OFDM, DS-UWB, and conventional pulse-position modulated pulses.

The second chapter will present results for gated-noise interference. In August 2004, the MB-OFDM Alliance Special Interest Group (MBOA-SIG) requested a waiver of FCC rules that require gating/hopping to be turned off during average power compliance measurements [6]. This group argues that UWB test procedures were developed specifically for pulse-based systems, putting them at an unfair disadvantage. MBOA-SIG also argues that the proposed MB-OFDM system, which hops between three bands, causes no greater interference to licensed services than pulsed UWB devices. Opponents argue that granting the waiver will put MB-OFDM at an unfair advantage as compared to other UWB systems and increase its interference to legacy receivers. To help resolve regulatory questions associated with non-continuous interference, gated Gaussian noise is included in this interference study. This signal eliminates the complexities associated with actual UWB modulations and provides a simple, parameterized theoretical basis for understanding the susceptibility characteristics.

1.2. Organization of this Book

This first chapter describes the methodology, test system, and procedures developed to characterize UWB emissions and measure susceptibility of C-band satellite DTV receivers. It is divided into five sections in the main body and eight appendices. The main body of this book is a comprehensive, high-level discussion providing orientation and motivation for the reader. Section 2 gives technical descriptions of the interference signals considered as well as discussion on related signal simulation and generation techniques. Section 3 examines interference signal characterization measurements and describes relevant metrics, test and measurement equipment, and procedures. Section 4 discusses technical aspects of the victim receiver. It summarizes relevant communication aspects of DTV systems, simulation of DTV satellite signals, and metrics to assess signal quality degradation. Section 5 describes a test to assess interference susceptibility of satellite DTV receivers to UWB signals.

Appendices to this chapter contain information supporting the main body. Appendix A describes the mathematics behind simulation of the interference signals. Appendices B and C discuss the operation and specific technical aspects of the vector signal generator used to generate the interference signals. Appendix D explains an efficient means for estimating the amplitude probability distribution of a complex-baseband signal. Appendix E discusses important issues related to signal characterization measurements using the vector signal analyzer. Appendix F provides detailed test system calibration measurements. Appendix G gives hardware specifications. Finally, Appendix H is a glossary of acronyms and units used in this chapter.

2. INTERFERENCE SIGNALS

This study evaluates DS-UWB, Dithered-Pulse Ultrawideband (DP-UWB), MB-OFDM, and gated Gaussian noise signals. DS-UWB and MB-OFDM prototypes existed during the evaluation period; however, since the prototype equipment was constrained by either availability or functionality, we chose to develop our own methods for generating these signals. As shown in figure 1, these methods required three principal tasks: signal description, computer simulation, and vector signal generation. This section summarizes our approach towards these tasks and associated challenges.

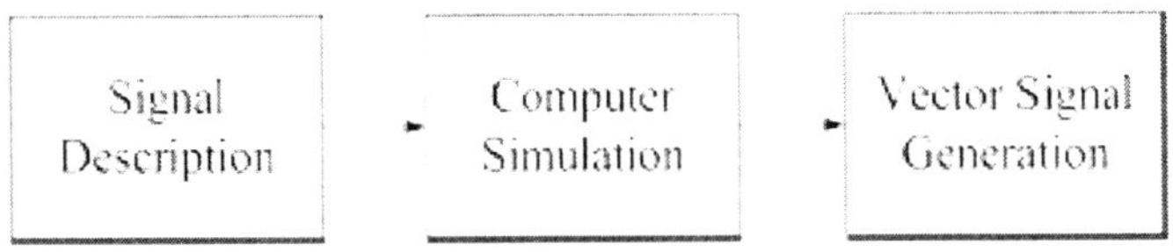

Figure 1. Signal generation tasks.

2.1. Signal Description

This subsection provides brief descriptions and signal parameters of the interfering signals evaluated in this study; more detailed mathematical descriptions are provided in Appendix A.

2.1.1. Direct-Sequence Ultrawideband

DS-UWB is currently being considered as an IEEE 802.15 standard [7]. DS-UWB transmits signals in two bands which occupy spectrum approximately from 3.1 to 4.8 GHz and 6.2 to 9.7 GHz. Our interest is chiefly in the lower band which overlays the C-band satellite spectrum.

DS-UWB devices transmit band-limited pulses in two operational modes – sparse mode and DS mode; both modes are illustrated in figure 2. Binary data modulates codewords composed of 1, 3, 6, 12, or 24 chips. A codeword with length 1 is the trivial case where chip and codeword are equivalent. Lengths 3, 6, and 12 use sparse codes where all but one chip have zero values. Lengths 12 and 24 have a small set of direct-sequence codewords, composed of chips with values of 1, -1, and 0.

Chips are generated at a nominal rate of 1320 million chips per second in the lower band. Slight variations of chip rate are used with length 12 and 24 direct sequence codewords to differentiate transmitters operating in co-located networks. Each codeword conveys 1 bit of information. Hence, shorter codewords have higher data rates. The data rate is reduced by the use of convolution codes with rates 1/2 and 3/4. As an example, a link using a 6 chip sparse codeword and 1/2 rate convolutional code can achieve 110 million bits per second (Mbps).

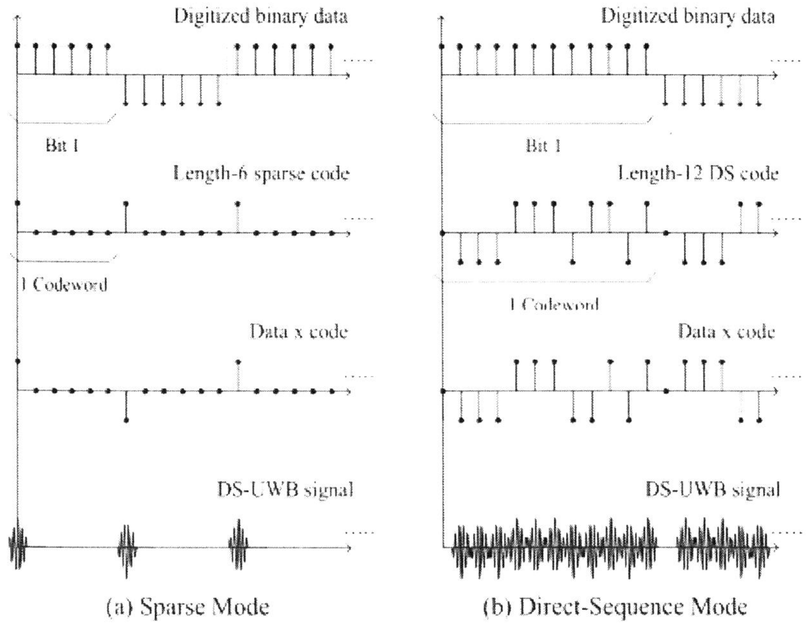

Figure 2. DS-UWB construction: (a) length-6 sparse codeword (1,0,0,0,0,0), and (b) length-12 DS codeword (0,-1,-1,-1,1,1,1,-1,1,1,-1,1).

A single simulated DS-UWB pulse is shown in figure 3. The root-raised cosine filter, with a roll-off factor of 0.3, limits the occupied spectrum to 1.716 GHz, which is 30% greater than the chip rate.

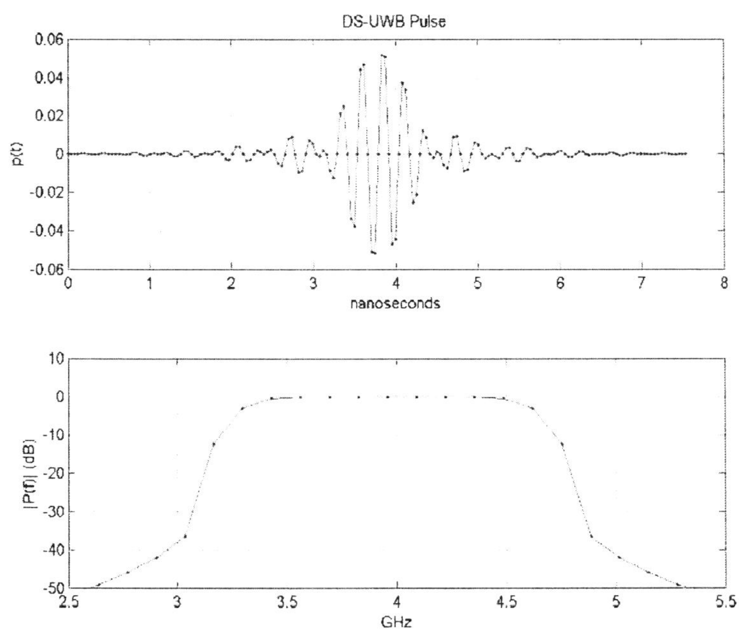

Figure 3. Simulated DS-UWB pulse sampled at 23,760 million samples per second.

Table 1 gives code lengths and codewords for the DS-UWB signals evaluated in this study. The convolutional code rate has no impact on signal characteristics and therefore is not included in the signal parameters that describe the permutations. Shaded rows highlight signals available from prototype equipment supplied by the manufacturer.

Table 1. DS-UWB Signals Considered in This Study

Signal	Code length	Codewords
DS-1	1	1
DS-2	3	0,1,0
DS-3	6	0,0,1,0,0,0
DS-4	12	0,-1,-1,-1, 1,1,1, -1,1,1, -1,1
DS-5	12	0,0,0,0,0,0,0,0,1,0,0,0
DS-6	24	-1,1,-1,-1,1,-1,-1,1,-1,0,-1,0, -1,-1,1,1,1,-1,1,1,1,-1,-1,-1,

2.1.2. Dithered-Pulse Ultrawideband

DP-UWB is not currently being considered as an IEEE 802.15 standard. However, it is included in our evaluation because it is a conventional pulsed UWB signal.

DP-UWB pulses are position modulated where pulse position is determined by the binary data and a dither factor. The dither factor is the fraction of a symbol period, T_{pulse}, to which the pulse position is bounded. This factor is often referred to as percentage dither by multiplying it by 100. For example, a 50% dithered pulse carrying a -1 value is bounded from the center of the symbol period to as much as $T_{pulse}/4$ behind the center. Similarly, the data value 1 positions the pulse ahead of the center of the symbol period by as much as $T_{pulse}/4$. Figure 4 depicts pulses that have been position modulated with 50% dithering. Pulse repetition frequencies (PRF) correspond directly to the data rate for binary data. This data rate would be reduced with the likely use of convolutional coding in a practical system.

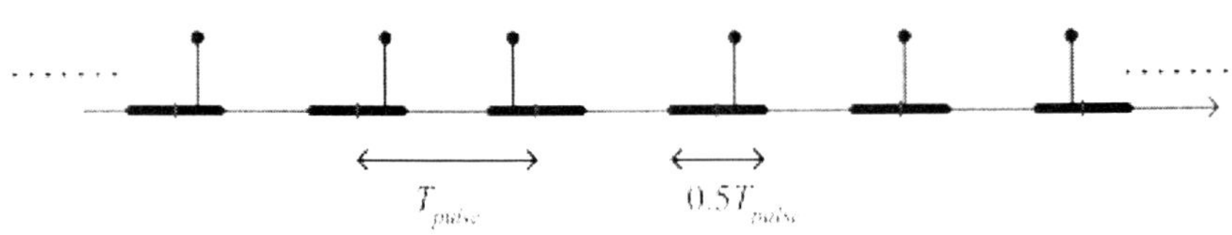

Figure 4. Example of 50% dithered pulses transmitting (...1, 1, -1, 1, 1, -1...).

A single simulated DP-UWB pulse is shown in figure 5. It is a negative unipolar pulse with an approximate 80-picosecond pulse width measured at a point corresponding to half the maximum pulse amplitude. The DP-UWB pulse frequency response is 10-dB down at 6.3 GHz.

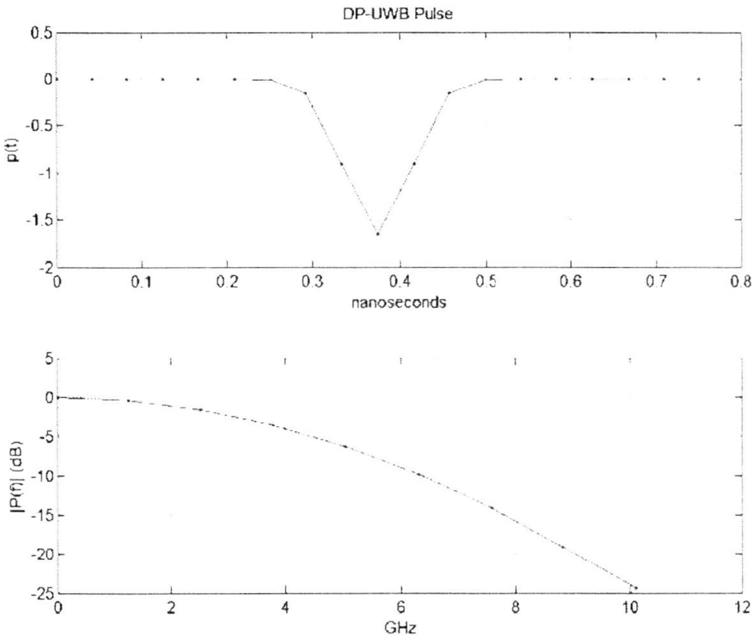

Figure 5. Simulated Gaussian pulse sampled at 24,000 million samples per second.

Table 2 gives PRFs of the DP-UWB signals evaluated in this study. Shaded rows indicate signals that could be generated with a hardware DP generator constructed in lieu of a prototype.

Table 2. 50%-Dithered-Pulse[2] UWB Signals Considered in This Study

Signal	PRF (MHz)
DP-1	0.1
DP-2	1
DP-3	10
DP-4	100

[2] Negative unipolar pulse with 80-picosecond pulse width.

2.1.3. Multi-Band OFDM Ultrawideband

MB-OFDM is currently being considered as an IEEE 802.15 standard [8]. MB-OFDM transmits in any one of fourteen bands whose center frequencies are spaced 528 MHz apart beginning at 3.432 GHz and ending at 10.296 GHz. In this study, we are interested in the band centered at 3960 MHz, which overlaps the C-band satellite spectrum.

A time-frequency code, used to minimize interference between collocated networks, determines the order and number of bands the transmitter hops to and the number of symbol periods the transmitter dwells in a band. For example, a 3/2 time-frequency code would hop over 3 bands and dwell at each band for 2 MB-OFDM symbol periods.

An MB-OFDM symbol consists of a 32-sample zero prefix, a 128-sample data block, and a 5-sample guard interval. These 165 samples, generated at a sampling frequency of 528

MHz, last 312.5 nanoseconds. An example of a single simulated MB-OFDM symbol is depicted in figure 6. Each MB-OFDM symbol conveys 100 QPSK symbols corresponding to 200 information bits. This throughput is reduced by the convolutional code rates 1/3, 11/32, 5/8, and 3/4. Redundant transmission of data on another frequency band for diversity purposes reduces the data rate further. A 110 Mbps rate corresponds to an 11/32 code rate with 2-band frequency diversity.

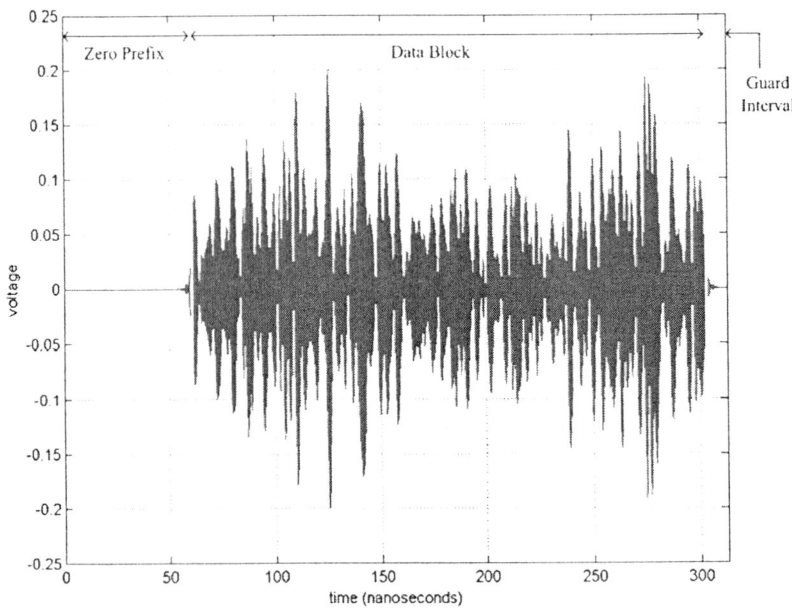

Figure 6. Simulated MB-OFDM symbol with 32-sample zero prefix, 128-sample data block, and 5-sample guard interval.

The 128-sample data block is created by inverse Fourier transforming the coefficients of 128 tones spaced 4.125 MHz apart. The tones include 100 QPSK-modulated data tones, 12 BPSK-modulated pilot tones evenly spaced throughout the data tones, one zeroed tone at the center frequency, 10 BPSK-modulated guard tones near the band edges, and five zero guard tones located at the band edges (which do not add to the bandwidth of the signal). Figure 7 illustrates the tone positions.

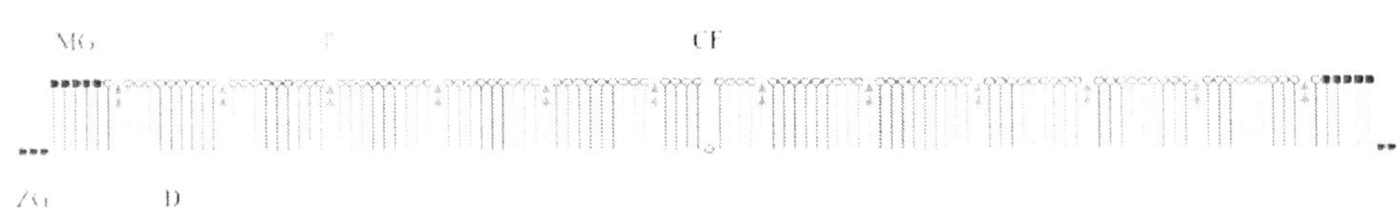

Figure 7. Position of zero guard tones (ZG), modulated guard tones (MG), pilot tones (P), zero center frequency tone (CF), and data tones (D).

Table 3 gives the number of bands and dwells for MB-OFDM signals considered in this study. Shaded rows indicate signals available from the prototype supplied by the manufacturer.

Table 3. MB-OFDM Signals Considered in This Study

Signal	Bands	Dwell
MB-1	1	1
MB-2	3	1
MB-3	3	2
MB-4	7	1
MB-5	7	2
MB-6	7	6
MB-7	13	1
MB-8	13	2
MB-9	13	12

2.1.4. Gated Gaussian Noise

Gaussian noise is a fundamental random process present in all communication systems. It has been studied extensively and provides a good reference to compare other forms of interference. In this study, we investigated how DS-UWB statistics compare with Gaussian noise statistics.

Gating is sometimes used in communication devices to increase signal power beyond the average power limits specified by the FCC. Given integration times that are large compared to a gating period, signal power can be amplified by the reciprocal of the duty cycle while maintaining the same average power. For example, if continuous Gaussian noise is gated with a 25% duty cycle and amplified by 6 dB, it has the same average power as the original continuous signal.

Frequency hopping of an MB-OFDM signal appears as gating to a fixed-frequency narrowband receiver. To help resolve compliance measurement issues for gated/hopped signals related to [6], gated Gaussian noise is considered as a general interference signal with ultra-wide bandwidth and without the complexities associated with actual UWB modulations. Table 4 gives on-times and duty cycles of periodically gated Gaussian noise signals considered in this study.

Table 4. Periodically Gated Gaussian Noise Signals Considered in This Study Duty cycle (percentage of time noise is gated on) GN-01 = Gaussian noise

GN-01 = Gaussian noise		Duty cycle (percentage of time noise is gated on)			
		50	25	12.5	6.25
On-time (nanoseconds)	10	GN-02	GN-03	GN-04	GN-05
	100	GN-06	GN-07	GN-08	GN-09
	1,000	GN-10	GN-11	GN-12	GN-13
	10,000	GN-14	GN-15	GN-16	GN-17

Table 5 gives the number of bands and dwells for gated-noise signals that emulate the MB-OFDM signals in table 3. Two distinct gating processes are used for the insertion of prefix and guard intervals and for the frequency hopping due to the time-frequency codes. Both gating processes contribute to the percentage of time that noise is actually on. All gated-noise permutations in tables 4 and 5 could be generated with a noise diode, an amplifier, and an electronic switch with a fast rise time.

Table 5. Gated Gaussian Noise Signals Emulating MB-OFDM Signals in table 3

Signal	Bands	Dwells	Percentage of time noise is actually on
GN(MB) – 1	1	1	77.6
GN(MB) – 2	3	1	25.9
GN(MB) – 3	3	2	25.9
GN(MB) – 4	7	1	11.1
GN(MB) – 5	7	2	11.1
GN(MB) – 6	7	6	11.1
GN(MB) – 7	13	1	6.0
GN(MB) – 8	13	2	6.0
GN(MB) – 9	13	12	6.0

2.2. Computer Simulation

Computer simulation translates the mathematical description of the signal to samples needed by the vector signal generator. Simulation of realistic, wide-bandwidth signals places a considerable burden on most computers with finite memory and processor resources. Consequently, steps were taken to reduce the complexity of the wireless network system and the number of samples needed to accurately represent the signal produced by it. Although a prototype UWB receiver will not be able to demodulate the simplified signal, from the perspective of the victim receiver, it is indistinguishable from a signal generated by a prototype UWB transmitter. This subsection describes the software implementation of the interference signals at a high level; specific details and further explanation of the computer simulation will be provided in a separate document.

2.2.1. Reduction of System Complexity

The operational standards for WPAN systems are still under development [7, 8], but the language and structure will be similar to wireless local area networks (WLAN) as described in the *IEEE 802.11 Handbook* [9]. As depicted in figure 8(a), wireless network systems are structured into two layers: medium access control (MAC) and physical (PHY). Briefly, the MAC layer supplies a reliable delivery mechanism for data transfer over noisy, unreliable wireless media. It reduces interference between systems by dictating when transmissions occur and how long they last. The PHY layer transmits and receives data over a shared wireless media; it is the interface between the MAC and wireless media. The PHY layer is divided into two sublayers: physical layer convergence procedure/protocol (PLCP) and physical medium dependent (PMD). The PLCP sublayer provides information needed by the receiver to demodulate the signal. The PMD establishes the mechanism for converting information bits into transmitted signals and from received signals.

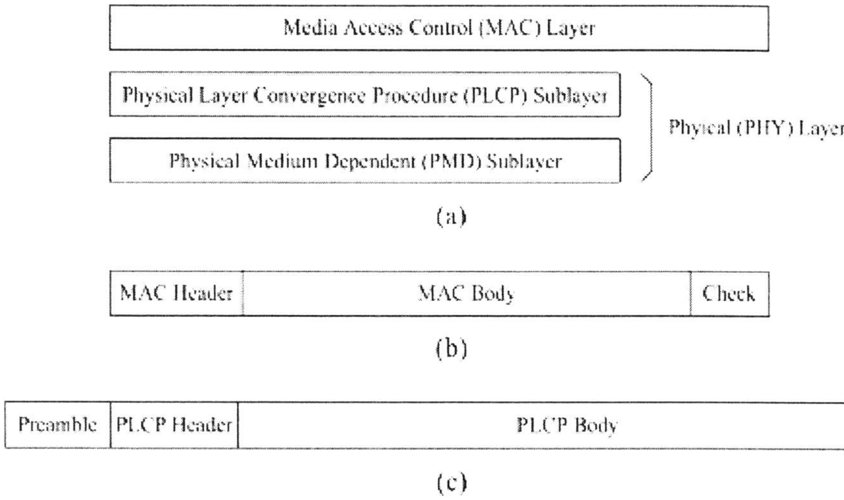

Figure 8. (a) Wireless network layers, (b) MAC frame, (c) PLCP frame.

MAC and PLCP information is organized into frames as shown in figures 8(b) and 8(c). The MAC frame includes a header, body, and a check sequence. The PLCP frame includes a preamble, header, and body; the PLCP body holds the MAC frame. Figure 9 depicts components of a typical wireless network transmitter where bits representing the PLCP frame are converted to signals for transmission under the control of the PMD sublayer.

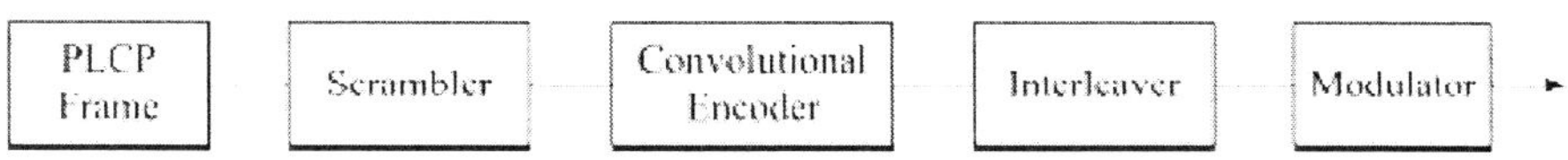

Figure 9. Transmitter PMD components.

Levels of simulation complexity are depicted in figure 10. The most sophisticated simulation includes MAC control of the PLCP frames. A less complex simulation transmits a steady stream of PLCP frames. The simplest simulation transmits a steady stream of scrambled MAC frames.

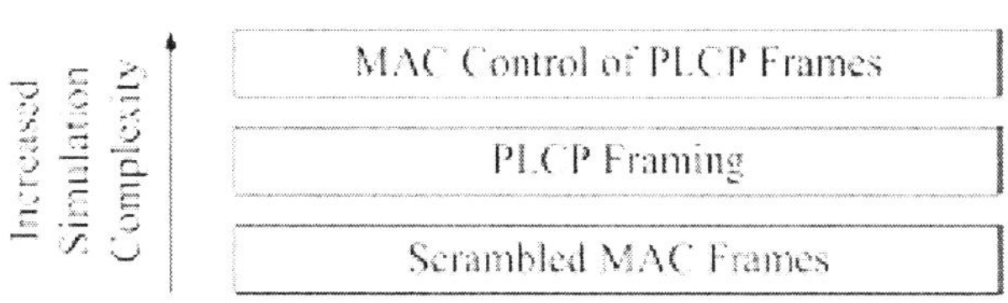

Figure 10. Levels of wireless network simulation complexity.

For the present scope of this study, simulation at the simplest level is adequate. As shown in figure 11, random data representing the scrambled, convolutionally encoded, and interleaved MAC frame modulates the UWB signals described in Section 2.1 and Appendix A. Justification for this approach is two-fold. First, the MAC frame is the most significant part of the PLCP frame. Second, MAC control of PLCP frames must include consideration of various traffic models, which would detract from questions related to the specific UWB modulators.

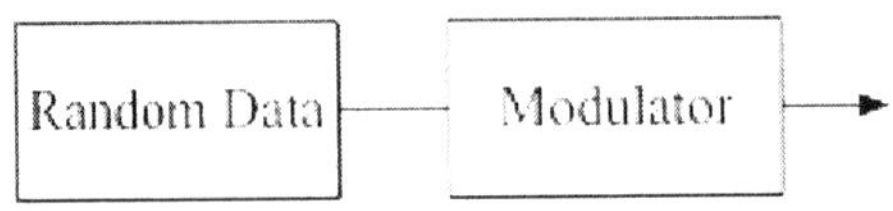

Figure 11. Simplified wireless network simulation. Figures A-1, A-2, and A-3 in Appendix A give specific details of DS-UWB, DP-UWB, and MB-OFDM modulators.

2.2.2. Reduction of Samples

For practical purposes, the number of samples and the corresponding sample rate need to be reduced so that reasonably long signal durations can be processed in a timely manner. For example, the sampling rate is often set to 4 or more times the highest frequency of interest. For real signals considered in this study, the highest frequency of interest exceeds 4 GHz, which requires more than 16 GSps. This produces far too many samples to process in a reasonable time for most computers.

Sample rates of signals can be reduced by bandpass filtering the real signal and converting the real-bandpass signal to complex-baseband (CBB) as depicted in figure 12. Conversion to CBB allows for the use of a sample rate that depends on the bandwidth, B, of the filter rather than highest frequency of the original signal. Simulation in CBB also has the benefit of generating waveforms readily implemented on the vector signal generator (VSG).

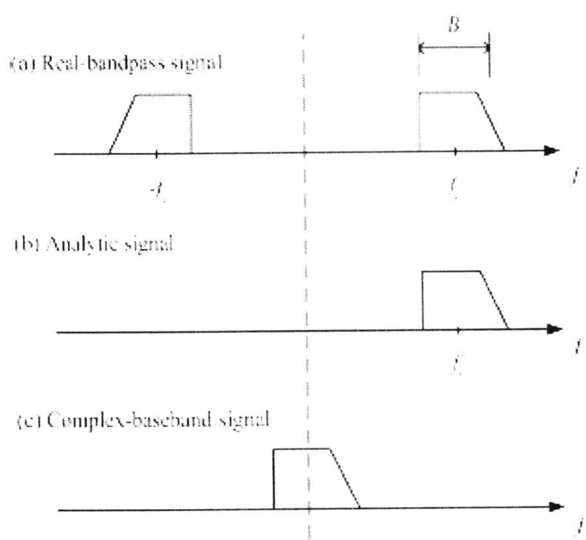

Figure 12. Spectrum amplitude of (a) real-bandpass, (b) analytic, and (c) CBB signals.

2.3. Implementation on the Vector Signal Generator

Simulated CBB interference waveforms were implemented on the VSG. A brief description of the VSG is given here with a more detailed description given in Appendix B. Figure 13 shows a highly simplified block diagram of the VSG. The simulated discrete-time CBB signal, stored in fixed-point format as a waveform file, is loaded into VSG memory. The VSG converts the digital in-phase (I) and quadrature-phase (Q) signals to I and Q analog signals using interpolation, digital-to-analog conversion, and analog reconstruction filtering. The analog I and Q signals are the inputs to the I/Q modulator, which converts the CBB signal to a real signal at the RF center frequency determined by the local oscillator (LO).

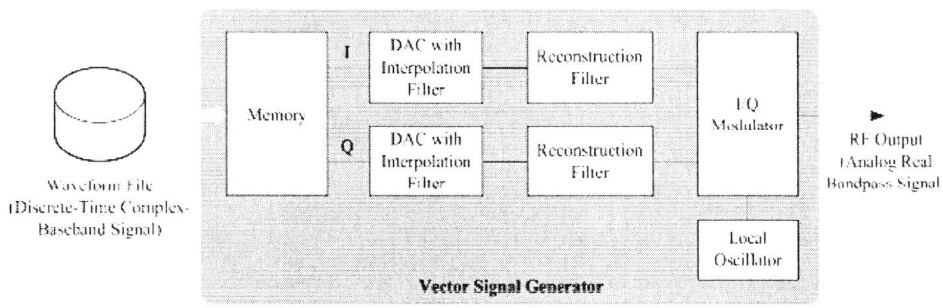

Figure 13. Vector signal generator block diagram. DAC stands for digital-to-analog converter.

The VSG produces an RF output that is somewhat degraded. Imperfect digital-to-analog conversion distorts the desired signal. This distortion is most severe at the band edges. Additionally, LO feed-through, intermodulation products, and other spurious responses from imperfect I/Q modulation can corrupt the desired signal. Appendix B discusses techniques to minimize the amplitude of these undesirable signals. Appendix C provides a detailed discussion of the most prominent VSG impairment, LO feed-through, and describes an experiment that demonstrates that LO feed-through is not a problem for the DTV interference susceptibility test.

3. Interference Characterization Measurements

Interference characterization measurements were made with the vector signal analyzer (VSA), spectrum analyzer (SA), digital oscilloscope (DO), and power meter. These measurements were made at C-band on signals directly out of the VSG. Measurements were also performed on prototypes when available. This section describes metrics used to characterize interference, data needed to compute the metrics, instruments used for collecting the data, and specific settings for the instruments.

3.1. Metrics and Data Needed to Compute Them

Interference signals can be characterized by statistics of the random variable representing the time-varying signal voltage. Measurement and statistical processing is made difficult by the high-frequency content of UWB signals. For example, DS-UWB and DP-UWB pulses shown in figures 3 and 5 require sample rates of 23.76 GSps and 24 GSps, respectively, to obtain adequate resolution. Currently, digitization at these sample rates requires state-of-the-art oscilloscopes. Fortunately, our interest is in the effect of interference on relatively narrowband receivers. Inside a superheterodyne victim receiver, interference is band-limited and downconverted to CBB along with the desired signal. Consequently, it makes sense to measure and statistically analyze the CBB signal, which can be sampled at slower rates.

Statistics of particular interest are peak voltage, root-mean-square (rms) voltage, amplitude probability distribution (*APD*), power spectral density (*PSD*), and other higher-order statistics [10]. *APD* characterizes a random process by displaying the percentages of time that a signal exceeds certain levels, which can be determined by histogramming or sorting the amplitudes. The sorting method for estimating *APD*s is described in Appendix D. *PSD* is a measure of the average power per unit bandwidth as a function of frequency. It can be determined from the Fourier transformed autocorrelation function or from averaged spectrum magnitudes.

3.2. Description of Test and Measurement Equipment

Previous UWB signal characterization studies [2, 3] relied mostly on the spectrum analyzer, power meter, and digital oscilloscope. In addition, this study utilizes the vector signal analyzer to acquire amplitude and phase samples of the CBB signal. These instruments differ in bandwidth, dynamic range, and the type of data they acquire. This subsection describes each instrument and explains tradeoffs associated with each. Measurement parameters are described for acquiring data necessary to compute metrics that characterize interference signals.

3.2.1. Vector Signal Analyzer

The Agilent 89461A VSA provides approximately 36-MHz maximum bandwidth with 50-dB dynamic range in the maximum bandwidth. In Appendix E, dynamic range is defined and measured. Figure 14 shows a simplified block diagram of the VSA. Signals at the RF input are attenuated to prevent overload and filtered to reject image frequencies. The RF signal is then downconverted to a final intermediate frequency (IF) of approximately 19 MHz. The final IF signal is converted to a digital IF signal by an analog-to-digital converter (ADC) operating at 95 million samples per second (MSps). The digital IF signal is converted to CBB using digital downconversion. The fastest sampling rate of the I and Q signals is 46.08 MHz corresponding to a 21.7-nanosecond sample period and a 36-MHz span. For the VSA, span refers to the bandwidth where amplitude and phase distortion are negligible. Decimation to slower "cardinal" sampling rates, each related by a factor of 1/2, can be used to generate other cardinal spans. For example, cardinal spans of 18, 9, 4.5, 2.25, 1.125, and 0.5625 MHz correspond to decimation factors of 1/2, 1/4, 1/8, 1/16, 1/32, and 1/64.

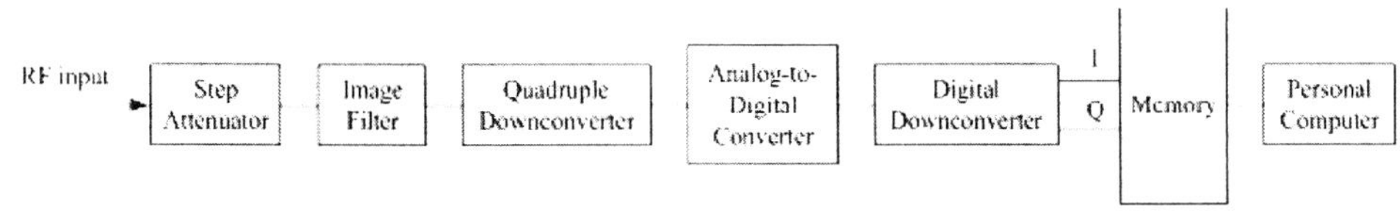

Figure 14. Conceptual block diagram of a vector signal analyzer.

VSA operation is controlled by vendor-supplied software running on a PC. VSA measurement settings include center frequency, span, input range (i.e., maximum signal amplitude), and number of samples. The software commands the VSA to acquire data at the smallest cardinal span greater than the requested span. Data is transferred to the PC and re-sampled to a rate corresponding to the requested span. The re-sampled data can be saved in memory for post-measurement analysis or processed for immediate display. For a more detailed discussion of the VSA refer to *Agilent Application Note 150-15* [11].

The VSA is good for characterizing interference to DTV receivers, since its bandwidth is commensurate to that of the satellite transponder. Furthermore, the VSA provides data with both amplitude and phase. This improves the versatility of the data and reduces the number of measurements required. For example, we are interested in illustrating signal characteristics in a variety of bandwidths corresponding to the satellite transponder, i.e., 36 MHz, the DTV signal used for interference susceptibility tests, i.e., 19.51 MHz, and other legacy systems (to investigate bandwidth scalability of statistics). Rather than remeasure at each bandwidth, digital filters are applied during post processing to achieve the specified processing bandwidths; this would not be possible if only magnitude data were available.

VSA settings for characterizing interference signals are given in table 6. Settings were chosen with three considerations in mind. First, the maximum VSA bandwidth is used to characterize interference to the satellite DTV channel. Second, it is preferable to use a cardinal span, rather than some arbitrary span, to avoid re-sampling by the vendor supplied software. Finally, at least 100,000 samples, free of correlations introduced by measurement equipment, are necessary to estimate statistics at the 0.01 percentile. Hence, at least 100,000 × 36/10 = 360,000 samples should be measured in the 36-MHz span, 100,000 × 4.5/1 = 450,000 samples should be measured in the 4.5-MHz span, and 100,000 × 0.5625/0.1 = 562,500 samples should be measured in the 0.5625-MHz span.

Table 6. VSA Measurement Settings for Various Post-Processing Bandwidths

Post-Processing Bandwidths (MHz)	Span (MHz)	Number of Samples	Center Frequency(MHz)	Input Range (dBm)
36, 19.51, 10	36	> 360,000		
1	4.5	> 360,000	3820	-30
0.1	0.5625	> 562,500		

Center frequency corresponds to the center of the DTV channel chosen for the interference susceptibility test. An input range of -30 dBm was chosen for these measurements. VSA measurement accuracy is only possible if the range of signal amplitudes resides within the dynamic range of the measurement device. Appendices E and F demonstrate how to set the average power of input signals to ensure accurate VSA measurements.

3.2.2. Spectrum Analyzer

The Agilent E4440A spectrum analyzer provides an 8-MHz maximum bandwidth with more dynamic range than the VSA. Figure 15 shows a simplified block diagram of the SA. The signal at the RF input is attenuated and passed through a preselection filter to reject strong out-of-band signals. The sweep generator instructs the SA to sweep over a specified frequency span, which shifts the RF input to a target IF for the two-stage downconverter. The final IF signal is centered at 7.5-MHz and converted to a digital IF signal by the analog-to-digital converter operating at 30 MSps. The digital IF signal is then converted to CBB using a digital downconversion and decimation process. At this point, the complex samples can be processed and displayed, or sent to a PC for post measurement analysis. For our purposes, digital signal processing inside the SA implements traditional swept measurements that involve application of a resolution bandwidth (*RBW*) filter, envelope detector, and video bandwidth (*VBW*) filter; conventional metrics, i.e., peak and root-mean-square (rms) amplitude, are displayed.

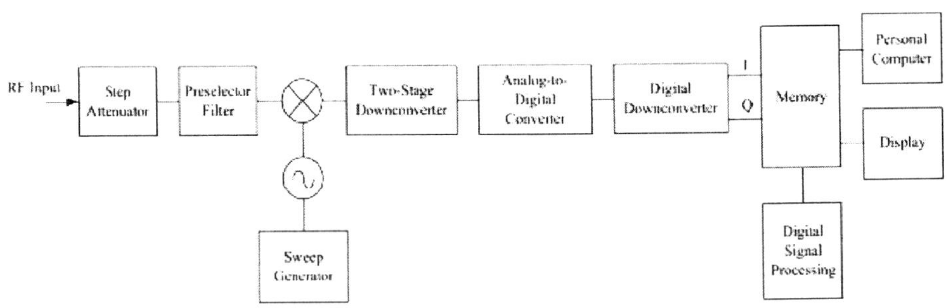

Figure 15. Conceptual block diagram of a digital spectrum analyzer.

SA measurement settings include center frequency, span, *RBW*, *VBW*, number of frequency steps (*Nf*), sweep time, and detector type. Center frequency and span define the frequency range the SA sweeps across. The sweep rate and resolution are determined by the span, number of frequency steps, and the length of time allotted for one sweep. For more in-depth discussion of spectrum analyzer operations, the reader is referred to *Agilent Application Note 150* [12].

The maximum SA bandwidth is smaller than that of the satellite DTV channel; however, SA measurements were used to provide *PSD* and peak amplitude estimates, verify VSA measurements, perform FCC compliance tests, and monitor signals. Table 7 gives SA settings for characterizing interference signals. Measurements at 3950 MHz are centered on the C-band satellite spectrum. Measurements at 3820 MHz are centered on the DTV channel used for susceptibility testing. Frequency step size, i.e., span divided by the number of frequency steps, is set to one sixth of the *RBW* to ensure that spectrum details are not missed. *VBW* is much wider than *RBW* to ensure peak amplitude accuracy. Sweep time is set to $1000 \times N_f \times k/RBW$ seconds to ensure rms amplitude accuracy. The Gaussian *RBW* filters require $k = 3$.

Table 7. Spectrum Analyzer Settings for Characterizing Interference Signals

Center Frequency (MHz)	Span (MHz)	*RBW* (MHz)	*VBW* (MHz)	*Nf*	Sweep Time (seconds)	Detector
3950	600	1	50	3600	10.8	Rms,
3820	40	0.1		2400	72.0	peak

3.2.3. Digital Oscilloscope

The Tektronix TDS7704B digital oscilloscope DO provides real signal samples with 7-GHz maximum bandwidth and less dynamic range than the VSA and SA. Figure 16 shows a block diagram of the DO. It performs anti-alias filtering and digitizes the signal at a maximum of 20 GSps. Trigger functions allow the user to control when digitization begins. DO measurement settings include sample rate, time scale, voltage scale, and number of samples.

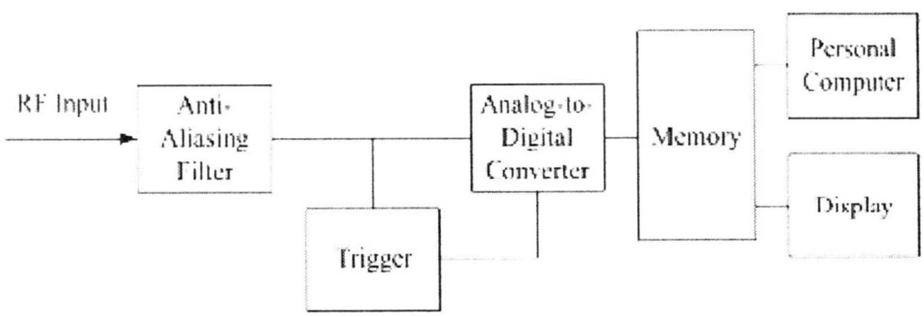

Figure 16. Conceptual block diagram of a digital oscilloscope.

The wide bandwidth of the DO allows for a qualitative "snapshot" of the full-band interference signals to reveal signal parameters such as pulse rise time, signal duration, and gating period.

Table 8 gives typical settings for DO measurements. Actual settings depend on the specific signal characteristics of interest. Maximum sample rate is used to ensure the highest resolution.

Table 8. Digital Oscilloscope Settings for Characterizing Interference Signals

Time Scale (microseconds/div)	Voltage Scale (millivolts/div)	Sample Rate (GSps)
40	50	20
4		

3.2.4. Power Meter

The Agilent 4419B power meter with the 9300A diode average power sensor provides average power measurements with 18-GHz maximum bandwidth and 80-dB dynamic range. Operating in a normal mode, the power meter provides 20 average power measurements per second.

This power meter provides calibrated average power measurements of both RF and IF signals. To get the most accurate reading, it is set to the center frequency of the measured signal and 1024 measurements are averaged; this takes approximately 60 seconds to complete.

4. Victim Receiver

C-band satellite television was chosen as the victim receiver for this interference susceptibility test because C-band receivers operate within the band allocated for UWB emissions, satellite signals are weak and vulnerable to interference, and a number of operational scenarios can be investigated. This section discusses relevant communication aspects of satellite DTV systems, simulation of the DTV satellite signal, and metrics to assess quality; a more general discussion of satellite communication techniques can be found in *Pratt, et al.* [13] and *Couch* [14].

4.1. Satellite Digital Television Description

The scope of this study is limited to C-band satellite DTV transmissions that comply with the commonly used Digital Video Broadcast (DVB) recommendations *ETS 300 421* [15] published by the European Telecommunications Standards Institute (ETSI). This subsection discusses the communication aspects of a single transponder channel for satellite DTV systems that comply with DVB recommendations.

4.1.1. MPEG-2 Encoder

DTV video and audio signals are digitized separately, compressed, and then multiplexed into a serial bit stream for transmission. Compression algorithms are applied to minimize the amount of data required to represent the video image or audio, while preserving the level of quality required for the given application. This is achieved by matching spectral characteristics of video and audio signals to the frequency response of human eyes and ears; further compression is achieved by not transmitting static information. DVB uses the MPEG-2 standard for compression.

There are two levels of multiplexing in DVB. At the lowest level, components that make up a program (e.g., video, audio, system-control information) are multiplexed. At a higher level, the programs themselves are multiplexing. The transport subsystem is based on fixed-length packets, each holding 188 bytes of information which represents one MPEG segment. The contents and timing of each packet are identified by packet headers.

4.1.2. Forward Error Correction and Modulation

DVB systems mitigate errors by reducing the baud rate and protecting the signal with forward error correction codes (FEC). Two FEC codes, Reed-Solomon (RS) and convolutional, are used. Briefly, the RS code has a block length of 204 bytes carrying 188 information bytes, or one MPEG segment, and can correct up to 8 byte errors per block. The convolutional code has length 7 and rates that include 1/2, 2/3, 3/4, 5/6, and 7/8. Interleaving (with a depth of 12 bytes) is used to shield the Reed-Solomon decoder from burst errors. FEC-coded bits are mapped into QPSK symbols for transmission. Symbols are band-limited by a root-raised-cosine (RRC) filter with a roll-off factor (α) of 0.35 and upconverted to the satellite channel radio frequency (RF).

Table 9 gives E_b/N_o and signal-to-noise ratio (*SNR*) performance requirements for DVB satellite television. E_b/N_o is the ratio of energy per bit over noise power spectral density. *SNR* requirements are 3 dB higher over E_b/N_o because QPSK maps two bits into one symbol. If the received signal is above the specified E_b/N_o, the specified FEC techniques are designed to provide a bit-error rate (BER) between 10^{-10} and 10^{-11} at the output of the RS decoder. This corresponds to less than one uncorrected error "event" per transmission hour. (Although *ETS 300 421* [15] is unclear as to the nature of an error "event," it is assumed to correspond to an MPEG segment decoding error.)

Table 9. Requirements for One Uncorrected Error "Event" per Transmission Hour [15]

Convolutional Inner Code Rate	E_b/N_o[3] Requirement (dB) for $10^{-11} \leq$ BER $< 10^{-10}$ After the RS decoder	Corresponding *SNR* Requirement (dB) for QPSK
1/2	4.5	7.5
2/3	5.0	8.0
3/4	5.5	8.5
5/6	6.0	9.0
7/8	6.4	9.4

[3] E_b/N_o values include a 0.8-dB modem implementation margin and the 0.36-dB noise bandwidth increase due to the RS code implementation.

4.1.3. Receiver

Figure 17 depicts the two-stage-downconversion superheterodyne architecture typically utilized by satellite DTV receivers. Immediately after the antenna feed, a low-noise block downconverter (LNB) amplifies the received signal and downconverts it to the first IF between 950 and 1450 MHz. The signal is carried to the receiver via coaxial cable, where it is downconverted once more, channel selected, and demodulated. The QPSK demodulator uses RRC filtering to minimize intersymbol interference and maximize *SNR*. The demodulated signal is applied to the Viterbi decoder, de-interleaved, and RS decoded. Finally, error-free MPEG segments are demultiplexed, decompressed, and delivered to the television set for viewing. Segments received in error are not passed on. Error concealment is used to hide the effects of their absence.

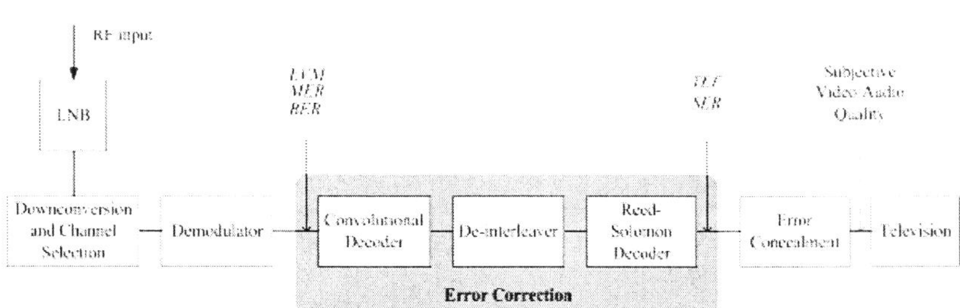

Figure 17. Signal quality metrics measured at different points in a DTV receiver.

4.2. DTV Quality Metrics

Video and audio quality is often subjectively evaluated by viewer and listener panels briefly exposed to received images and sounds. Acquisition of subjective quality metrics is qualitative, time consuming, and prone to perceptual error because there are a wide variety of techniques used by receivers to conceal the effects of MPEG segment errors.

Alternatively, quantitative signal quality metrics from various points in the receiver are available from DTV signal monitoring instruments. In this study, signal quality metrics from the Tektronix MTM400 DVB signal monitor are recorded to evaluate degradation of the victim receiver. Available metrics are modulation error rate (*MER*), error vector magnitude (*EVM*), pre-Viterbi bit-error rate (*BER*), and cumulative transport error flag (*TEF*) from which MPEG segment error rate (*SER*) can be computed. Figure 17 shows where the metrics are taken from within the receiver signal processing chain.

MER and *EVM* are taken after I/Q demodulation. An example of a demodulated QPSK constellation and its I and Q components are shown in figure 18.

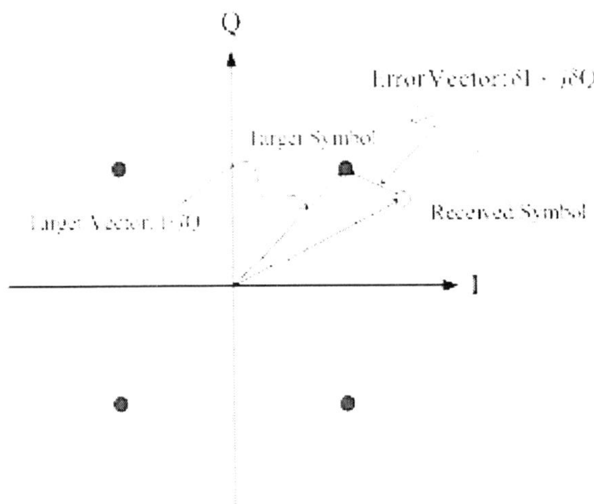

Figure 18. QPSK constellation with both target and received symbols.

Modulation error ratio in dB is defined from these components by

$$MER = 10\log\left(\frac{\sum_{n=1}^{N}\left(I_n^2 - Q_n^2\right)}{\sum_{n=1}^{N}\left(\delta I_n^2 + \delta Q_n^2\right)}\right).$$

where N data points are taken to achieve sufficient accuracy. For QPSK, *EVM* can be expressed in terms of *MER* as $EVM = 10^{2\text{-}MER/20}$.

SER is post-Reed-Solomon MPEG segment error rate and is calculated from *TEF* and Coordinated Universal Time (UTC) as

$$SER = \frac{TEF(t_2) - TEF(t_1)}{(t_2 - t_1)R_{seg}}$$

where R_{seg} is the MPEG segment rate with units of MPEG segments per second.

4.3. Operational Scenario Chosen for DTV Interference Susceptibility Tests

The operational scenario of the victim receiver chosen for this study is shown in table 10; it corresponds to DTV channel 581 from geostationary satellite Satcom C3 positioned approximately 35,786 km above the equator at 131.0° W longitude. Modulation, compression, and error correction schemes comply with DVB recommendations described in Section 4.1.

Table 10. Satellite Simulation Scenarios for DTV Interference Susceptibility Tests

Transponder Center Frequency	3820 MHz
Modulation	QPSK with root-raised cosine filter ($\alpha = 0.35$)
Compression	MPEG-2
Data Rate	26.970353 Mbps
Symbol Rate	$R_S = 19.510468$ Mbaud
Reed-Solomon Error Correction	$R_{RS} = 188/204$
Convolutional Error Correction	$R_{conv} = 3/4$, $K = 7$
Interleaving	Depth = 12 bytes
Signal-to-Noise Ratio	{8.5, 11.5, 14.5} dB

Calculation of the operational segment rate begins at the byte level. 188 information bytes are packed into a Reed-Solomon codeword, interleaved, and convolutionally encoded. Overhead associated with the error correction scheme is quantified by the rates $R_{RS} = 188/204$ and $R_{conv} = 3/4$. The length in bits of one MPEG segment is

$$\ell_{seg} = \frac{188}{R_{RS} R_{conv}} \times 8 \frac{\text{bits}}{\text{byte}} = 2176 \frac{\text{bits}}{\text{segment}}$$

which segments are transmitted is

$$R_{seg} = \frac{m}{\ell_{seg}} R_S$$

where 2^m is the modulation alphabet size and R_S is the symbol rate. For the operational scenario, $m = 2$ for the QPSK alphabet, $R_S = 19.51$ Mbaud, and $R_{seg} = 17{,}932$ segments per second corresponding to $T_{seg} = 55.76$ microseconds per segment. The symbol rate and RRC roll-off factor correspond to an occupied bandwidth of approximately $19.51 \times (1+0.35) = 26.34$ MHz.

SNR is determined primarily by the size of the receiving dish, the noise figure of the LNB, and the path length of the signal. Three values, separated by 3 dB, were chosen to include a reasonable set of satellite transmission scenarios; 8.5-dB *SNR* corresponds to the minimum requirement for one uncorrected error event per hour [15].

4.4. Satellite Signal Simulation

Complexity is built into satellite television systems to increase capacity and improve quality of the broadcast. Transponders onboard geostationary satellites capture uplink signals and retransmit the signals back down to Earth. A typical satellite has twelve 36-MHz transponders. Some satellites double the number of channels by using horizontal and vertical polarizations. The transmitted satellite signal propagates down through the atmosphere and is received by dual-polarized, high-directivity dish antennas.

For the purpose of quantifying interference susceptibility, receiving a "live" satellite signal presents significant disadvantages. Interference can be introduced by the cross-polarized channel on the same frequency, the co-polarized signal on the adjacent channel, and devices located near the receiving antenna. The signal is weakened by attenuation and depolarization effects of the atmosphere. Distortion is caused by nearby ground reflections. Also, "live" video content can be inconsistent and unpredictable. These problems are alleviated by simulating the satellite signal and coupling that signal, along with the simulated interference signal, to the victim receiver with a coaxial cable and a directional coupler rather than a pair of antennas.

Figure 19 shows a block diagram for the satellite signal generator used in this study. The digital video stream is SMPTE-259M, the native video format for MPEG-2, and therefore provides the highest quality input signal. The video clip used is "Flower Garden" from the International Telecommunication Union Recommendation *ITU-R BT.802.1* [16]. This clip is 20 seconds long and repeats itself forward and then backward to minimize discontinuity. The steady scan and detailed video content of "Flower Garden" provides a scene with lots of spatial and temporal information that constantly forces the encoder to generate new digital video. The encoder creates a 26.97-Mbps MPEG-2 bit stream, the modulator adds FEC and creates a 19.51-Mbaud QPSK-modulated signal centered at 70 MHz, and two upconversion stages shift the signal to 3820 MHz. Details on the upconverter are provided in Appendix F.

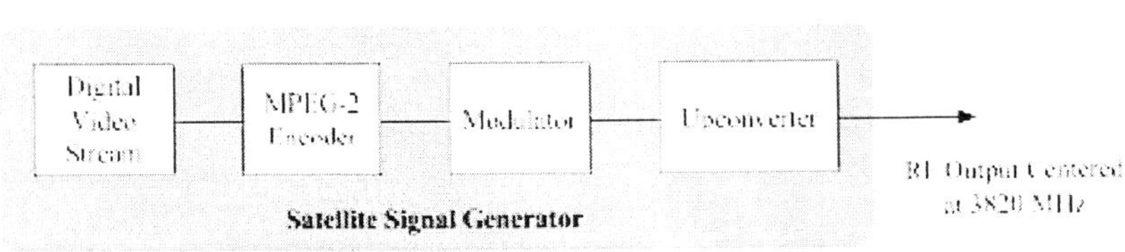

Figure 19. Block diagram for the satellite signal generator. Hardware specifications are in Appendix G.

5. INTERFERENCE SUSCEPTIBILITY TESTS

This section provides technical details and justification for tests to assess ultrawideband interference susceptibility of C-band satellite DTV receivers.

5.1. Methodology

DVB error correction and concealment schemes cause steep receiver performance curves where a 1-dB difference in *SNR* can cause abrupt changes in picture quality. For our operational scenario, the received picture is seemingly perfect at an *SNR* of 6 dB, yet totally blanked by errors at 4 dB. To resolve this abrupt change in performance, the test system and procedure were designed for precision and repeatability.

The developed test system:

- simulates interference signals in software, so the widest range of UWB signal parameters are available for testing; generates interference signals with a VSG from simulated waveforms, so interference amplitude is precisely controlled; characterizes interference signals with a VSA, which provides both signal amplitude and phase information for post-measurement filtering and analyses; simulates the satellite signal, so that scene content and *SNR* are controlled;
- conductively couples signals to the satellite receiver, so uncertainties introduced by radio channel interference, attenuation, and distortion are eliminated; evaluates performance degradation quantitatively with signal quality metrics from an MPEG stream monitor rather than subjective quality metrics; is completely characterized, so that its effects on the test are known. Also, the test procedure: is under software control to eliminate procedural errors; calibrates the measurement system at the beginning of each test to ensure consistency and system integrity.

5.2. Test System

A block diagram of the measurement system is shown in figure 20. Characterization measurements and component specifications are provided in Appendices F and G. The system is comprised of five subsystems: (1) interference signal generation, (2) satellite signal generation, (3) coupling, (4) victim receiver, and (5) calibration.

Interference signals are generated from software simulated waveforms with a VSG. The VSG provides 0.01-dB amplitude control and is capable of completely turning off the interference signal and terminating the port in 50 ohms. Satellite signals are simulated and step variable attenuator VA provides 1-dB control of the satellite *SNR*. A switch was provided that can completely shut the satellite signal off and terminate the port in 50 ohms during calibration.

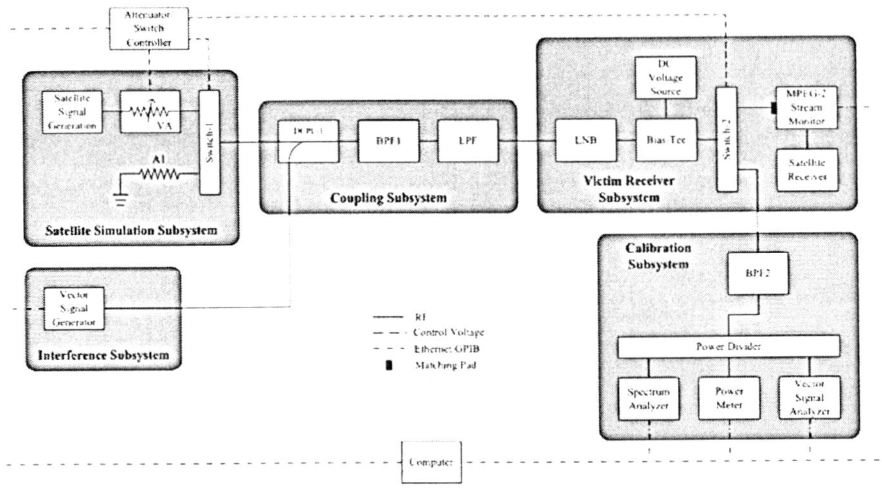

Figure 20. Block diagram for DTV interference susceptibility test. System calibration measurements are in Appendix F. Hardware specifications are in Appendix G.

Satellite and interference signals are added via the coupling subsystem. The interfering signal is applied to the port of directional coupler DCPL-1 that attenuates the signal by 10 dB. The bandwidth of bandpass filter BPF1 corresponds to that of the satellite transponder channel. Lowpass filter LPF assists BPF1 in removing unwanted mixing products introduced by the satellite simulator upconverter. These filters were not included in the satellite simulation subsystem because they also band-limit the interfering signal. Limiting the bandwidth of the interfering signals ensures that performance degradation is due to co-channel interference rather than LNB compression due to out-of-band interference. It also ensures that signals generated by other VSGs or prototype devices will have the same interference bandwidth.

In the receiver subsystem, the LNB downconverts signals to L-band. A bias-tee provides the LNB with the necessary dc voltage and impedance match. The MPEG stream monitor provides a pass-through, L-band signal for the satellite receiver. Scenes transmitted by the satellite simulator can be viewed on a video monitor attached to the satellite receiver. Switch-2 enables the signal to be directed to either the MPEG stream monitor or the calibration subsystem.

In the calibration subsystem, measurements made during each test are performed with the VSA and power meter. The SA is used for signal monitoring and diagnostics. Bandpass filter BPF2 is used to limit the bandwidth of the noise amplified by the LNB and measured by the power meter. BPF2 bandwidth is much wider than that of BPF1 and the VSA, so it does not influence VSA calibration measurements. Care was taken to ensure that VSA measurements are made within the dynamic range of the system and measurement instrument (see Appendix E).

5.3. Procedures

A single DTV interference susceptibility test, for any one interference signal given in Section 2.1, involves system calibration followed by quality-metric acquisition.

5.3.1. Calibration

At the beginning of each test, the system is calibrated to ensure that all hardware is operating properly and the satellite signal and interference amplitudes are set properly. The calibration procedure consists of setting the amplitude of the satellite simulator to a level that corresponds to the test *SNR* and the amplitude of the interference signal to a level that affords an accurate VSA measurement. Measurements of the satellite signal plus system noise, interference plus system noise, and system noise alone are then made with the VSA and power meter. These measurements are used in post-measurement processing to precisely determine *SNR* and interference-to-noise ratio (*INR*) conditions present during the test.

5.3.2. DTV Quality Metric Acquisition

DTV quality metrics are acquired with the following procedure at discrete interference levels:

1. Interference is turned off and the receiver is allowed to acquire and demodulate the desired satellite signal in the absence of interference for 20 seconds.
2. Interference at the specified amplitude is applied for approximately 200 seconds.
3. During the time of interference exposure, *MER*, *EVM*, *BER*, *TEF*, and UTC time are sampled at about 3 samples per second and written as a single record to a data file.

During coarse-resolution acquisition, this cycle is repeated at increasing interference amplitudes separated by 2-dB intervals, beginning at an interference level that has negligible effect on signal quality and ending where *SER* is in excess of 10^{-1}. The fine-resolution acquisition begins 2 dB below the interference level where the first *TEF* was counted in the coarse-resolution acquisition. Amplitude is incremented in 0.1-dB steps until 10^{-1} *SER* is once again achieved.

Data from the fine-resolution acquisition is used to determine the "threshold of visibility," i.e., *SER* where degradation is first perceived. Video quality studies have shown that this threshold correlates to approximately $SER \approx 10^{-4}$ [17]. From the standpoint of a Bernoulli experiment, where independent interference samples degrade reception of independent MPEG segments, 200 seconds of applied interference is adequate for estimating the *SER*.

6. SUMMARY

This chapter gives procedures for statistical characterization of UWB interference signals and assessment of interference susceptibility of C-band satellite television receivers. Our approach is to inject carefully characterized interference into an operating narrowband receiver and measure the susceptibility of the receiver with precisely defined signal quality metrics.

The victim receiver chosen for this interference susceptibility test is C-band satellite digital television. This receiver demodulates signals transmitted in the 3.7 to 4.2 GHz frequency range, which lies within the band allocated for UWB devices. We chose to generate the satellite signal in the laboratory rather than use an unpredictable live signal, to generate interference signals with a vector signal generator rather than rely solely on prototype devices, and to perform signal and system characterization measurements primarily with the vector signal analyzer to provide comprehensive data capable of being post-processed in many ways.

Specifically, interference signals are simulated in software (see Section 2.2 and Appendix A), so the widest range of UWB signal parameters are available for testing, and generated with a VSG (see Section 2.3 and Appendix B), so interference amplitude is precisely controlled. Characterization of the interference signals with a VSA (see Section 3.2.1) provides both signal amplitude and phase information for post-measurement filtering and analyses. Satellite signal simulation (see Section 4.4) provides control over the scene content and *SNR*. Signals are conductively coupled to the satellite receiver to eliminate uncertainties introduced by radio channel interference, attenuation, and distortion. Signal quality metrics, i.e., *MER, EVM, BER*, and *SER*, from an MPEG stream monitor (see Section 4.2) are used to evaluate performance degradation quantitatively. Finally, the test procedure is under software control to eliminate procedural errors and calibrates the measurement system at the beginning of each test to ensure consistency and system integrity.

Comprehensive data acquired will be analyzed to determine if there are common signal characteristics that predict interference potential. Results will be useful for regulatory agencies that are currently charged with defining UWB emission limits and corresponding compliance measurement procedures. Results will also be useful to the communications industry that will only thrive if compliance measurement procedures fairly evaluate interference potential of all UWB signals.

7. Acknowledgments

The authors recognize John McCorkle of Freescale, Inc. for his leadership, insight, and insistence on the highest quality. We learned a great deal from working with John in all relevant technical areas. Dr. Paul Runkle of Signal Innovations provided a wealth of expertise in the field of digital signal processing. Given that signal processing was the crux of this project, Paul made significant contributions by ensuring the integrity and efficient execution of our simulations.

This investigation required a number of ITS engineers with a broad range of expertise, some of whom are not included in the list of authors. The authors recognize J. Randy Hoffman, Richard Statz, and Steven Engelking for developing much of the low level software routines needed to automate the test; Stephen Wolf for his expertise in video quality assessment; Frank Sanders for sharing knowledge learned in previous interference tests on C-band earth station receivers [18]; Dr. Roger Dalke for his knowledge of random processes and experimental uncertainty analyses, and his critical eye; Andrew Rogers and Michael Mitchell for their attention to detail in hardware construction and signal characterization measurements; and Dr. William Kissick for his support and leadership.

8. REFERENCES

[1] "First Report and Order in the matter of ET Docket 98-153 (Revision of Part 15 of the Commission's Rules Regarding Ultra-Wideband Transmission Systems)," *FCC 02-48*, adopted Feb. 14, 2002, released Apr. 22, 2002.

[2] W.A. Kissick, Ed., "The temporal and spectral characteristics of ultrawideband signals," *NTIA Report.* 01-383, Jan. 2001.

[3] J.R. Hoffman, M.G. Cotton, R.J. Achatz, R.N. Statz, and R.A. Dalke, "Measurements to determine potential interference to GPS receivers from ultrawideband transmission systems," *NTIA Report.* 01-384, Feb. 2001.

[4] J.R. Hoffman, M.G. Cotton, R.J. Achatz, and R.N. Statz, "Addendum to NTIA Report 01-384: Measurements to determine potential interference to GPS receivers from ultrawideband transmission systems," *NTIA Report.* 01-389, Sep. 2001.

[5] J.R. Hoffman, E.J. Haakinson, and Y. Lo, "Measurements to determine potential interference to public safety radio receivers from ultrawideband transmission systems," *NTIA Report.* 03-396, Apr. 2003.

[6] "Office of Engineering and Technology Declares MBOA-SIG Request for a Waiver of Part 15 for an Ultra-Wideband System to be a "Permit-but-Disclose" Proceedings for Ex Parte Purposes," *Public Notice, ET Docket No. 04-352, DA 04-2793*, released Sep. 3, 2004.

[7] R. Kohno, M. Welborn, and M. McLaughlin, "DS-UWB Physical Layer Submission to 802.15 Task Group 3a," *IEEE P802.15-04/0137r0*, Mar. 2004.

[8] Batra, et al., "Multi-band OFDM Physical Layer Proposal for IEEE 802.15 Task Group 3a," *IEEE P802.15-03/268r3*, Mar. 2004.

[9] O'Hara and A. Petrick, *IEEE 802.11 Handbook*, New York, NY: IEEE Press, 1999.

[10] C.L. Nikias and A.P. Petrotulu, *Higher-Order Spectra Analysis*, Upper Saddle River, NJ: PTR Prentice Hall, 1993.

[11] Agilent, "Vector Signal Analysis Basics," *Application Note. 150-15.*

[12] Agilent, "Spectrum Analyzer Basics," *Application Note. 150.*

[13] T. Pratt, C. Bostian, and J. Allnutt, *Satellite Communications*, John Wiley and Sons, 2003.

[14] L.W. Couch, *Digital and Analog Communication Systems*, New York, NY: Macmillan Publishing Company, 1987.

[15] ETS 300 421: "Digital broadcasting systems for television, sound, and data services: Framing structure, channel coding, and modulation for 11/12 GHz satellite services," *European Telecommunications Standards Institute*, Dec. 1994.

[16] ITU-R BT.802.1, "Test Pictures in Sequences for Subjective Assessments of Digital Codecs Conveying Signals Produced According to Recommendation ITU-R BT.601," *International Telecommunication Union*, Jul. 1994.

[17] J.C. Whitaker, "Digital television transmission systems," Ch. 13 in *Standard Handbook of Video and Television Engineering*, J.C. Whitaker, Ed., New York, NY: McGraw-Hill, 2000.

[18] F. Sanders, "Measurements of pulsed co-channel interference in a 4-GHz digital earth station receiver," *NTIA Report.* 02-393, May 2002.

[19] ETS TR 101 290: "Digital video broadcasting (DVB); Measurement guidelines for DVB systems," *European Telecommunications Standards Institute*, v1.2.1, 2001.

[20] J.W. Modestino and B. Sankur, "Modelling and analysis of impulsive noise," Part VII, Section 6 in *Communication Systems and Random Process Theory*, J.K. Skwirzynski, Ed., Alphen aan den Rijn, The Netherlands: Sijthoff and Noordhoff, 1978.

[21] Agilent Technologies, Inc., Application Note 57-2, "Noise figure Measurement Accuracy – The Y-Factor Method," Mar. 19, 2004.

Appendix A: Mathematical Signal Description

This section provides mathematical expressions for $u(t)$, i.e., the real-valued representation of interference signals pertinent to this study.

A.1. Direct-Sequence Ultrawideband

A block diagram of the simulated direct-sequence ultrawideband (DS-UWB) transmitter is shown in figure A-1.

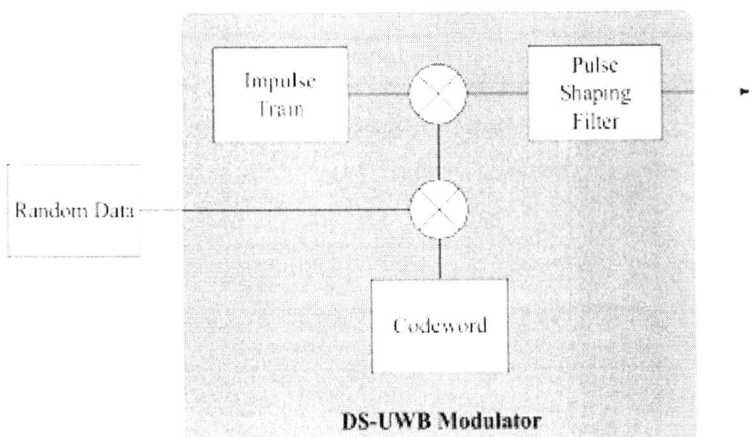

Figure A-1. Simulated DS-UWB transmitter block diagram.

The simulated DS-UWB signal is mathematically described by

$$u(t) = \sum_{k=-\infty}^{\infty} \sum_{l=1}^{L} d_k c_l p_{DS}(t - kLT_{chip} - (l-1)T_{chip}) .$$

where d_k is the kth random data element with values {1, -1}, L is the codeword length with values of {1, 3, 6, 12, and 24}, is the lth codeword chip with values {1, 0, -1}, and T_{chip} is the chip period.

The impulse response of the pulse shaping filter is

$$p_{DS}(t) = h_{rrc}(t)\sin(6\pi f_{chip} t) \quad .$$

where $h_{rrc}(t)$ is the impulse response of the root-raised-cosine (RRC) filter and f_{chip} is the chip rate. RRC filters are often specified by their frequency response

$$H_{rrc}(f) = \begin{cases} 1 & |f| \le f_{co}(1-\alpha) \\ 0 & |f| > f_{co}(1+\alpha) \\ \sqrt{\dfrac{1+\cos\left[\pi(|f| - f_{co}(1-\alpha))/2\alpha f_{co}\right]}{2}} & f_{co}(1-\alpha) < |f| \le f_{co}(1-\alpha) \end{cases}$$

where f_{co} is the filter cutoff frequency equal to $f_{chip}/2$ and α is the roll-off factor.

A.2. Dithered-Pulse Ultrawideband

A block diagram of the simulated dithered-pulse ultrawideband (DP-UWB) transmitter is shown in figure A-2.

The simulated DP-UWB signal is mathematically described by

$$u(t) = \sum_{k=-\infty}^{\infty} p_{DP}\left(t - kT_{pulse} - \frac{T_{pulse}}{2} - \frac{d_k \beta_k T_{pulse}}{2}\right) \quad .$$

where p_{DP} is the pulse-shaping filter impulse response, T_{pulse} is the average pulse repetition period, and d_k is the kth random data element with values {1, -1}. β_k is the kth dithered position relative to T_{pulse}, which is uniformly distributed with a probability density

$$g(\beta) = \begin{cases} 1/\beta_{max} & 0 < \beta < \beta_{max} \\ 0 & \text{otherwise} \end{cases} \quad .$$

where β_{max} is the dither fraction, which is a constant bounded by $0 < \beta_{max} \le 1$.

The pulse-shaping filter has a Gaussian frequency response

$$P_{DP}(f) = \exp\left(-\sigma^2 f^2\right) \quad .$$

where

$$\sigma = \sqrt{\frac{\ln 2}{2B^2}}$$

and B is the 3-dB bandwidth.

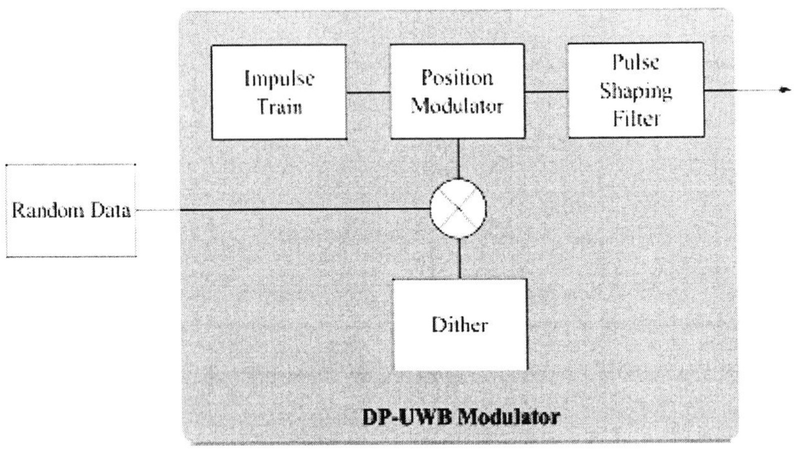

Figure A-2. Simulated DP-UWB block diagram.

A.3. Multi-Band OFDM Ultrawideband

A block diagram of the simulated multi-band OFDM (MB-OFDM) transmitter is shown in figure A-3.

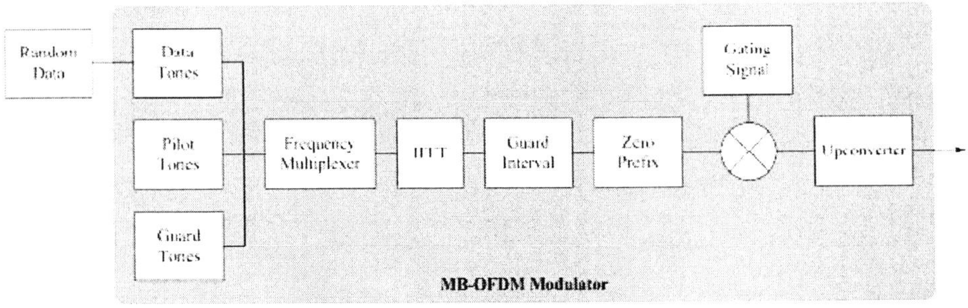

Figure A-3. Simulated MB-OFDM transmitter block diagram.

Mathematical description of the MB-OFDM signal begins with defining the 128-tone coefficient vector

$$D_n = \begin{cases} 0 & n = \{1\} & \text{center} \\ 0 & n = \{63, 64\}, \{65, 66, 67\} & \text{zero guard} \\ \exp\left[j\pi\left(a_n - \frac{1}{4}\right)\right] & n = \{58, 59, 60, 61, 62\}, \{68, 69, 70, 71, 72\} & \text{modulated guard} \\ \exp\left[j\pi\left(a_n - \frac{1}{4}\right)\right] & n = \{6, 16, 26, 36, 46, 56\}, \{74, 84, 94, 104, 114, 124\} & \text{pilot} \\ \exp\left[j\pi\left(\frac{a_n}{2} + \frac{1}{4}\right)\right] & \text{otherwise} & \text{data} \end{cases}$$

where a_n is the nth coefficient value that modulates data, pilot, and modulated guard tones. These coefficients take on values of {0, 1, 2, 3}, {0, 1}, and {0, 1} for data, pilot, and modulated guard tones respectively. Indices for the tone coefficient vector can be translated to actual frequencies by the equation

$$f = \begin{cases} f_c - (n-1)\Delta f & 1 \le n \le 64 \\ f_c - (n-129)\Delta f & 65 \le n \le 128 \end{cases}$$

where f_c is the center frequency and f is the 4.125-MHz tone spacing.

Random coefficients QPSK modulate the data tones. A fixed pattern of coefficients {0, 1, 0, 0, 1, 0, 0, 1, 0, 0, 1, 0} randomized by a bit from a 127-bit pseudorandom sequence, BPSK modulates the pilot tones. The same 127-bit pseudorandom sequence BPSK modulates the modulated guard tones between symbols. Guard tones within the same symbol are differentiated by a distinct offset into the 127-bit pseudorandom sequence.

The tone coefficients are inverse Fourier transformed to data block samples by

$$d_m = \frac{1}{N}\sum_{n=0}^{N-1} D_n \exp\left(\frac{j2\pi mn}{N}\right) \qquad m = 0, 1, 2, \ldots, N-1$$

where D_n is the nth QPSK symbol and N = 128. The MB-OFDM symbol (see figure 6) is formed by placing the data block between the 32-sample zero prefix and 5-sample guard interval suffix making the total symbol length, L, 165 samples in duration. The kth MB-OFDM symbol is

$$u_{k,l} = \begin{cases} 0 & 0 < l \le 32 \\ d_{l-33} & 32 < l \le 160 \\ 0 & 160 < l \le 165 \end{cases}$$

The simulated MB-OFDM signal is mathematically described as

$$u(t) = \Re\left\{ \left(\sum_{k=-\infty}^{\infty} \sum_{l=1}^{L} u_{k,l}\, h_{rect}\left(t - kT_{OFDM} - (l-1)\Delta t\right) \right) \exp(-j2\pi f_c t) \right\} .$$

where h_{rect} is a filter with a rectangular frequency response bounded by $\pm 1/(2\ t)$, $T_{OFDM} = L\ \Delta\ t$ = 312.5 nanoseconds is the MB-OFDM symbol period, $\Delta\ t = 1/(N\ \Delta\ f)$ 1.9 nanoseconds, and denotes the real part of a complex number.

Finally, MB-OFDM signals hop between 14 band center frequencies according to the specified time-frequency code. The effect the time-frequency code has on a single frequency is simulated by multiplying the signal by the gate function $g(t;\ T = bdT_{OFDM},\ \tau = dT_{OFDM})$, where b is the number of bands, d is the number of periods the signal dwells in each band. The periodic gating function shown in figure A-4 is defined mathematically as

$$g(t;T,\tau) = \sum_{k=-\infty}^{\infty} R(t-kT;\tau) \quad \text{where} \quad R(t;\tau) = \begin{cases} 1 & 0 \le t < \tau \\ 0 & \text{otherwise} \end{cases} \quad \text{and} \quad T > \tau .$$

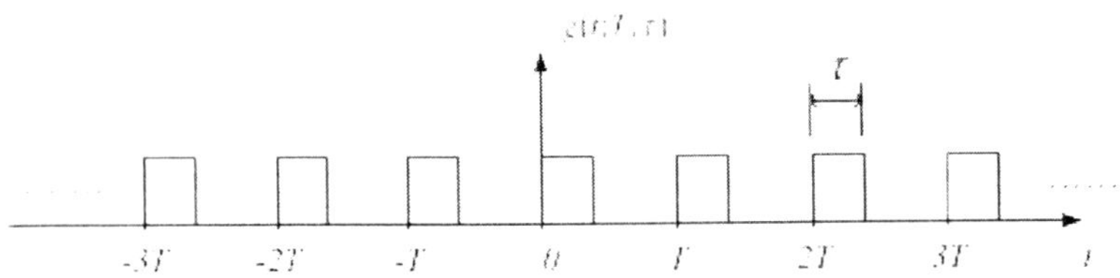

Figure A-4. Periodic gating function.

A.4. Gated Gaussian Noise

A block diagram of the simulated gated Gaussian-noise source is shown in figure A-5.

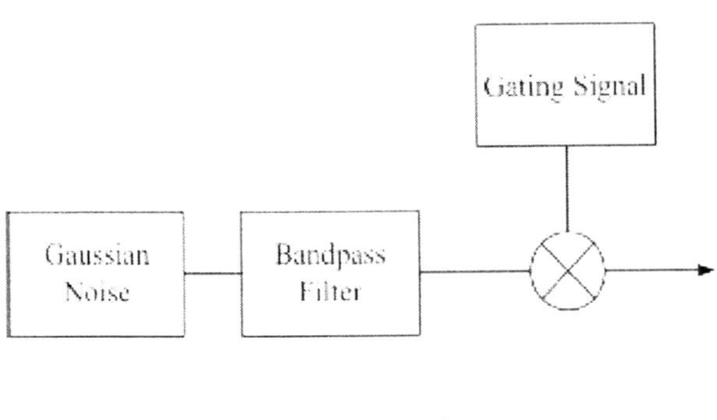

Figure A-5. Simulated gated Gaussian noise transmitter block diagram.

Gated Gaussian noise is described by

$$n(t;T,\tau) = (n(t) * h_{BPF}(t))g(t;T,\tau) \quad .$$

where * is the convolutional operator and $n(t)$ is a zero-mean Gaussian random process with the probability density

$$g(n) = \frac{1}{\sqrt{2\pi}\sigma_n} \exp\left(\frac{-n^2}{2\sigma_n^2}\right) \quad .$$

where σ_n is the standard deviation of $n(t)$.

APPENDIX B: VECTOR SIGNAL GENERATOR OPERATION

The dual arbitrary waveform generator (AWG) inside the Agilent E8267C Vector Signal Generator (VSG) is used in this study to generate various types of interference signals. Complex-baseband (CBB) waveforms are simulated in software, loaded into VSG memory, and generated by the dual AWG. This appendix describes instrument operation, techniques to optimize RF performance, relevance of these techniques to our study, conditioning of simulated waveforms, and VSG operating procedures.

B.1. Operation in the Dual AWG Mode

A block diagram of the VSG operating in the dual AWG mode is shown in figure B-1. Digital in-phase (I) and quadrature-phase (Q) signals, previously loaded into memory, are converted to analog I and Q signals by digital-to-analog converters (DACs) with digital interpolation filters and analog reconstruction (lowpass) filters. Interpolating filters increase the data rate by a factor of four and ease requirements of the reconstruction filters. Reconstruction filters smooth the stair-step responses that are present in the analog I and Q signals at the output of the DACs.

The smoothed analog I signal at the output of the reconstruction filter modulates an LO while the corresponding Q signal modulates a 90° phase-shifted LO. These modulated signals are summed to produce a real bandpass signal centered at the LO frequency. After this, only the amplitude of the bandpass signal is adjusted.

Signal amplitude is dependent upon the settings for the automatic level control (ALC) and the RF output power. When the ALC is on, the ALC circuitry senses the RF output power and uses this level in a feedback loop to maintain the RF output power at a constant calibrated level. The ALC causes unwanted amplitude fluctuations in signals with variable amplitudes. Since the interference signals have amplitude variations, we chose to turn off the ALC.

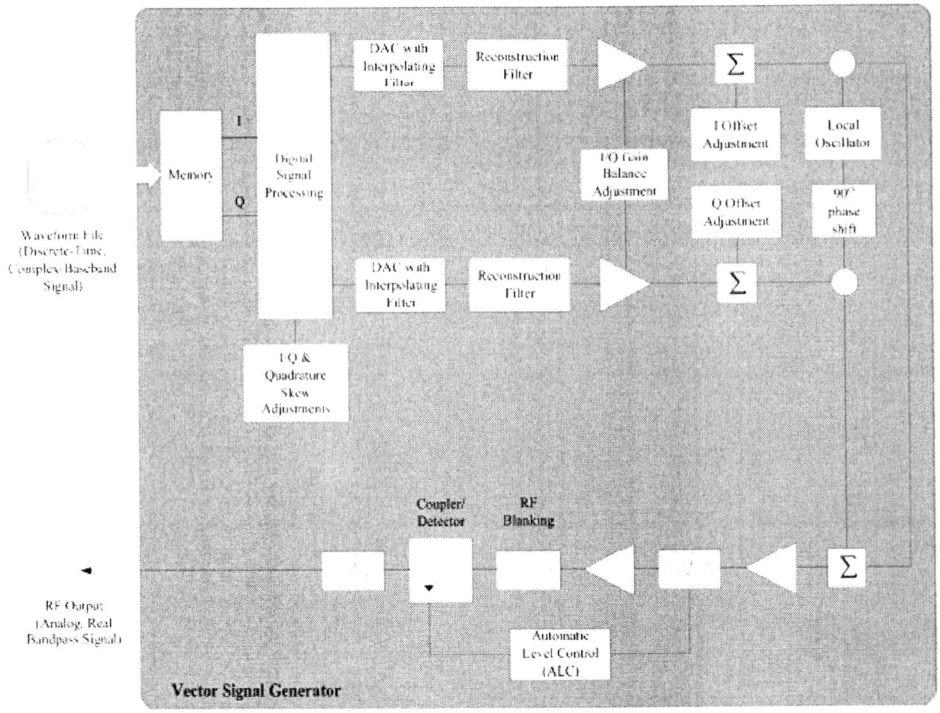

Figure B-1. Block diagram of the Vector Signal Generator.

B.2. Techniques to Optimize RF Output Performance

Hardware I/Q modulation within the VSG generates undesirable signals as well as the desired waveform. Undesirable signals include local oscillator (LO) feed-through, intermodulation products, and other spurious responses. LO feed-through, produced by I and Q DC offset voltages within the I/Q modulator, is perhaps the most prominent unwanted signal. While generation of the undesirable signals is unavoidable, the amplitude of the undesirable signals can be minimized using various techniques including I/Q calibration, pre-distortion, RF blanking, and frequency shifting.

B.2.1. I/Q Calibration and Adjustments

Effects of imperfect I/Q modulation are compensated for by adjustments to parameters that include I and Q DC offsets, I/Q gain balance, I/Q timing skew, and quadrature skew. I and Q DC offset adjustments vary the DC voltage of I and Q signals, respectively. I/Q gain balance adjustment varies the gain of the I signal relative to the gain of the Q signal. I/Q skew adjustment delays I relative to Q or vice versa. Quadrature skew changes the phase angle between I and Q signals.

I/Q adjustments can be made with custom or vendor-supplied internal I/Q adjustment routines. Custom routines consist of playing a test waveform through the VSG and monitoring its output on a spectrum analyzer (SA). Optimum I/Q modulation settings are found by repetitively adjusting the parameters and observing the effects of adjustment on the

SA. For example, a two-tone test signal, played through the VSG, can be used to minimize LO feed-through. Optimum DC offset settings are found by repetitive adjustment of these settings and monitoring the amplitude of the LO feed-through on the SA. In contrast, vendor-supplied routines use internal test waveforms and monitoring functions.

Since both custom and vendor-supplied routines provided similar results, we chose the vendor-supplied routine to reduce development time and test-system complexity. To maintain the best-possible performance, we perform I/Q calibration before each DTV interference susceptibility test.

B.2.2. Waveform Pre-Distortion

Pre-distortion software is available from the manufacturer that can be used to further minimize the effects of I/Q modulator imperfections in addition to performing an I/Q calibration. This software requires that the VSG be connected to an Agilent PSA or ESA spectrum analyzer. The software then applies corrections to I and Q data to minimize undesirable signals as determined by automated measurements by the SA.

While this technique can provide superb results in minimizing undesirable signals, it has several limitations that preclude its use in our measurements. First, this technique requires monitoring the RF output of the VSG on a SA any time data correction is desired. This would add more complexity to the measurement system. Second, changing the settings on the VSG requires new corrections to be made on I and Q waveform data. Perhaps an even more serious limitation is that this technique can only be used with waveforms that use a sampling rate of 100 million samples per second (MSps). Many of the waveforms that we use have other sampling rates. For these reasons, we chose not to use pre-distortion.

B.2.3. RF Blanking

RF blanking can be used to mitigate LO feed-through when the signal is known to be off. For our tests, this includes gating periods. RF blanking uses the pulse modulation circuitry in the VSG to turn off the RF output. The user generates a separate marker file that contains Boolean values corresponding to each data sample in the waveform file. Each Boolean value in the marker file instructs the pulse modulation circuitry to turn the RF output on or off while the corresponding data sample in the waveform is being played on the VSG. Using a marker file, the user can specify turning off the RF output whenever a waveform data sample is zero, thus squelching the LO feed-through that would normally be present in the RF output. The pulse-modulation circuitry used for RF blanking typically provides 80 dB of attenuation for the RF-off state.

Practical limitations associated with RF blanking prevented us from using it. Most importantly, there is a finite rise and fall time associated with the pulse modulation circuitry. Unfortunately, the rise and fall times can be as long as 10 nanoseconds, which is as long as the sample period. Consequently, RF blanking can modulate the desired signal. This modulation is most evident when there are numerous transitions between zero and non-zero voltages. Additionally, the VSG introduces a delay between the time a waveform file sample is played and when RF blanking occurs. This delay is on the order of 40 to 50 nanoseconds and varies from one VSG to another. The delay must be experimentally determined and incorporated into the marker file.

B.2.4. Frequency Shifting the Complex-Baseband Signal

Another method of minimizing the effects of LO feed-through is to frequency shift the CBB signal by f_0 during computer signal simulation so that the LO is outside the bandwidth of the desired signal as shown in figure B-2. f_0 should be chosen so that all significant spectral content of the desired signal is undistorted by the band edges of the VSG frequency response. If this distortion is unavoidable, it can be mitigated to some extent by applying an inverse filter, which compensates for the VSG frequency response, to the waveforms during simulation. As demonstrated in Appendix C, LO feed-through was not a problem for the DTV interference susceptibility tests; hence, we chose not to frequency shift.

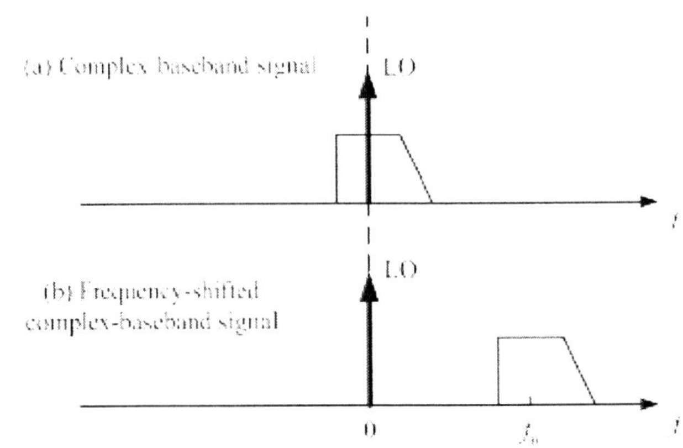

Figure B-2. Frequency shifting the complex-baseband signal by f_0

B.3. Waveform Conditioning

In this study, the simulated CBB waveforms of various interference signals are created and stored sequentially in a binary waveform file as a sequence of discrete-time, double-precision I and Q values. To be useful to the VSG, these values must be converted to a fixed-point format. This conversion process is performed in three steps: normalization, scaling, and quantization.

Normalization is accomplished by finding the maximum of the absolute values of all I and Q data values and dividing all the I and Q values by this maximum value. After normalization, all I and Q values are within the range $-1 \leq \{I, Q\} \leq 1$.

Scaling compensates for the effects of sample interpolation performed by the VSG prior to digital-to-analog conversion and reconstruction filtering. The interpolator, which runs at four times the AWG sample rate, often causes samples with large voltage transitions to exceed the range of the digital-to-analog converter (DAC). As recommended by the manufacturer and to ensure that DAC over-range errors do not occur, normalized I and Q data values are scaled by a factor of 0.7. After scaling, all I and Q values are within the range $-0.7 \leq \{I, Q\} \leq 0.7$. A less conservative scaling factor can be determined experimentally (on a signal by signal basis) by monitoring VSG DAC over-range error indicators.

Finally, I and Q floating point values are converted to fixed-point numbers corresponding to the DAC resolution. For a 16-bit DAC, normalized and scaled floating-point values are multiplied by $(2^{15} - 1)$ and converted to 16-bit signed integer I and Q values. After quantization, all I and Q values are within the range $-22{,}937 \leq \{I, Q\} \leq 22{,}937$. Prior to loading the waveform into VSG RAM, it is important to resolve discrepancies between fixed-point formats that may exist between the computer, which processed the samples, and the VSG, which uses them. In our case, samples are processed on a PC using an Intel processor and played on a VSG using a Motorola processor. This requires a conversion from little endian fixed-point format to big endian.

B.4. Operating Procedures

In this study, VSG operational procedures are controlled via software using SCPI commands. This permits highly repeatable operation. The sequence of commands used to setup and operate the VSG is provided below:

1. System Preset: Perform the same operation as the front panel Preset button. Set the VSG to the factory defined state.
2. Clear Status Register: Clear the status byte register including the error queue and all event registers.
3. ALC Off: Turn off automatic level control.
4. Set RF Frequency: Set LO frequency that the I and Q samples modulate.
5. Clear Waveform Memory: Delete WFM1 waveform memory (RAM) and wait 2 seconds before issuing the next command to allow deletion to complete.
6. Load Waveform File: Load normalized waveform file from personal computer into WFM1 waveform memory using FTP.
7. Select Waveform Segment: Select the waveform segment corresponding to file loaded into WFM1 waveform memory.
8. Set Sample Rate: Set sample rate according to that specified in simulation.
9. AWG On: Turn on baseband AWG and wait 10 seconds before issuing the next command to allow baseband reconfiguration.
10. Modulation On: Enable waveform modulation of LO.
11. Set Run-Time Scaling to 100%: The factory default is 70% and is only appropriate when loaded waveforms are not already properly normalized.
12. I/Q Calibration: Perform I/Q calibration.
13. Set RF Output Power Level.
14. RF On: Turn on RF output.

Appendix C: Effects of VSG LO Feed-through on DTV Interference Susceptibility Tests

This appendix provides a quantitative analysis of the effects of VSG LO feed-through on digital television (DTV) interference susceptibility tests.

C.1. LO Feed-through Inherent to DTV Systems

As described in *ETSI Measurement Guidelines for DVB Systems* [19] and shown in figure A-1, unwanted LO feed-through is inherent to quadrature-amplitude modulated (QAM) DTV systems. Note that QPSK receivers being tested in this effort are a subset of QAM. LO feed-through is often characterized by a carrier suppression specification defined as

$$CS = \frac{P_{DTV}}{P_{LO}} \quad . \tag{C-1}$$

where P_{DTV} is the total power of the QAM signal without LO feed-through and P_{LO} is the power of the LO feed-through.

The signal has a noise equivalent bandwidth of B_{DTV}, which we assume is equal to the QPSK symbol rate, *RS*. We also assume that the spectrum analyzer (SA) uses Gaussian-shaped filters whose noise equivalent bandwidth is approximately the resolution bandwidth (*RBW*). Since the noise and the DTV signal are zero-mean random processes, the mean powers of each signal can be separated arithmetically. As a result, the noise, signal, and LO feed-through power as well as the carrier suppression ratio can be computed from SA measurements.

Various terms in the computation are used. $PSD(f)$ refers to power spectral density in watts/Hz, which is, in general, frequency dependent. *P* refers to power in watts. Finally, *M* refers to a SA power measurement in watts. For this appendix, *M* is measured in 1-kHz *RBW* with root-mean-square (rms) detection and 0.3-second integration time (i.e., sweep time divided by number of frequency steps). Subscripts *nSYS*, *DTV*, and *LO* refer to contributions from system noise, satellite DTV signal, and local-oscillator feed-through respectively.

M_{nSYS} is average power of the system noise measured with the system signal path cascaded with the SA, as shown in figure 20, but with the DTV signal turned off. Noise from the SA and LNB are included in the measurement and we assume it to be flat across the satellite channel. From this measurement, power density of the system noise is computed as

$$PSD_{nSYS}(f) = \frac{M_{nSYS}}{RBW} \quad .$$

$M_{DTV+nSYS}$ is the average power of the DTV signal plus system noise measured by the SA at frequency f_1 slightly offset from the center frequency of the DTV signal but still in the region where the DTV power spectral density is constant. The corresponding power spectral density without system noise is

$$PSD_{DTV}(f_1) = \frac{M_{DTV+nSYS}}{RBW} - PSD_{nSYS}(f) \quad .$$

Average power of the DTV signal is

$$P_{DTV} = PSD_{DTV}(f_1) \cdot B_{DTV} = PSD_{DTV}(f_1) \cdot R_s \quad .$$

$M_{LO+DTV+nSYS}$ is the average power of the LO feed-through plus DTV signal plus system noise measured by the SA centered on the DTV signal. From this measurement, power of the LO feed-through is computed as

$$P_{LO} = M_{LO+DTV+nSYS} - M_{DTV+nSYS} \quad .$$

Carrier suppression is computed from P_{LO} and P_{DTV} with equation C-1.

Figure C-1 shows LO feed-through generated by the satellite simulation subsystem, described in Section 4.4, and measured by the spectrum analyzer. Satellite simulator power was set so that SNR was approximately 8.5 dB – the lowest SNR tested and therefore the most susceptible to LO feed-through from the VSG. The following measurements were made: $M_{nSYS} \approx$ -90.5 dBm, $M_{DTV+nSYS} \approx$ -82.0 dBm, $M_{LO+DTV+nSYS} \approx$ -71.6 dBm. Calculations from these measurements give $P_{LO} \approx$ -72.2 dBm, $P_{DTV} \approx$ -38.6 dBm, and CS $\approx$ 34 dB. Note that these results are given in dB, but the calculations were performed in linear units.

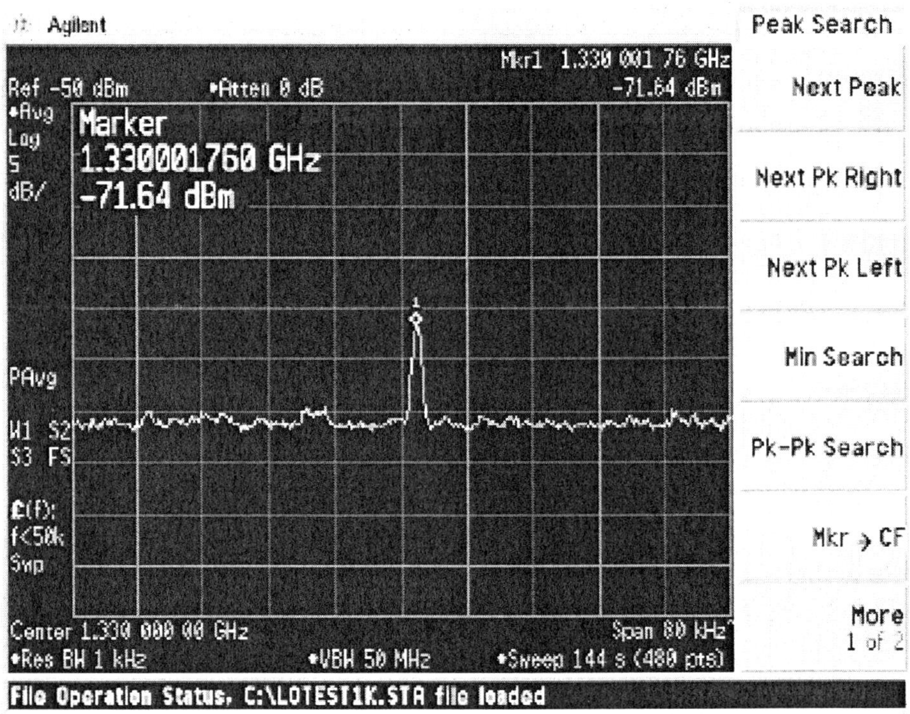

Figure C-1. LO feed-through from satellite signal generator.

C.2. Experiment

Satellite receivers that demodulate QPSK signals are designed to tolerate reasonable levels of LO feed-through. However, it is not clear how much additional LO feed-through, introduced by the VSG-generated interference signals in our DTV interference susceptibility test, can be tolerated.

Figure C-2 depicts hardware used to emulate the VSG as it is used during DTV interference susceptibility testing to generate gated-noise signals from simulated waveforms. The noise diode, amplifier, and high-speed switch generate a gated-noise signal, while the signal generator adds a CW signal that emulates the VSG LO feed-through.

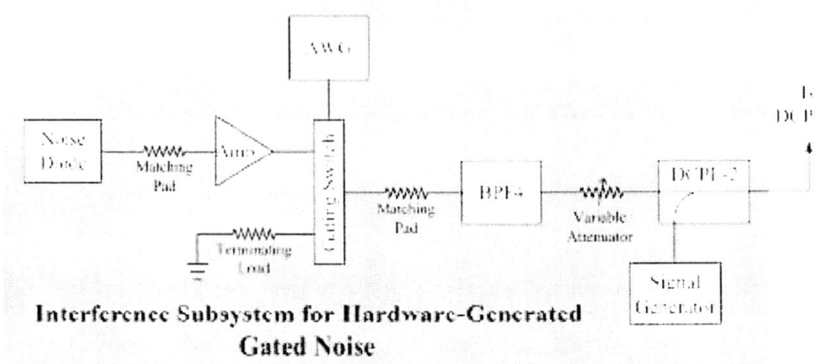

Figure C-2. Interference subsystem for testing the effects of LO feed-through.

The measurement procedure consisted of the following four steps:

1. Baseline signal-quality measurement: The satellite simulation subsystem was set to 8.5-dB *SNR*. Baseline interference, i.e., hardware-generated gated noise, was coupled into the test system at directional coupler DCPL-1 (see figure 20) at a level that caused the segment error rate (*SER*) to be approximately 10^{-4}. Baseline signal quality was recorded.
2. Baseline-plus-CW signal-quality measurement: A CW signal was introduced at directional coupler DCPL-2 and signal quality was recorded with the DTV receiver exposed to baseline-plus-CW interference. The CW signal was centered in the DTV channel at 3820 MHz and increased in 5-dB steps. The baseline-plus-CW measurements were stopped at a CW level that caused *SER* to exceed 10^{-2}.
3. Baseline and baseline-plus-CW power measurement: The CW amplitude was set to the highest level that did not cause signal quality to deviate from the baseline signal quality. SA measurements at the CW frequency and offset from the CW frequency were made at the output of the LNB in 1-kHz *RBW*. Calculations using these measurements determined the amount of CW power alone. Figure C-3(a) shows SA measurements of hardware-generated baseline-plus-CW Gaussian noise interference (GN-01).
4. VSG interference and LO feed-through power measurement: The hardware-generated interference subsystem was replaced by the VSG loaded with the same

corresponding gated-noise waveform. VSG amplitude was adjusted so that the gated-noise power, away from the LO feed-through, matched the baseline power measured previously. SA measurements at the LO frequency and offset from the LO frequency were made at the output of the LNB in 1-kHz *RBW*. Calculations using these measurements were used to determine the amount of LO feed-through power alone. Figure C-3(b) shows SA measurements of VSG-generated Gaussian noise interference (GN-01).

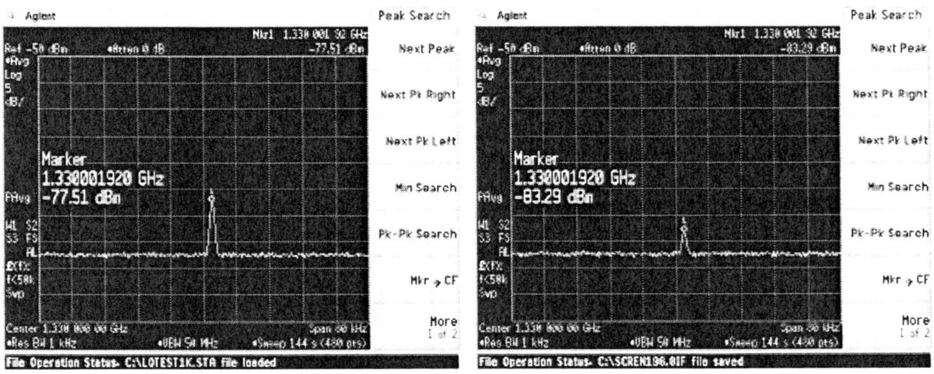

(a) (b)

Figure C-3. SA measurements of (a) hardware-generated Gaussian-noise plus CW and (b) VSG-generated Gaussian noise with LO feed-through.

C.3. Results

Interference permutations considered for this experiment were continuous Gaussian noise (GN-01), gated noise with 100-nanosecond on time and 50% duty cycle (GN-07), and gated noise with 100-nanosecond on time and 6.25% duty cycle (GN-09). VSG-generated GN-09 has the strongest LO feed-through relative to the intended-interference power. This is because gating attenuates the noise average power but has no effect on the LO feed-through power. In fact, GN-09 LO feed-through will be the strongest encountered out of all gated-noise interference permutations listed in Section 2.1.5 because the VSG amplitude has to be set higher to create an $SER = 10^{-4}$ condition.

Figures C-4 – C-6 display best estimates of four DTV signal quality metrics, i.e., segment error rate (*SER*), pre-Viterbi bit error rate (*BER*), modulation error rate (*MER*), and error vector magnitude (*EVM*). These metrics are plotted against increasing levels of CW power representing VSG LO feed-through. Systematic error, introduced by the measurement system, has been mostly removed from these results. LO feed-through of the satellite simulator (black) and VSG (red) are plotted as vertical lines. Signal quality metrics began to degrade at CW powers of -60 dBm, -62 dBm, and -65 dBm for GN-01, GN-07, and GN-09 respectively.

C.4. Effect of VSG LO Feed-through on Carrier-Suppression Ratio

The vertical red and black lines in figure C-4 – C-6 show that LO feed-through of the satellite simulator and VSG were lower than the threshold where DTV quality metrics began to degrade. However, it is the combination of these residual carriers that we are concerned with. Figure C-7 shows how the combination of LO feed-through powers degrade *CS* from 34 dB.

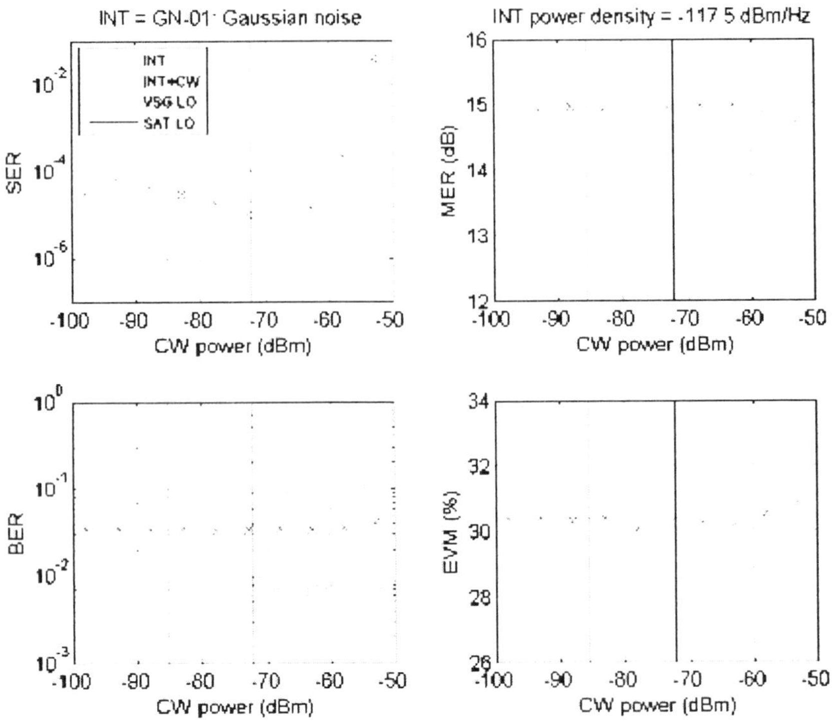

Figure C-4. LO feed-through measurements for Gaussian noise interference.

The highest VSG LO feed-through for the gated-noise signals listed in Section 2.1.5 is approximately -74 dBm = -77 + 3 dBm, where -77 dBm is the LO feed-through power of the VSG generating GN-09 at a level that causes *SER* to be 10^{-4} (illustrated in figure C-6) and 3 dB is a conservative estimate of the additional interference power needed to bring that *SER* to 10^{-1}. Figure C-7 shows that this degrades carrier suppression to approximately 32 dB; this is the minimum *CS* for the gated-noise interference potential study. In other words, all gated-noise interference signals generated in this study will have a *CS* 32 dB.

Results in Section C.3 show that, in the worst-case, DTV signal quality metrics began to degrade at an LO feed-through level of -65 dBm, which degraded carrier suppression to approximately 26 dB. Therefore, as shown in figure C-7 there is a 6-dB margin between the measured *CS* threshold and the minimum *CS* for the gated-noise interference potential study.

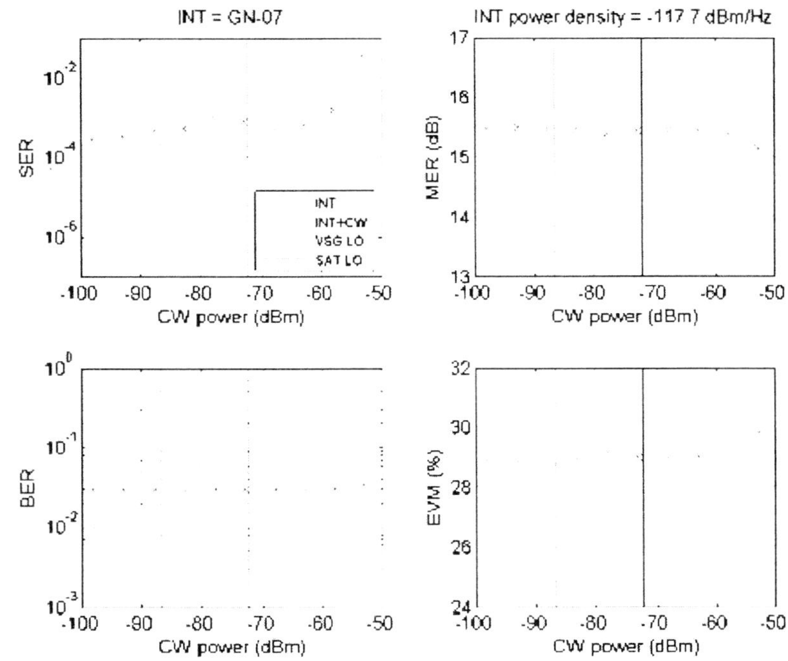

Figure C-5. LO feed-through measurements for gated noise interference with a 100-nanosecond on-time and 25% duty cycle.

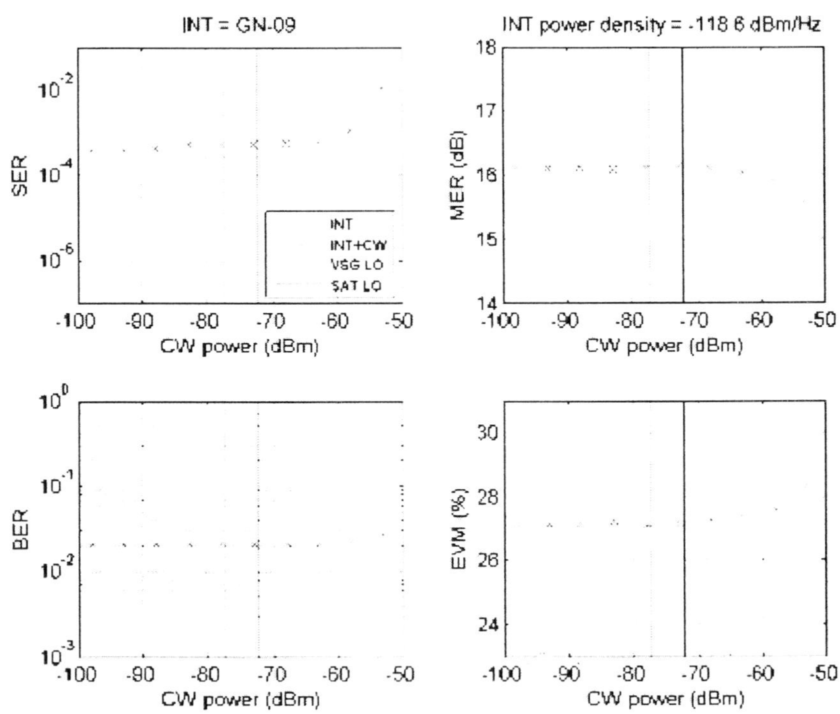

Figure C-6. LO feed-through measurements for gated noise interference with a 100-nanosecond on-time and 6.25% duty cycle.

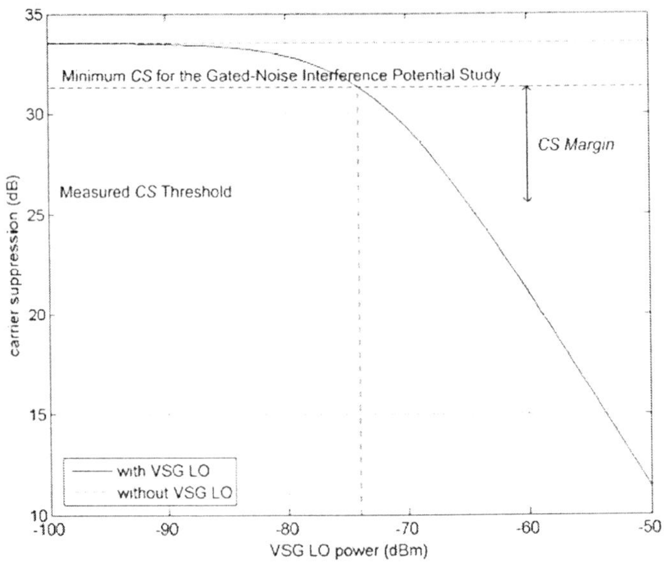

Figure C-7. Effects of VSG LO on carrier suppression.

C.5. Conclusion

This experiment shows that the highest VSG LO feed-through power, -74 dBm, for all gated-noise interference signals generated in this study degrades the carrier suppression ratio from 34 to 32 dB. This is approximately 6 dB above the carrier suppression ratio that degraded signal quality metrics for a DTV receiver operating at $SER \approx 10^{-4}$. Hence, VSG LO feed-through is not a problem for our DTV interference susceptibility tests.

Appendix D: Estimating and Graphing the Amplitude Probability Distribution of Complex-Baseband Signals

The amplitude probability distribution (*APD*) function has been found to be useful in characterizing signals and evaluating effects of interference on victim receivers. The following discussion defines the *APD* mathematically and demonstrates how it can be estimated and plotted with the MATLAB® language.

D.1. Estimating the *APD*

The mathematics and terminology of the *APD* are based on the concept of random variables. A random variable is a function, X, which assigns a real number, $X(v)$, to every element, v, of a sample space. In our case, the random variable assigns a real number representing the amplitude of a complex-baseband signal drawn from a sample space via signal simulation or measurement. The *APD* is expressed as

$$F(a) = \wp\{X(v) > a\}$$

or more commonly

$$F(a) = \wp(X > a) \quad .$$

where a is an amplitude value and $\wp$ ( ) means probability.

A discrete estimate of the *APD* can be obtained from a finite set of samples. This is accomplished by sampling the complex-baseband signal N times, converting these samples to the amplitudes

$$a(n) \quad , \qquad n = 1, 2, \cdots, N \quad ,$$

ordering the amplitudes from smallest to largest

$$a[n] \quad , \qquad n = 1, 2, \cdots, N \quad ,$$

where the square brackets distinguish the ordered amplitudes from the unordered amplitudes, and computing

$$\wp(X > a[n]) = 1 - \frac{n}{N} \quad .$$

The routine "[a, p] = apd(s)" (given in Appendix D.5) estimates the *APD* from N amplitude samples. Elements in the vector "s" are checked to make sure they are amplitudes, i.e., that they are real and positive. "s" is ordered from smallest to largest via "a = sort(s)", where "sort" is a MATLAB function, and corresponding probabilities are computed via "p = 1-(1 : N)/N".

D.2. Plotting the *APD*

*APD*s are typically displayed on a Rayleigh graph whose axes are transformed by functions that linearize the *APD* of the Rayleigh-distributed amplitudes of complex Gaussian noise. Derivation of these functions begins with the expression of the *APD* of a Rayleigh-distributed random variable

$$\wp(X > a) = \exp\left(\frac{-a^2}{\sigma^2}\right) .$$

where σ represents the standard deviation of zero-mean complex noise. This function is linearized by the natural logarithm function and converted to decibels by

$$10\log(a^2) = 10\log(\sigma^2) - 10\log(-\ln(\wp(X > a))) .$$

Graphical coordinates, x and y, are introduced by

$$y = \log(a^2)$$

and

$$x = 10\log(-\ln(P_0)) - 10\log[-\ln(\wp(X > a))] ,$$

where P_0 is the probability at the graph origin.

The x-axis is in units of percentage and labeled "percentage of time ordinate is exceeded" or "percent exceeding ordinate." The x-axis probabilities generally range from 0.0001 to 99 percent for problems associated with modern receiver design.

Signals are often normalized. Common normalization factors are the root-mean-square (rms) voltage of the signal or the rms voltage of white Gaussian noise. If normalized by the rms voltage of the signal, the y-axis is labeled "dB relative to rms voltage" or "dB relative to average power." If normalized by the rms voltage of white Gaussian noise, the y-axis is labeled "dB relative to measurement system average noise power" or "dB relative to *kTB*." Otherwise, units appropriate to the samples such as dBV or dBW are used.

The routine "plotapd(a,p)" plots the *APD* on a Rayleigh graph. The routine begins by checking for zero amplitudes and probabilities that will generate log function error messages. Zero amplitudes are possible for signals simulated without added noise. The largest or peak amplitude of the *APD* estimate is assigned zero probability. "Plotapd" circumvents these problems by replacing the zero value with the MATLAB constant "not a number" (NaN). The MATLAB "log" or "log10" functions will not generate error messages when operating on NaNs and the MATLAB "plot" routine will not plot points in which the abscissa or ordinate is NaN. The remainder of the routine transforms data and axes to the graphical coordinate system using the equations for x and y above. y-axis labeling is dependent on how or if the samples were normalized and therefore is left outside the plotting routine.

D.3. Testing the Routines

The routine "testapd" demonstrates the use of the "apd" and "plotapd" routines by checking the median and mean values of two test signals (corresponding plots are shown in figures D-1 and D-2). The first test signal is a ten-thousand element vector of linearly increasing voltages ranging from 0 to 9999. The median (i.e., the amplitude exceeded by 50 percent of the samples) of this signal is 5000 V or 74.0 dBV. The second test signal is a one-million element vector of complex Gaussian noise voltages with zero mean and a variance of 2 V^2. The rms voltage is equal to the square root of the variance or 3.0 dBV. The corresponding probability is equal to 1/*e* or 0.367 and expressed on the graph as 36.7%.

D.4. Additional Comments

Several important amplitude statistics, including rms voltage or average power, can be computed with the information found in the *APD*. However, the APD cannot provide information concerning the signal time-domain behavior without other a priori information such as knowing that it is a periodically pulsed signal.

Additionally, the *APD* is strongly affected by the type of band-limiting filter; the *APD* estimate accuracy and precision are dependent on the sample interval and number of samples used. Consequently, these parameters should be documented with the *APD* graph.

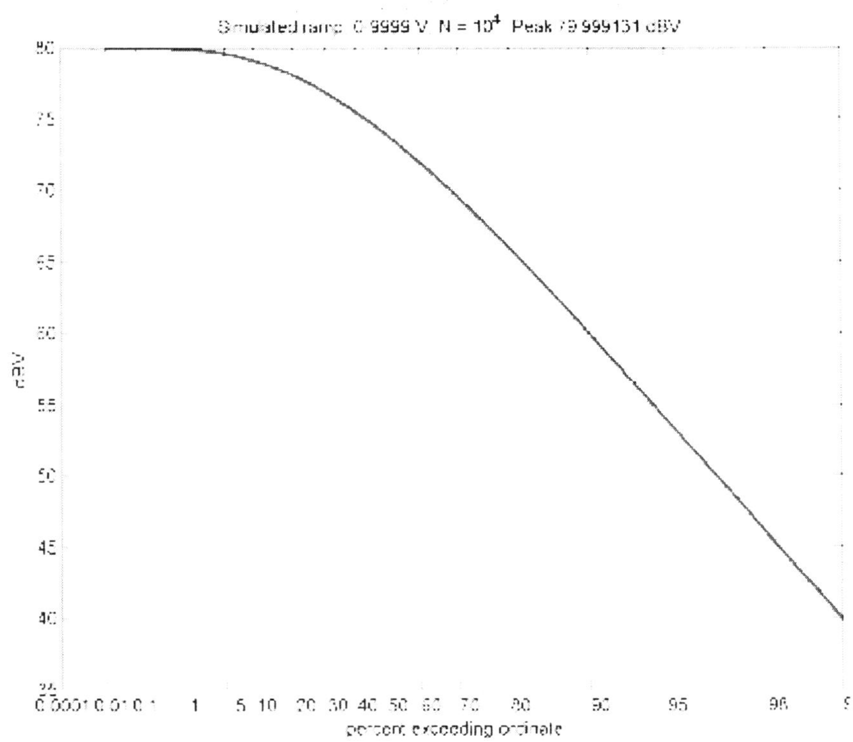

Figure D-1. APD of a ramp function increasing from 0 to 9999 V.

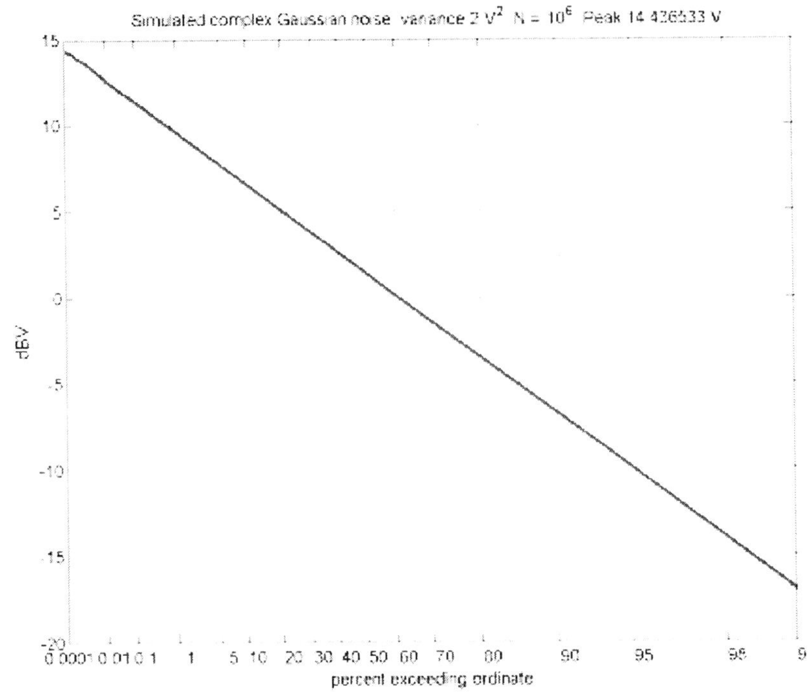

Figure D-2. *APD* of complex Gaussian noise with mean = 0 V and variance = 2 V^2.

D.5. MATLAB Code

D.5.1. testAPD.m

```
% testAPD
% demonstrates how to use apd(s) and plotapd(a,p).

% ramp
N = 10.0^4;
v = [0:N-1];
[a,p] = apd(abs(v));
plotapd(a,p);
title(sprintf('Simulated ramp, 0-9999 V, N = 10^4, Peak %f dBV',20*log10(a(N))))
ylabel('dBV')
pause

%complex Gaussian noise
N = 10.0^6;
v = randn(1,N) + i*randn(1,N);
[a,p] = apd(abs(v));
plotapd(a,p);
title(sprintf('Simulated complex Gaussian noise, variance 2 V^2, N = 10^6, Peak %f
V',20*log10(a(N))))
```

```
ylabel('dBV')
```

D.5.2. apd.m

```
function [a, p] = apd(s);
% [a, p] = apd(s) estimates the amplitude probability distribution function.
%
% input parameters:
% s = amplitude samples
%
% return variables:
% a = ordered amplitudes
% p = probability that the ordered ampitude is exceeded

if isreal(s) and min(s)>=0
a = sort(s);
N = length(a);
p = 1 - [1:N]/N;
else
disp('Input values must be amplitudes i.e. real and positive values.');
end
```

D.5.3. plotapd.m

```
function plotapd(a,p)
% plots (a,p) pairs representing the APD on Rayleigh graph.
%
% input variables:
% a = ordered amplitudes
% p = probability of ordered amplitude

% x-axis labels (labels must be same length)
xticklabel = ['0.0001';' 0.01';'0.1';'1';'5';'10';'20';'30';'40';'50';'60';'70';'80';' 90';' 95';'98';'
99'];
ptick = [0.0001 .01 0.1 1 5 10 20 30 40 50 60 70 80 90 95 98 99]/100;
porigin = ptick(1);

% replace 0 valued amplitudes with NaN to avoid logarithm error messages
if min(a) == 0
idx = find(a==0);
a(idx) = NaN;
end

% replace peak amplitude 0 probability with NaN to avoid logarithmic error messages
p(length(a)) = NaN;

% map a and p and plot
x = 10.0*log10(-log(porigin)) - 10.0*log10(-log(p));
```

```
y = 20.0*log10(a);
plot(x,y,'LineWidth',2);
grid

% customize x axis
xlabel('percent exceeding ordinate');
xtick = 10.0*log10(-log(porigin))-10.0*log10(-log(ptick));
xLim([min(xtick) max(xtick)]);
set(gca,'XTick',xtick);
set(gca,'XTickLabel',xticklabel);

return
```

Appendix E: Dynamic Range of the Vector Signal Analyzer

Statistics such as average power, amplitude probability distribution (*APD*), and power spectral density are measured with the vector signal analyzer (VSA) in this study. Accurate estimates of these statistics require the amplitudes of measured signals to lie within the dynamic range of the VSA. This appendix describes how average powers of measured signals are set to ensure accurate VSA measurements. First, a continuous wave (CW) dynamic range measurement is described. Next, average power measurements of four fundamental signals with differing peak-to-average power ratios (P/A) are described, and *APD* measurements of these signals are provided. Finally, guidelines for setting the average power of measured signals are given.

Four fundamental signals are considered in this appendix: CW, Gaussian noise, power-Rayleigh noise [20] with an exponent of 4, and a periodic pulse train with a 1-MHz pulse repetition frequency (PRF). All signals, including CW, were simulated in complex baseband (CBB) at a sample rate of 100 million samples per second (MSps), loaded into the VSG memory, upconverted to 3820 MHz, and applied to the input of the VSA. A power meter was used to verify average signal power applied to the input of the VSA. The VSA center frequency and input range were set to 3820 MHz and -30 dBm, respectively, and measurements were made in 36- and 3-MHz VSA spans.

E.1. Dynamic Range Measurement

Dynamic range was measured by applying incrementally stronger CW signals to the VSA and recording amplitude estimates. The CW signal was generated with the vector signal generator (VSG) and connected to the VSA with a coaxial cable. VSA measurements were performed at increasing power levels in 2-dB steps. The initial input level was well below the noise floor of the VSA and the highest amplitude level was above its 1-dB compression point. Power levels where VSA overload occurred were noted.

Blue curves in figures E-1 and E-2 show results from the dynamic range measurements. Average power computed from VSA samples is reported as "measured power" and average power measured by the power meter at the VSA input is reported as "input power." VSA overdrive occurred at the same input power for both spans. The noise floor in the 36-MHz span, i.e., -77 dBm, is approximately 11 dB higher than the noise floor in the 3-MHz span, i.e. -88 dBm, which corresponds to the difference in bandwidths.

For the purposes of this study, dynamic range is defined as the straight portion of the CW curves in figures E-1 and E-2. It is approximately 50 dB and 60 dB for the 36- and 3-MHz VSA spans.

E.2. Average Power Measurements of Fundamental Signals

The dynamic range measurement procedure was repeated for Gaussian noise, power-Rayleigh noise, and a periodic pulse train. Results are plotted in figures E-1 and E-2; the periodic pulse train is denoted by the acronym UPS, which stands for uniform pulse spacing. As shown in table E-1, VSA overload occurred at the lowest average power for power-Rayleigh noise, next for the periodic pulse train, and last for Gaussian noise. All occurred well below the CW overload. Interestingly, the measured power departed from the straight, top-portion of the CW curve at varying input powers. All measured powers departed from the straight, bottom-portion of the curve at the same input power, i.e., approximately -70 dBm and -80 dBm for the 36- and 3-MHz VSA spans, respectively.

Table E-1. Minimum Input Power of Signals Causing VSA Overload (dBm)

VSA Span	36 MHz	3 MHz
CW	-20	-20
Gaussian Noise	-32	-42
Power-Rayleigh	-48	-57
1-MHz Pulse Train	-32	-43

E.3. *APD* Measurements of Fundamental Signals

Next, *APD*s were calculated from the VSA measurements performed previously and plotted in figures E-3 through E-10. These plots show how the choice of average power affects *APD* measurement accuracy, which is made evident by accurate replication of *APD* features. For example, peak features were poorly replicated when strong amplitudes caused amplifier compression. Table E-2 shows the range of average powers where *APD*s were measured accurately. Each signal has a distinct P/A, defined as the power present 0.01% of the time divided by the average power. Gaussian noise P/A is approximately 10 dB, and CW P/A is 0 dB. Unlike Gaussian-noise and CW P/A, power-Rayleigh and periodic-pulse P/A are dependent on the measurement bandwidth. Table E-2 gives P/A measured in 36- and 3-MHz VSA spans.

Table E-2. Information from *APD*s Measured by the VSA

	Input Power Range where *APD*s were Measured Accurately (dBm)		P/A (dB)	
VSA Span	36 MHz	3 MHz	36 MHz	3 MHz
CW	-40/-20	-50/-20	0	0
Gaussian Noise	-63/-33	-73/-43	10	10
Power-Rayleigh	-63/-43	-84/-54	21	15
1-MHz Pulse Train	-53/-33	-64/-34	15	4

E.4. Conclusion

Results show that the average input power of signals measured by the VSA should be determined from careful consideration of VSA dynamic range and characteristics of the signal. Failure to do so will result in inaccurate average power and *APD* estimates. As a guideline for VSA signal characterization measurements, the average power of the measured signal should be set 20 - 25 dB above the bottom of the VSA dynamic range. Using this guideline, average powers of measured signals were set to -45 dBm and -55 dBm for 36- and 3-MHz VSA spans, respectively. Signals with higher P/A than those used in this appendix can be accurately measured by reducing the average power accordingly.

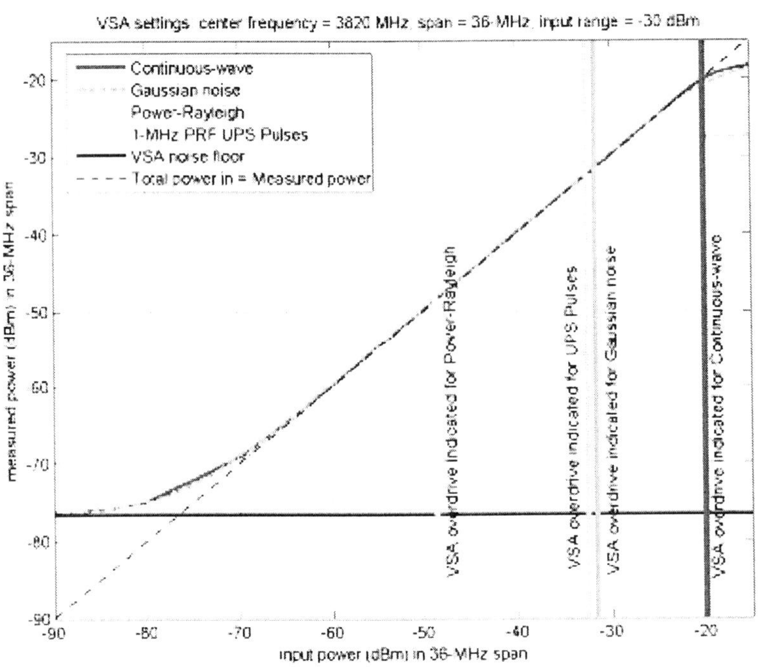

Figure E-1. Measured average power versus input average power of the VSA at span = 36 MHz.

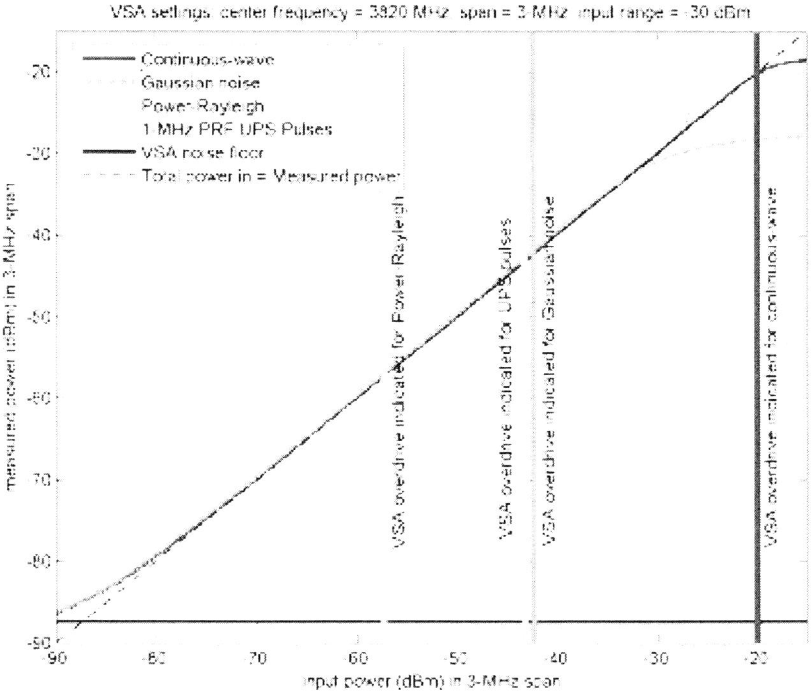

Figure E-2. Measured average power versus input average power of the VSA at span = 3 MHz.

Figure E-3. *APD*s of continuous-wave signals at decreasing power levels. Average power at VSA input is shown at right edge.

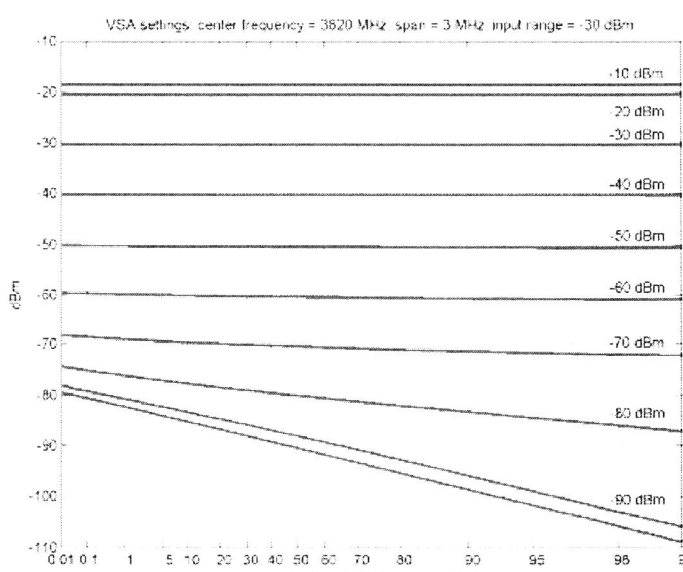

Figure E-4. *APD*s of continuous-wave signals at decreasing power levels. Average power at VSA input is shown at right edge.

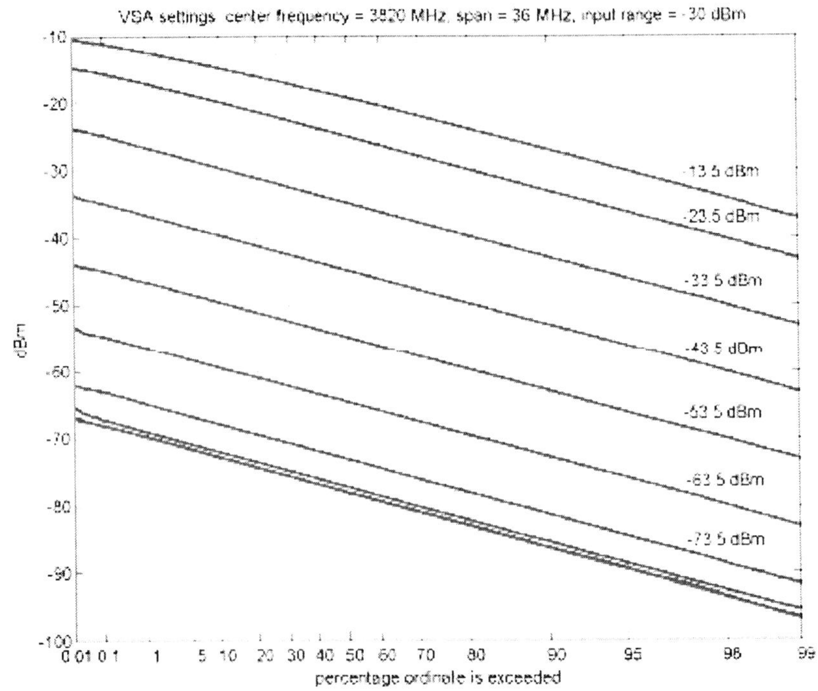

Figure E-5. *APDs* of Gaussian noise at decreasing power levels. Average power at VSA input in 36-MHz span is shown at right edge.

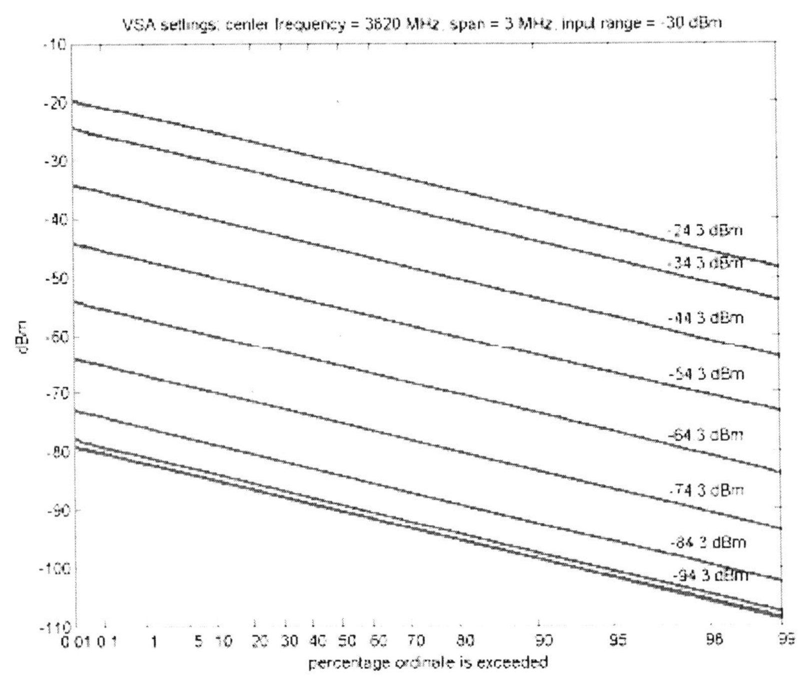

Figure E-6. *APDs* of Gaussian noise at decreasing power levels. Average power at VSA input in 3-MHz span is shown at right edge.

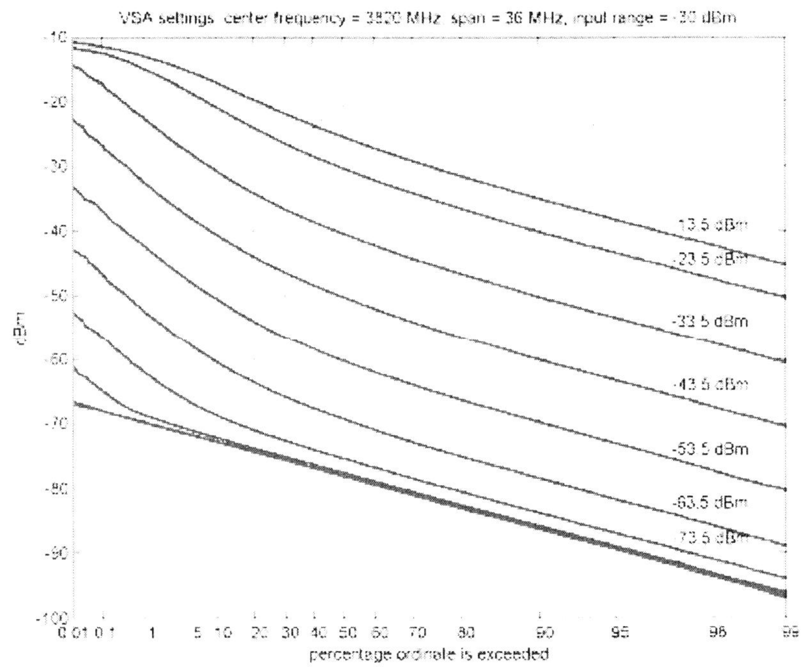

Figure E-7. *APD*s of power-Rayleigh signals at decreasing power levels. Average power at VSA input in 36-MHz span is shown at right edge.

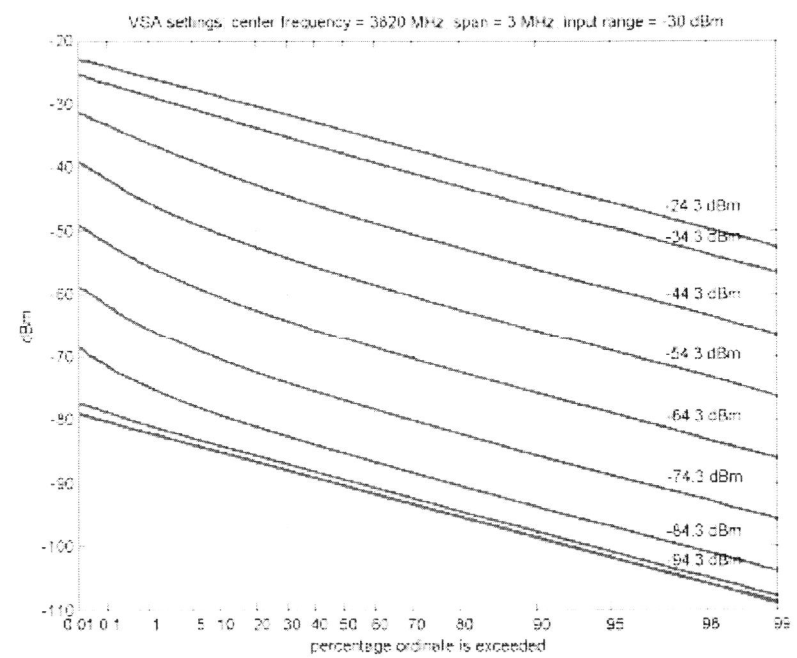

Figure E-8. *APD*s of power-Rayleigh signals at decreasing power levels. Average power at VSA input in 3-MHz span is shown at right edge.

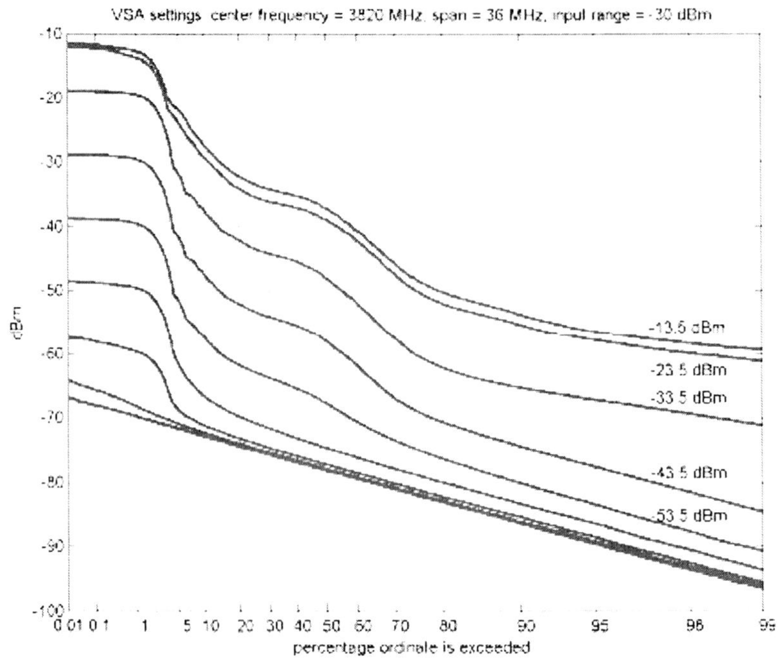

Figure E-9. *APD*s of 1-MHz PRF, uniform-pulse spaced pulses at decreasing power levels. Average power at VSA input in 36-MHz span is shown at right edge.

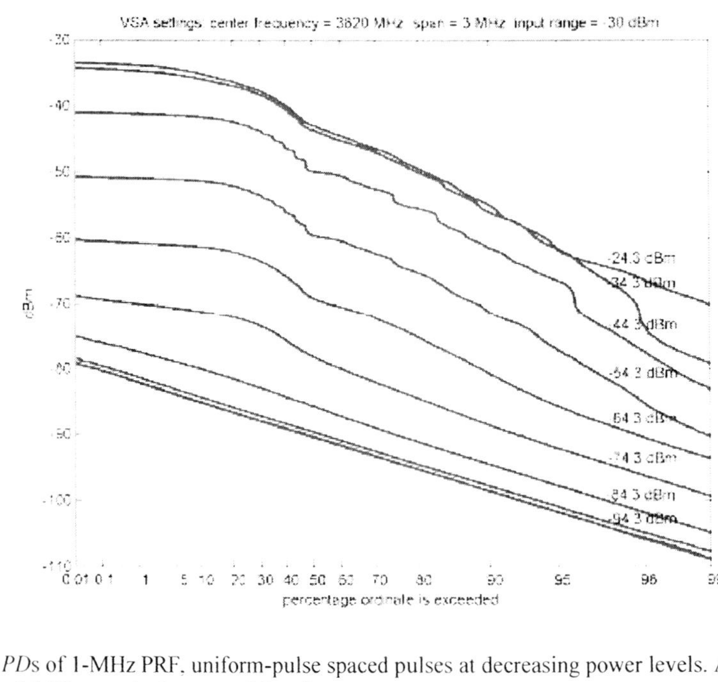

Figure E-10. *APD*s of 1-MHz PRF, uniform-pulse spaced pulses at decreasing power levels. Average power at VSA input in 3-MHz span is shown at right edge.

APPENDIX F: TEST SYSTEM CHARACTERIZATION

This appendix provides characterization measurements and analyses of various parts of the DTV interference susceptibility test system.

F.1. Characterization of the Signal Paths

The interference and satellite signal paths of the DTV interference susceptibility test system, shown in figure 20, were characterized by gain, 1-dB compression point, noise figure, and frequency response. The interference signal path is defined from the interference signal input of directional coupler DCPL-1 to the input of the vector signal analyzer (VSA), and the satellite signal path is defined from the satellite signal input of DCPL-1 to the input of the VSA.

Gain and 1-dB compression point were measured using a continuous-wave (CW) input signal at 3820 MHz. The input compression point is expressed as an average power at the input of DCPL-1. The output compression point is defined at the VSA input and expressed as an average power and corresponding peak voltage. Peak voltage, expressed as

$$V_{peak} = \sqrt{2 \times 50 \times 10^{(P_{dBm}-30)/10}} .$$

is commonly used by the VSA and can be measured with an oscilloscope at the VSA input. Noise figures of both the interference and satellite signal paths were measured using a noise diode and spectrum analyzer according to the Y-Factor Method as described in [21].

Table F-1 summarizes the results for both interference and satellite signal paths. Note that the measurements of the output 1-dB compression point for both signal paths are within 0.1 dB. Within the precision of the measurement, these values should be considered to be the same.

Table F-1. System Parameters for the Victim Receiver and Calibration Subsystems

	Noise Figure (dB)	Gain (dB)	Input 1-dB Compression Point (Average Power; dBm)	Output 1-dB Compression Point (Average Power; dBm)	Output 1-dB Compression Point (mVpeak)
Interference Signal Path	18.7	34.9	-35.3	-1.4	269.2
Satellite Signal Path	9.3	44.5	-44.8	-1.3	272.3

Magnitude of the system frequency response was measured by injecting a CW input signal that is incrementally increased in frequency in discrete steps (from 3770 MHz to 3870 MHz) into directional coupler DCPL-1 and measuring the output (at the input to the VSA) with a spectrum analyzer (SA), which tracked the CW input signal in zero-span mode. Both the CW signal generator and the SA operated under computer control to perform this measurement. Measurements of both the interference and satellite signal paths are shown in figures F-1 and F-2. Figure F-1 shows system gain over a 100-MHz bandwidth, which is dominated by the 40-MHz, 1-dB bandpass filter (BPF1 in figure 20). Note that the interference signal path has approximately 10 dB less gain than the satellite signal path due to attenuation in the directional coupler. Figure F-2 shows system gain over the 40-MHz passband of BPF1 at higher resolution. Note that the flatness over the passband for the entire signal path is approximately ±0.75 dB.

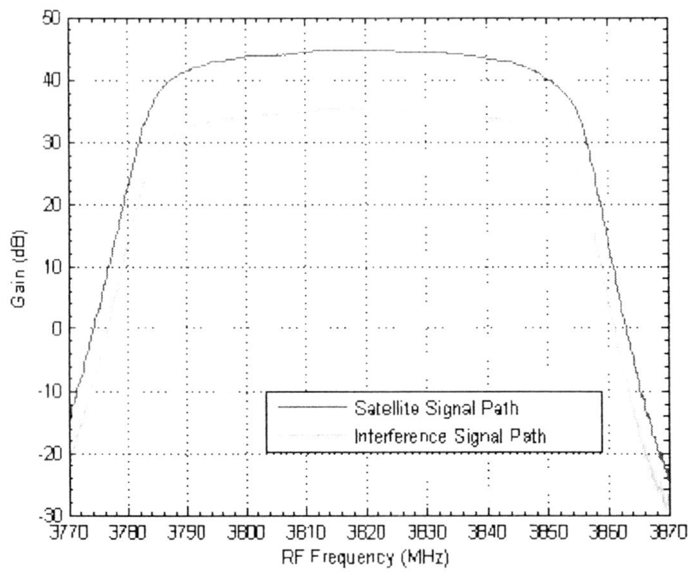

Figure F-1. Frequency response of signal paths.

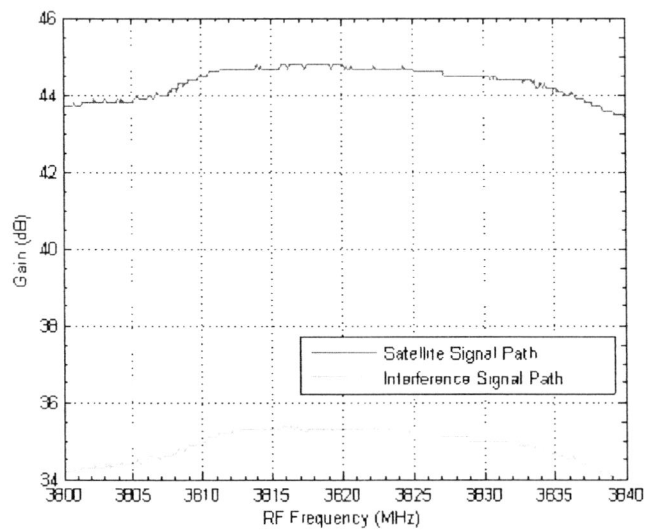

Figure F-2. Passband response of signal paths.

F.2. Dynamic Range of Interference Signal Path Cascaded with the VSA

During calibration measurements in the DTV interference susceptibility test, the VSA is used to measure interference signals at the output of the interference signal path. It is important to ensure that these measurements are performed at signal levels that do not compress the LNB or the VSA, but are high enough above the noise floor so that lower amplitudes are accurately measured.

Dynamic range measurements, similar to those described in Section E.1, are displayed in figure F-3. The input range of the VSA was selected to be -10 dBm. Dynamic range of the interference signal path cascaded with the VSA, defined by the straight portion of the CW curve, is approximately 34 dB. In the bandwidth of the VSA, peak-to-average ratios of interference signals do not exceed 25 dB. Setting the average power of these signals to 25 dB below the top of the dynamic range eliminated compression effects and provided a 9-dB margin to accurately measure low amplitudes within the signal.

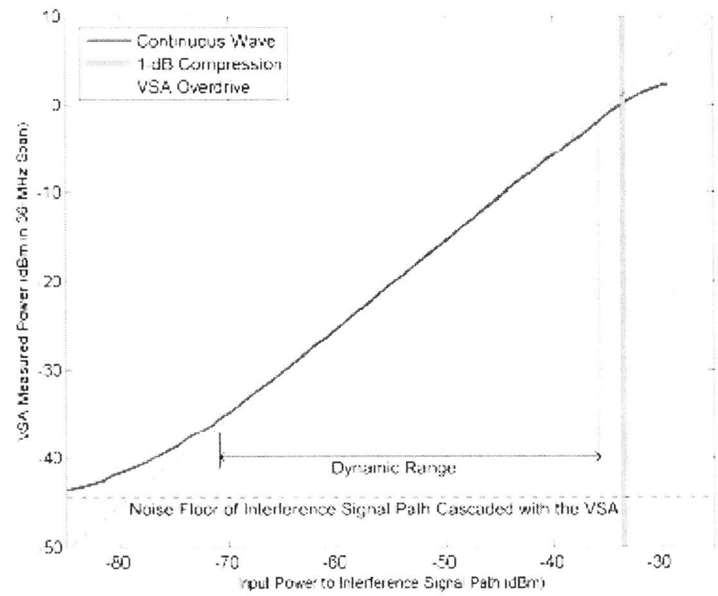

Figure F-3. Dynamic range of the interference signal path cascaded with the VSA.

F.3. Intermodulation Products in Satellite Signal Generator

As a part of the satellite signal simulator shown in figure 17, the upconverter frequency shifts the satellite signal from the modulator intermediate frequency of 70 MHz to the satellite channel center frequency of 3820 MHz. Shifting is done in two "mixer" stages. In the first stage, the 1260-MHz local oscillator (LO) frequency produces lower and upper sideband signals centered at 1190 and 1330 MHz. Bandpass filter BPF3 passes the upper sideband signal to the second stage whose 5150-MHz LO frequency produces lower and upper sidebands centered at 3820 and 6480 MHz. Bandpass filter BPF1 in the coupling subsystem (see figure 20) passes the lower sideband to the LNB. Practical mixer components, represented by mixer-1 and mixer-2 in figure F-4, create higher- order intermodulation (IM) products along with the desired frequency conversion products. Figures F-5 and F-6 show the theoretical frequency and corresponding order of these products for the two mixing stages. The 70-MHz intermediate frequency (IF) signal at the input of mixer-1 is depicted by the only first-order, black-horizontal line in figure F-5. It is 36-MHz wide, which is large compared to the 19.51-MHz wide DTV signal actually being used for testing. Mixing the IF signal with a 1260-MHz LO (shown as a red circle) produces multiple mixing products depicted by the black-horizontal lines at IM orders greater than one. The desired frequency conversion product of the first stage is the upper-sideband signal centered at 1330 MHz, which lies in the passband of bandpass filter BPF3 and is the IF for the second mixing stage (illustrated by the only first-order, black-horizontal line in figure F-6). High-order mixing products of the second mixing stage are attenuated by BPF1 (shown in figure 20). Notice that for both mixing stages there are no mixing products below tenth order that lie within the 36-MHz passband of the satellite signal path (bounded by the blue-vertical-dashed lines) – i.e. no higher-order mixing products overlap the desired frequency conversion products. Therefore, the simulated satellite signal centered at 3820 MHz is uncorrupted by higher-order IM products.

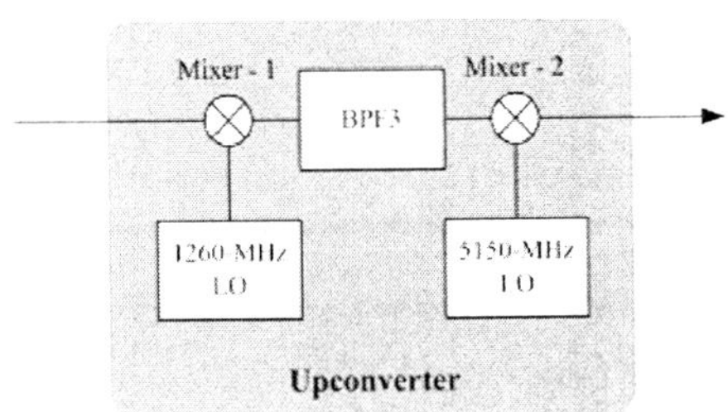

Figure F-4. Upconverter block diagram. Hardware is given in Appendix G.

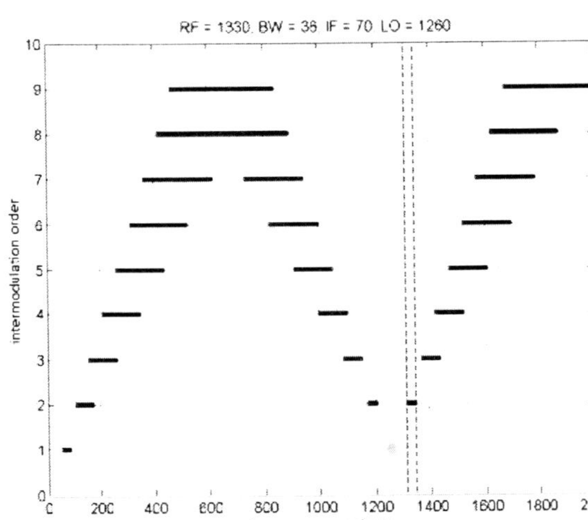

Figure F-5. IM products due to the first mixing stage of the satellite signal generator.

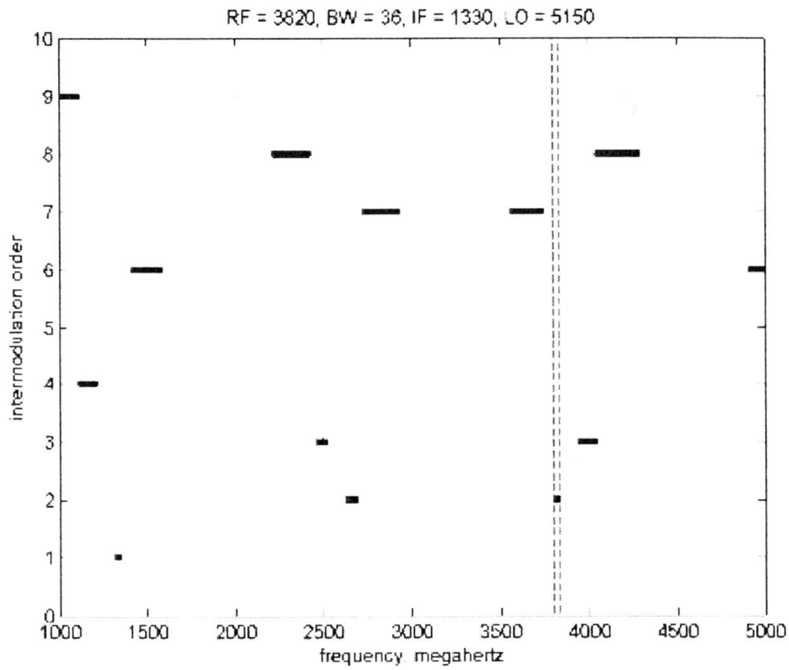

Figure F-6. IM products due to the second mixing stage of the satellite signal generator.

Appendix G: Hardware Specification

G.1. Interference Generation Subsystem

Amp (figure C-2) = Amplifier
Manufacturer: Unknown
Model: 3742-3-35
Serial: 1108
Gain: 38 dB
Frequency Range: 3.7 – 4.2 GHz

AWG (figure C-2) = Arbitrary Waveform Generator
Manufacturer: Agilent
Model: 33250A

BPF4 (figure C-2) = Bandpass Filter
Manufacturer: Microwave Filter Co.
Model: 7892-SMA-A Frequency
Range: 3.7 – 4.2 GHz
DCPL - 2 (figure C-2) = Directional Coupler
Manufacturer: Norsal Ind.
Model: 4871-10 Frequency

Range: 0.2 – 18 GHz

Gating Switch (figure C-2)
Manufacturer: Mini-Circuits
Model: ZASWA-2-50DR Frequency
Range: dc – 5 GHz
Switching Time: 5 ns (typical), 15 ns (maximum)

Noise Diode (figure C-2)
Manufacturer: Noise/Com, Inc.
Model: NC3208A
Serial: E804
Frequency Range: 0.1-18 GHz
ENR = 27 dB

Signal Generator (figure C-2)
Manufacturer: Agilent
Model: E8267C option 520 (VSG)
Frequency Range: 250 kHz – 20 GHz
Resolution: 0.001 Hz CW, 0.001 Hz all sweep modes
Switching Speed: < 12 ms typical

Variable Attenuator (figure C-2)
Manufacturer: Hewlett Packard
Model: HP 8494H and HP 8496B
Frequency Range: 0 – 18 GHz
Attenuation Range: 0 – 11 dB at 1-dB step and 0 – 110 dB at 10-dB step

Variable Attenuator (figure C-2)
Manufacturer: Narda Microwave
Model: 791 FM
Frequency Range: 2 – 12.4 GHz
Attenuation Range: 0 – 40 dB

VSG (figures 13, 20, and B-1) = Vector Signal Generator
Manufacturer: Agilent
Model: E8267C option 520
Frequency Range: 250 kHz – 20 GHz
Resolution: 0.001 Hz CW, 0.001 Hz all sweep modes
Switching Speed: < 12 ms typical

G.2. Calibration Subsystem/Test and Measurement Equipment

BPF2 (figure 20) = Bandpass Filter (tunable)
Manufacturer: KandL Microwave

Model: 5BT-750/1500-5N/N
Serial: 82-9839
Frequency Range: 750 - 1500 MHz
Percentage Bandwidth: 5%

Oscilloscope (figure 16) – Digital Phosphor Oscilloscope
Manufacturer: Tektronix
Model: TDS7704B
Hardware Analog Bandwidth: 7 GHz
Rise Time: 62 ps 10% to 90% (typical)
Memory: 1 MByte (standard)

Power Divider (figure 20)
Manufacturer: Midwest Microwave
Model: PWD-5520-03-SMA-79
Frequency Range: 0.5 – 2.0 GHz
Isolation: 15 dB typical
VSWR: 1.5:1

Power Meter (figure 20)
Manufacturer: Agilent
Model: E4419B
Frequency Range: 9 kHz – 110 GHz, sensor dependent
Power Range: -70 to +44 dBm, sensor dependent

Power Meter Sensor (figure 20)
Manufacturer: Agilent
Model: E9300A
Frequency Range: 10 MHz – 18 GHz
Power Range: -60 to +20 dBm

Spectrum Analyzer (figures 15 and 20)
Manufacturer: Agilent
Model: E4440A
Frequency Range: DC – 26.5 MHz
Maximum Resolution Bandwidth: 8 MHz

Vector Signal Analyzer (figures 14 and 20)
Manufacturer: Agilent
Model: 89641A
Frequency Range: DC – 6 GHz
IF Sample Rate: 95 MS/s

G.3. Satellite Simulation Subsystem

1260-MHz LO (figure F-4) = Signal Generator
Manufacturer: Hewlett-Packard
Model: HP83413
Frequency Range: 10 MHz – 20 GHz

5150-MHz LO (figure F-4) = Signal Generator
Manufacturer: Hewlett-Packard
Model: HP 8672A
Frequency Range: 2 GHz – 18 MHz

A1 (figure 20) – Attenuator

Attenuator/switch driver (figure 20)
Manufacturer: Agilent
Model: 11713A

BPF3 (figure F-4) = Bandpass Filter (tunable)
Manufacturer: KandL Microwave
Model: 5BT-750/1500-5N
Serial: 8913-1
Frequency Range: 750 – 1500 MHz
Percentage Bandwidth: 5%

Digital Video Stream (figure 19) – PCI-Bus Board for Uncompressed Audio/Video Recording
Manufacturer: DVS
Model: ClipStationPRO

Mixer – 1 (figure F-4) = Frequency Mixer
Manufacturer: Mini-Circuits
Model: ZFM-2000 LO/RF
Frequency Range: 100 – 2000 MHz IF
Frequency Range: DC – 600 MHz
Maximum Conversion Loss: 9.5 dB
Minimum LO/RF Isolation: 20 dB

Mixer – 2 (figure F-4) = Frequency Mixer
Manufacturer: Mini-Circuits
Model: ZX05-C60 LO/RF
Frequency Range: 1600 – 6000 MHz IF
Frequency Range: DC – 2000 MHz
Maximum Conversion Loss: 8.3 dB
Minimum LO/RF Isolation: 23 dB (typical), 17 dB (minimum)

Modulator (figure 19)
Manufacturer: ComTech EF Data
Model: SDM-2020

MPEG-2 Encoder (figure 19)
Manufacturer: Motorola
Model: SE-1000
Input: analog or digital
Output: DVB-ASI, DHEI, or DS3

Switch – 1 (figure 20)
Manufacturer: Dow-Key Microwave
Model: 401-2308
Frequency Range: 0 – 26.5 GHz RF
Circuit: SPDT

VA (figure 20) = Step Variable Attenuator
Manufacturer: Hewlett Packard
Model: HP 8494H and HP 8496B
Frequency Range: 0 – 18 GHz
Attenuation Range: 0 – 11 dB at 1-dB step and 0 – 110 dB at 10-dB step

G.4. Coupling Subsystem

BPF1 (figure 20) = Bandpass Filter
Manufacturer: TTE
Model: 310-3820M-40M-A
Center Frequency: 3820 MHz
1-dB Bandwidth: 40 MHz
50-dB Bandwidth: 110 MHz
Insertion Loss: 4 dB
VSWR: 1.5:1

DCPL – 1 (figure 20) = Directional Coupler
Manufacturer: Narda.
Model: 4202B-10
Frequency Range: 1 – 12.4 GHz
Coupling: 10 dB
Insertion Loss: 1.3 dB (maximum)
Directivity: 15 dB for 1 – 8 GHz, 12 dB for 8 – 12 GHz
VSWR: 1.35

LPF (figure 20) = Low-pass Filter
Manufacturer: TTE
Model: LMC11-7G-A

3-dB Maximum Frequency: 7 GHz
50-dB Minimum Frequency: 13 – 20 GHz
Insertion Loss: 0.2 dB
VSWR: 1.5:1

G.5. Receiver Subsystem

Bias Tee (figure 20)
Manufacturer: Orbital
Model: BT1–V2-F-S-B-B
Frequency Range: 900 – 2100 MHz
Insertion Loss: 0.5 dB
Passband Ripple: 0.3 dB (maximum)
VSWR: 1.3:1
LNB Interface: 75Ω F-type
Receiver Interface: 50Ω SMA

LNB (figure 20) – LNA plus Block Downconverter
Manufacturer: Norsat International, Inc.
Model: 3120 C-band PLL LNB
Input Frequency Range: 3.4 – 4.2 GHz
LO Freq: 5.15 GHz
Stability: +/- 5 kHz
Noise Temperature = 20 K
Gain: 60 dB

MPEG-2 Stream Monitor (figure 20)
Manufacturer: Tektronix
Model: MTM400

Satellite Receiver (figure 20)
Manufacturer: Motorola/General Instrument
Model: DSR4500 NTSC

Switch - 2 (figure 20)
Manufacturer: Dow-Key Microwave
Model: 401-2308
Frequency Range: 0 – 26.5 GHz
RF Circuit: SPDT

ZM (figure 20) = 50 /75 Impedance Matching Pad
Model: 9086-50/75
Power: 1 W
Frequency: 3 GHz
Maximum VSWR: 1.25:1

Insertion Loss: 5.7 dB

APPENDIX H: GLOSSARY

ADC	analog-to-digital converter
ALC	automatic level control
APD	amplitude probability distribution
AWG	arbitrary waveform generator
BER	bit error rate (italicized *BER* denotes measured pre-Viterbi bit error rate)
BPF	band-pass filter
BPSK	binary phase-shift keying
CBB	complex baseband
CRADA	cooperative research and development agreement
CS	carrier suppression
CW	continuous wave
DAC	digital-to-analog converter
dB	decibel
dBm	decibel referenced to 1 milliwatt
dBV	decibel referenced to 1 volt
dBW	decibel referenced to 1 watt
DCPL	directional coupler
DO	digital oscilloscope
DP-UWB	dithered-pulse ultrawideband
DS-UWB	direct-sequence ultrawideband
DTV	digital television
DVB	digital video broadcast
ETSI	European Telecommunications Standards Institute
EVM	error vector magnitude
FEC	forward error correction
FCC	Federal Communications Commission
FIR	finite impulse response
FTP	file transfer protocol
GHz	gigahertz = 10^9 cycles per second
GN	Gaussian noise
GSps	10^9 samples per second
I	in-phase
IF	intermediate frequency
IEEE	Institute of Electrical and Electronics Engineers, Inc.
IM	intermodulation
IRD	integrated receiver decoder
ITS	Institute for Telecommunication Sciences
ITU	International Telecommunication Union
LNB	low-noise block downconverter
ln	natural logarithm
LO	local oscillator
log	base-10 logarithm
LPF	low-pass filter
LSB	least significant byte

Appendix H. (Continued).

MAC	medium access control
MB-OFDM	multi-bandorthogonalfrequency-domain multiplex
MBOA-SIG	MB-OFDM Alliance Special Interest Group
Mbps	million bits per second
MER	modulation error rate
MHz	megahertz = million cycles per second
MPEG	Motion Picture Experts Group
MSB	most significant byte
MSps	million samples per second
ns	nanosecond = 10^{-9} seconds
NTIA	National Telecommunications and Information Administration
P/A	peak-to-average ratio
PHY	physical layer
PLCP	physical layer convergence procedure/protocol
PMD	physical medium dependent
PRF	pulse repetition frequency
PSD	power spectral density
Q	quadrature phase
QAM	quadrature amplitude modulating
QPSK	quadrature phase-shift keying
RBW	resolution bandwidth
RF	radio frequency
Rms	root-mean-square
RRC	root-raised cosine
RS	Reed-Solomon
SA	spectrum analyzer
SCPI	Standard Commands for Programmable Instrumentation
SER	segment error rate
SMPTE	Society of Motion Picture and Television Engineers
SNR	signal-to-noise ratio
TEF	transport error flag
UPS	uniform pulse spacing
UTC	coordinated universal time
UWB	ultrawideband
V	volt
VA	variable attenuator
VBW	video bandwidth
VSA	vector signal analyzer
VSG	vector signal generator
WLAN	wireless local area network
WPAN	wireless personal area network

In: Interference Potential of Ultrawideband Signals
Editors: M. Cotton, R. Achatz, pp. 69-179

ISBN: 1-60021-359-6

PART 2: MEASUREMENTS OF GATED-NOISE INTERFERENCE TO C-BAND SATELLITE DIGITAL TELEVISION RECEIVERS*

ABSTRACT

This chapter demonstrates that digital television (DTV) susceptibility to gated-noise interference cannot be predicted by interference power characteristics alone. It was found that DTV susceptibility is also dependent on temporal characteristics of the interfering signal and the bandwidth of the DTV receiver. A test system was developed to inject interference with known characteristics into a victim receiver and quantitatively measure susceptibility. In this experiment, a C-band satellite DTV victim receiver was exposed to gated-noise interference, whose temporal characteristics are defined by gating parameters such as on-time, fractional on-time, and off-time. The specific gating parameters considered in this chapter include on-times of 0.01, 0.10, 1.00, and 10.00 μs and fractional on-times of 1.00, 0.50, 0.25, 0.125, and 0.0625. Results showed that DTV susceptibility was strictly dependent on average power of the interfering signal only when off-times were less than the reciprocal bandwidth of the victim receiver. For longer off-times, however, susceptibility was dependent on the temporal characteristics of the interfering signal. Moreover, high correlation was observed between susceptibility and forward error correction performance of the receiver.

Key words: digital television; interference; satellite communications; ultrawideband.

1. INTRODUCTION

In April 2002, the Federal Communications Commission (FCC) released *FCC 02-48* [1] legalizing intentional, low-power ultrawideband (UWB) emissions between 3.1 GHz and 10.6 GHz for communications devices operated indoors. UWB emissions were limited to -41 dBm average power in 1 -MHz bandwidth and 0 dBm peak power in 50-MHz bandwidth, where average power is measured over a 1 -millisecond integration time and peak power measurement duration is unspecified. The rules define a UWB device as one that emits signals with 10-dB bandwidth greater than 500 MHz or greater than 20% of the center frequency.

---

* Excerpted from Report TR-05-429, dated August 2005.

The FCC rules do not specify how the bandwidth requirement is achieved, consequently allowing industry considerable breadth in choosing a modulation. This breadth is exemplified by the development of Direct-Sequence Ultrawideband (DS-UWB) and Multi-band Orthogonal Frequency-Domain Multiplexing (MB-OFDM) ultrawideband technologies. Proponents of DS-UWB and MB-OFDM both seek standardization from IEEE (Institute of Electrical and Electronics Engineers) 802.15 working group 3a on high-rate (greater than 20 million bits per second) Wireless Personal Area Networks (WPAN). As the name implies, WPAN is intended for short-distance (nominally less than 10 meters) wireless networking of devices such as PCs, personal digital assistants, and mobile phones.

Both DS-UWB and MB-OFDM transmitters are based on state-of-the-art integrated circuitry. DS-UWB modulation controls pulse polarity and hence supports phase shifting modulations. It achieves its ultra-wide bandwidth by transmitting sufficiently narrow pulses. MB-OFDM simultaneously modulates 122 carriers spaced 4.125 MHz apart to achieve its ultra-wide bandwidth and frequency hops the modulated carriers between non-overlapping bands.

Since previous work performed at the Institute for Telecommunication Sciences (ITS) [2 – 5] did not specifically look at susceptibility of receivers to interference from DS-UWB or MB-OFDM signals, and since there is little published information on this subject, ITS entered into a Cooperative Research and Development Agreement (CRADA) with the Freescale subsidiary of Motorola, Inc. to study how susceptibility could be quantified in terms of interference signal characteristics.

1.1. Experiment

Interference potential is a general concept where performance degradation of a victim receiver is predicted from interference signal characteristics. Interference potential is derived from numerous susceptibility tests on receivers with a variety of bandwidths and signal demodulation techniques. This chapter describes one such test where C-band satellite digital television (DTV) is exposed to gated-noise interference. Development of the test system and procedure was described in Part 1 of this book [6].

In this chapter, only continuous- and gated-noise signals are considered as interferers. While it is unlikely that these signals will be used in communications, their similarities to DS-UWB and MB-OFDM, respectively, are unmistakable and their analytic tractability is profoundly useful. More specifically, these signals emulate the noisy nature of DS-UWB and MB-OFDM as well as the gating characteristics of MB-OFDM caused by frequency hopping in and out of a victim receiver operational band. This emulation is accomplished without the mathematical complexities associated with actual UWB modulations. Comparisons between continuous noise and DS-UWB, and gated noise and MB-OFDM, will be discussed in Part 3 of this book.

C-band satellite television was chosen as the victim receiver for the susceptibility tests because it demodulates signals transmitted in the 3.7 to 4.2 GHz frequency range, which lies within the band allocated for UWB operation. It also uses a variety of bandwidths and signal demodulation techniques, which makes it an ideal victim receiver for the study of interference potential. Additionally, instruments capable of providing quantitative signal quality data from various receiver subsystems are readily available.

1.2. Organization of Book

The main body of this book presents DTV susceptibility data and characterization measurements for gated-noise interference with the following gating parameters: on-time (τ_{on}) = {10, 100, 1000, 10000} ns and fractional on time (ζ) = {1.0, 0.5, 0.25, 0.125, 0.0625}. These signals are denoted by the acronym GN.

Section 2 provides DTV susceptibility test results plotted two different ways. First, post-Reed-Solomon segment error rate *(SER)* and pre-Viterbi bit error rate *(BER)* are plotted as a function of interference-to-noise ratio *(INR)* to demonstrate how susceptibility depends on average power of the interfering signals. Second, *INR* and *BER* at the threshold of visibility (TOV), i.e., INR_{TOV} and BER_{TOV}, are plotted as a function of ζ to demonstrate how DTV susceptibility and forward error correction (FEC) performance depend on the gating parameters of the interfering signals.

Section 3 provides temporal and amplitude analyses of measured GN signals to demonstrate the effects of band-limiting. The temporal analysis is based on crossing statistics. Burst duration *(BD)* and burst interval *(BI)* estimates of band-limited GN signals are compared to on- and off-times of the corresponding gating functions. The amplitude analysis is based on amplitude probability distributions (*APD*s). Peak-to-average ratio *(P/A)* estimates of the band-limited GN signals are compared to corresponding ultra-wide-bandwidth *P/A* limits. Also provided in Section 3 is a spectral analysis based on power spectral densities (*PSD*s) of measured GN signals.

Section 4 discusses how DTV susceptibility depends on the characteristics of interfering signals. It summarizes the important points of the test, describes findings, and evaluates the significance and scope of the findings.

Appendices to this chapter contain information supporting the main body. Appendix A provides DTV susceptibility test results and signal characterization measurements for gated-noise signals that emulate MB-OFDM. Appendix B gives an overview of the test system, provides precise measurement definitions for signal-to-noise ratio *(SNR)*, *INR*, *SER*, and *BER*, and provides some degree of validation via comparison of theoretical and measured DTV signal quality metrics for Gaussian noise degradation. Appendix C develops theoretical expressions for the amplitude and spectral properties of gated Gaussian noise in an ultra-wide transmission bandwidth. Finally, Appendix D provides *APD*s of the gated-noise signals in a variety of bandwidths measured at the satellite radio frequency (RF) as well as *APD*s of the signals at the first intermediate frequency (IF) in the presence of low-noise block downconverter (LNB) noise for interference powers corresponding to INR_{TOV}.

2. DTV Susceptibility to GN Interference

Results of the DTV susceptibility tests for GN interference are described in this section. It presents gating parameters, discusses relevant signal powers and DTV signal quality metrics, and provides SER and BER versus INR plots and INR_{TOV} and BER_{TOV} versus ζ plots.

2.1. Gating Parameters

Table 1 gives on-time, fractional on-time, and off-time (τ_{of}) gating parameters of the periodically gated-noise interference signals considered in the main body of this chapter. The set of gated-noise signals in table 1 is denoted by the acronym GN, and each individual signal is specified by identifiers in the first column. An illustration of the gating parameters is given in figure 1. Appendix A extends the scope of this experiment to GN(MB) interference signals, which are gated-noise signals that emulate MB-OFDM.

Table 1. GN Gating Parameters

GN	τ_{on} (μs)	τ_{of} (μs)	ζ
01	~	0.00	1.0000
02	0.01	0.01	0.5000
03	0.01	0.03	0.2500
04	0.01	0.07	0.1250
05	0.01	0.15	0.0625
06	0.10	0.10	0.5000
07	0.10	0.30	0.2500
08	0.10	0.70	0.1250
09	0.10	1.50	0.0625
10	1.00	1.00	0.5000
11	1.00	3.00	0.2500
12	1.00	7.00	0.1250
13	1.00	15.00	0.0625
14	10.00	10.00	0.5000
15	10.00	30.00	0.2500
16	10.00	70.00	0.1250
17	10.00	150.00	0.0625

2.2. Signal Powers and DTV Signal Quality Metrics

Figure 2 illustrates reference points for the signal powers and DTV signal quality metrics acquired during the DTV susceptibility tests. *SNR* and *INR* define the average power of the DTV and interfering signals. These average power ratios were derived from vector signal analyzer data measured at IF and band-limited to B_{DTV} = 19.51 MHz in post-measurement processing by the same root-raised-cosine (RRC) filter found in the demodulator of the victim receiver. Figure B-1 provides a block diagram of the test hardware configuration and Section B.2 mathematically describes the power ratio measurements. DTV signal quality metrics *SER* and *BER* were acquired with the MPEG-2 transport stream monitor.[1] Section B.3 provides mathematical definitions of the DTV signal quality metrics.

[1] Modulation error ratio (MER) was also acquired with the MPEG-2 stream monitor. MER is an important DTV signal quality metric referenced to the output of the demodulator. Measured MER, however, was inconsistent with theory which prevented its use in this report.

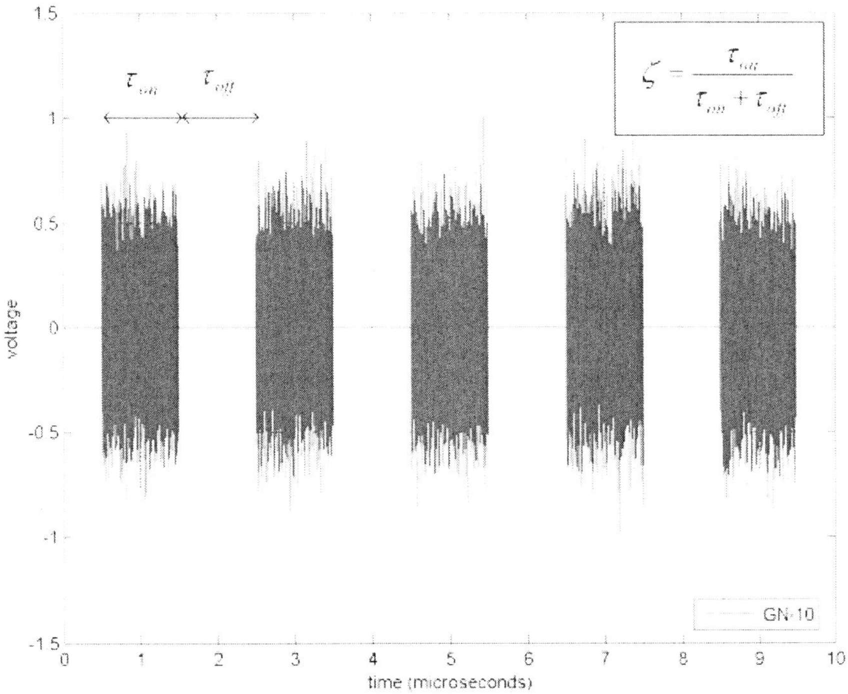

Figure 1. Simulated gated-noise signal, GN-10.

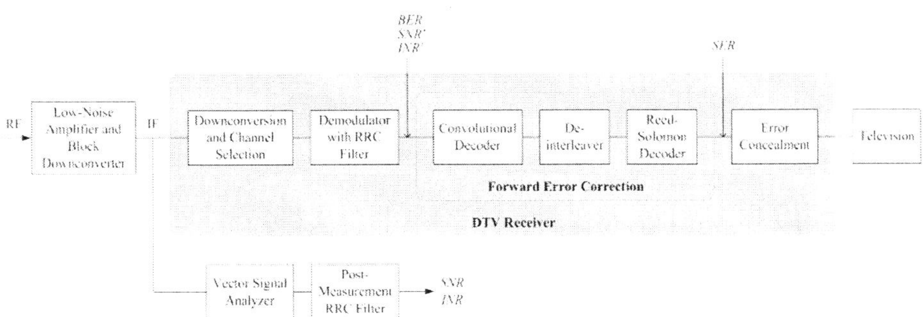

Figure 2. *SNR, SNR', INR, INR', BER,* and *SER* reference points.

It is important to note that *SNR* and *INR* were not measured at the same point as *BER*. Theoretical expressions for DTV signal quality metrics for Gaussian noise degradation are provided in Section B.4; these expressions use signal-to-noise ratio at the output of the demodulator *(SNR')* as the dependent variable. Equation (B-1) was derived empirically to account for imperfect demodulation effects and compute *SNR'* from *SNR*. Using this model, *SNR'* = {8.2, 10.5, 12.5} dB was calculated from *SNR* = {9, 12, 15} dB. Note that this model was derived under Gaussian noise assumptions and cannot be generalized for gated-noise interference to obtain *INR'* from *INR* without further research. Therefore, test results are displayed as a function of the measurable dependent variables *SNR* and *INR*.

2.3. DTV Signal Quality as a Function of Interference Average Power

Figures 3 – 8 provide composite plots of measured *SER* and *BER* as a function of *INR* for *SNR* = {9, 12, 15} dB. Each page is dedicated to a single *SNR*, each plot is dedicated to a specific τon, and each curve represents a single ζ. Figures 3, 5, and 7 provide composite plots of *SER* while figures 4, 6, and 8 provide composite plots of *BER*. The following are general comments regarding the shift, separation, and slope of these curves.

Both *SER* and *BER* shifted toward greater *INR* with increasing *SNR*. This occurred because more interference was needed to degrade stronger satellite signals.

SER curves with fixed *SNR* and τon show increased separation (from the continuous-noise case) with decreasing ζ. Likewise, all curves with fixed *SNR* and ζ exhibit increased separation with increasing τ_{on}. Hence, *SER* curves with τ_{on} = 10 ns and $\zeta \sim 0.25$ lay close to the continuous-noise case, while those with τon = 10,000 ns and $\zeta = 0.0625$ had the greatest separation. A notable anomaly to this trend occurred for GN-09 (τ_{on} = 100 ns, $\zeta = 0.0625$), where the shallow slope of the *SER* curve caused it to cross higher-ζ *SER* curves.

SER curves were generally steeper than corresponding *BER* curves, which indicates that *SER* was more sensitive than *BER* to changes in average interference power. *SER* steepness was due to two stages of forward error correction (FEC), which were applied to post-Reed-Solomon segments but not to pre-Viterbi bits. Another trend was observed for sufficiently small ζ, *i.e.*, $\zeta = 0.0625$, where slopes of the *SER* curves were flatter compared to higher-ζ cases. This may be attributed to conditions more favorable to FEC performance or to longer off-times where errors were unlikely. In either case, DTV susceptibility to small-ζ gated-noise interference seemed to be more dependent on the temporal characteristics than on the power characteristics of the interference signal.

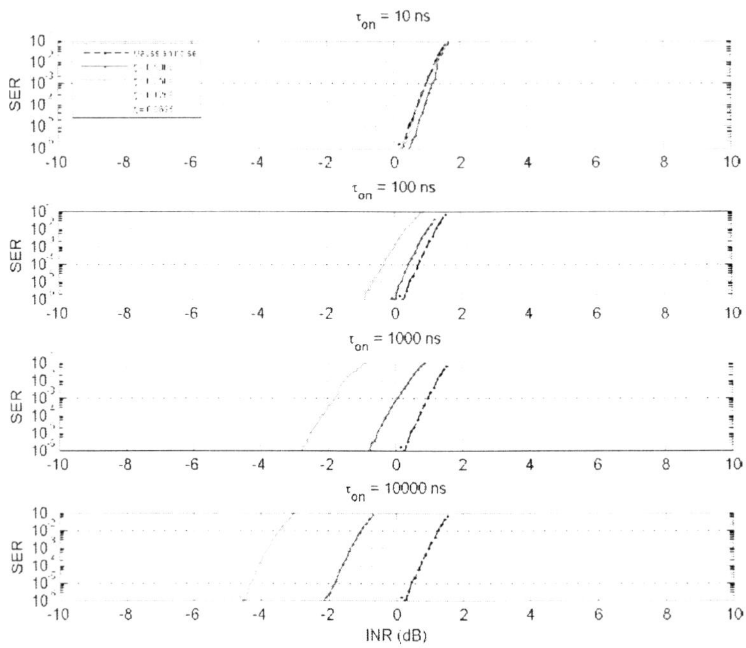

Figure 3. SER versus INR for a DTV receiver operating at SNR = 9 dB and exposed to GN interference.

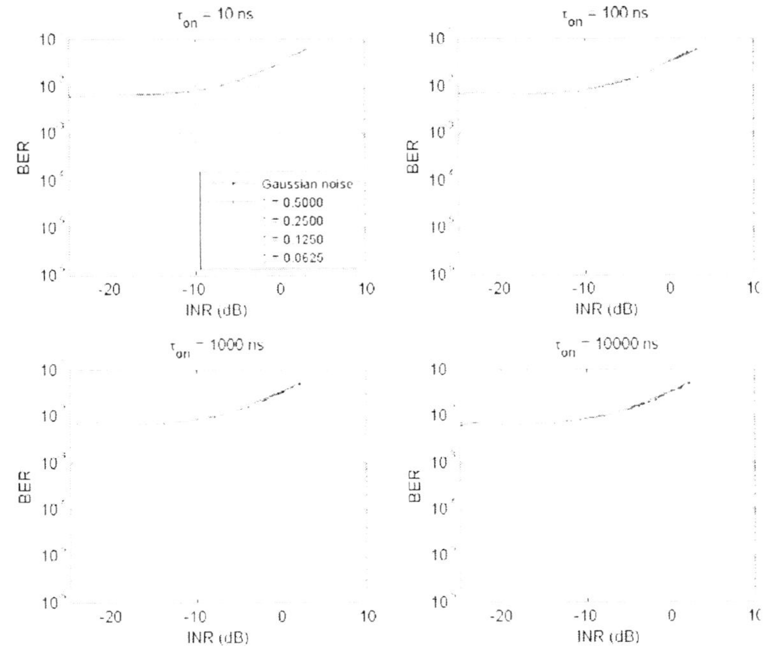

Figure 4. BER versus INR for a DTV receiver operating at SNR = 9 dB and exposed to GN interference.

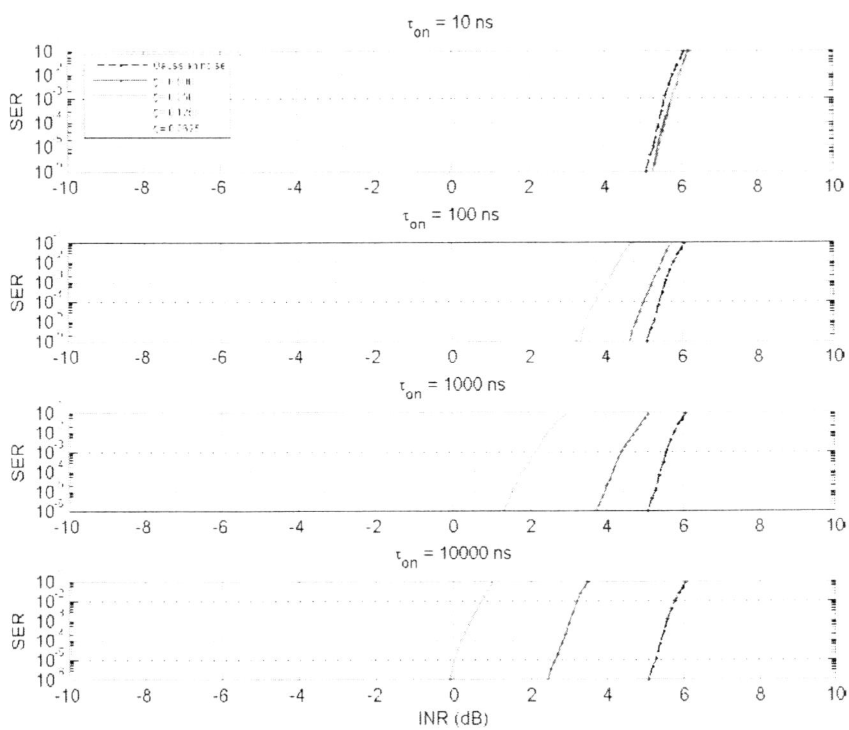

Figure 5. SER versus INR for a DTV receiver operating at SNR = 12 dB and exposed to GN interference.

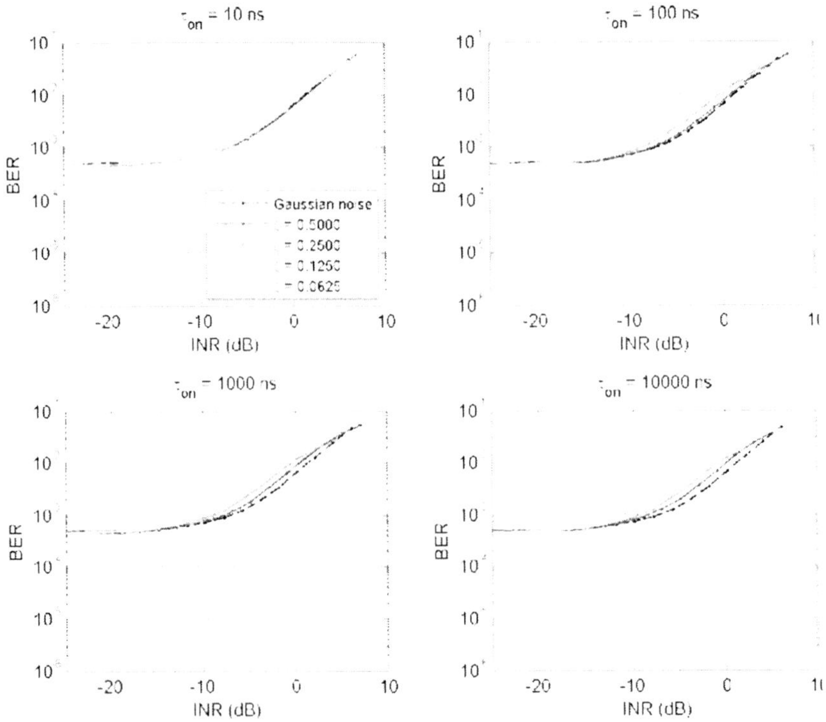

Figure 6. BER versus INR for a DTV receiver operating at SNR = 12 dB and exposed to GN interference.

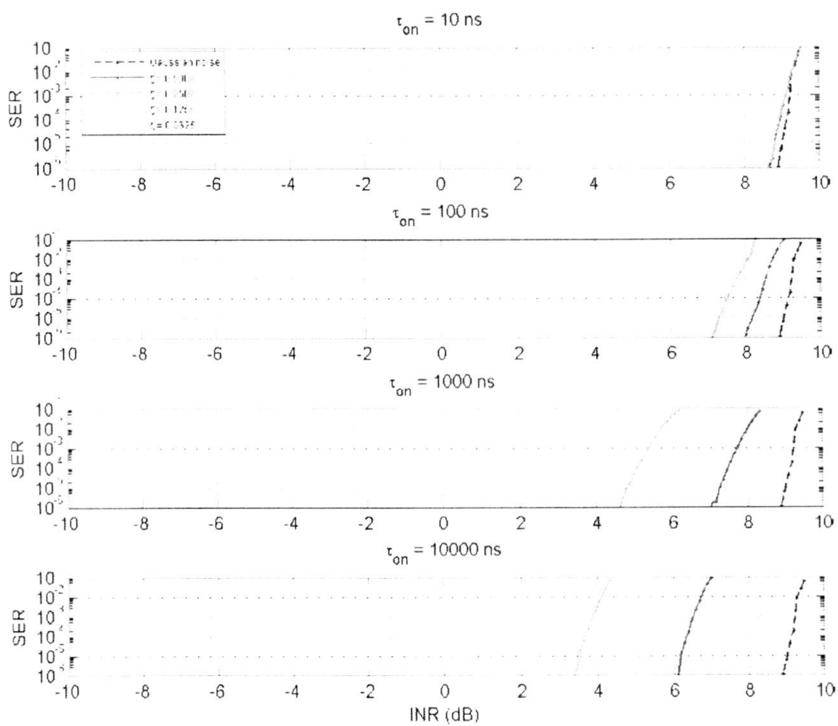

Figure 7. SER versus INR for a DTV receiver operating at SNR = 15 dB and exposed to GN interference.

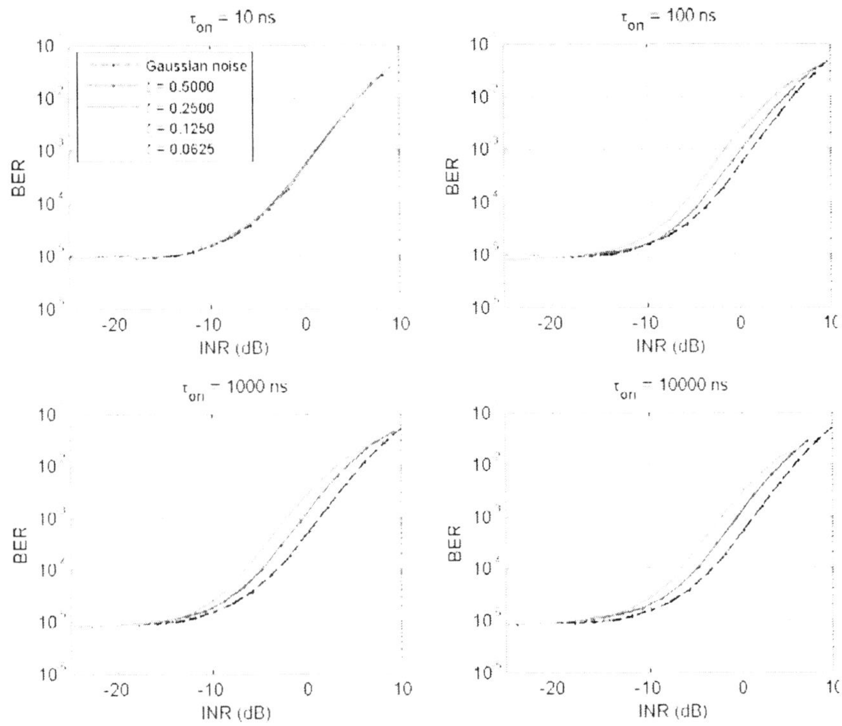

Figure 8. BER versus INR for a DTV receiver operating at SNR = 15 dB and exposed to GN interference.

2.4. DTV Susceptibility and FEC Performance as a Function of Interference Fractional On-Time

Figure 9 displays measured *SER* and *BER* curves for a DTV victim receiver operating at SNR = 15 dB and exposed to increasing levels of continuous-noise interference. This figure also identifies important DTV susceptibility and FEC performance metrics. Importantly, SER_{TOV} = 1 0-4 is a level identified by video quality studies [7] as the threshold of visibility where video quality degradation is first evident.

INR that corresponds to SER_{TOV}, INR_{TOV}, is our primary DTV susceptibility metric. That is, receiver susceptibility increases as INR_{TOV} decreases. In figure 9, INR_{TOV} is found vertically from the intersection of the measured *SER* curve and the SER_{TOV} horizontal line.

BER that corresponds to SER_{TOV}, BER_{TOV}, is our FEC performance metric. This metric quantifies the ability of the FEC to overcome bit errors caused by interference to achieve SER_{TOV}. In other words, larger BER_{TOV} indicates more corrected bits and better FEC performance. In figure 9, BER_{TOV} is found horizontally from the intersection of INR_{TOV} and the measured *BER* curve.

Table 2 gives INR_{TOV} and BER_{TOV} metrics for all GN interference tests. A distinctive trend was the positive correlation between INR_{TOV} and BER_{TOV}. This demonstrates that DTV susceptibility was dependent on how effective the FEC was at overcoming bit errors caused by the interference. For example, the FEC was most effective at mitigating continuous-noise degradation and the receiver was least susceptible to continuous-noise interference.

Table 2. Measured DTV Susceptibility and FEC Performance for GN Interference

GN	Gating Parameters			SNR = 9 dB		SNR = 12 dB		SNR = 15 dB	
	τon (μs)	τoff (μs)	ζ	$INRTOV$ (dB)	$BERTOV$	$INRTOV$ (dB)	$BERTOV$	$INRTOV$ (dB)	$BERTOV$
01	~	0.00	1.0000	0.7	0.038	5.4	0.038	9.2	0.038
02	0.01	0.01	0.5000	0.9	0.038	5.5	0.038	9.0	0.038
03	0.01	0.03	0.2500	0.7	0.038	5.6	0.038	9.0	0.038
04	0.01	0.07	0.1250	0.1	0.036	5.1	0.034	8.5	0.034
05	0.01	0.15	0.0625	-0.3	0.031	4.1	0.029	7.5	0.029
06	0.10	0.10	0.5000	0.4	0.035	5.0	0.033	8.4	0.033
07	0.10	0.30	0.2500	-0.3	0.030	3.8	0.028	7.5	0.028
08	0.10	0.70	0.1250	-1.2	0.025	3.2	0.023	6.7	0.023
09	0.10	1.50	0.0625	-1.5	0.021	3.4	0.018	7.0	0.018
10	1.00	1.00	0.5000	-0.3	0.032	4.2	0.030	7.5	0.030
11	1.00	3.00	0.2500	-2.2	0.022	1.8	0.018	5.1	0.018
12	1.00	7.00	0.1250	-4.6	0.015	-0.4	0.011	2.7	0.011
13	1.00	15.00	0.0625	-6.2	0.012	-2.5	0.007	1.1	0.007
14	10.00	10.00	0.5000	-1.6	0.025	2.8	0.022	6.3	0.022
15	10.00	30.00	0.2500	-4.1	0.017	0.2	0.012	3.7	0.012
16	10.00	70.00	0.1250	-6.5	0.013	-2.2	0.007	1.1	0.007
17	10.00	150.00	0.0625	-8.6	0.010	-4.5	0.004	-1.2	0.004

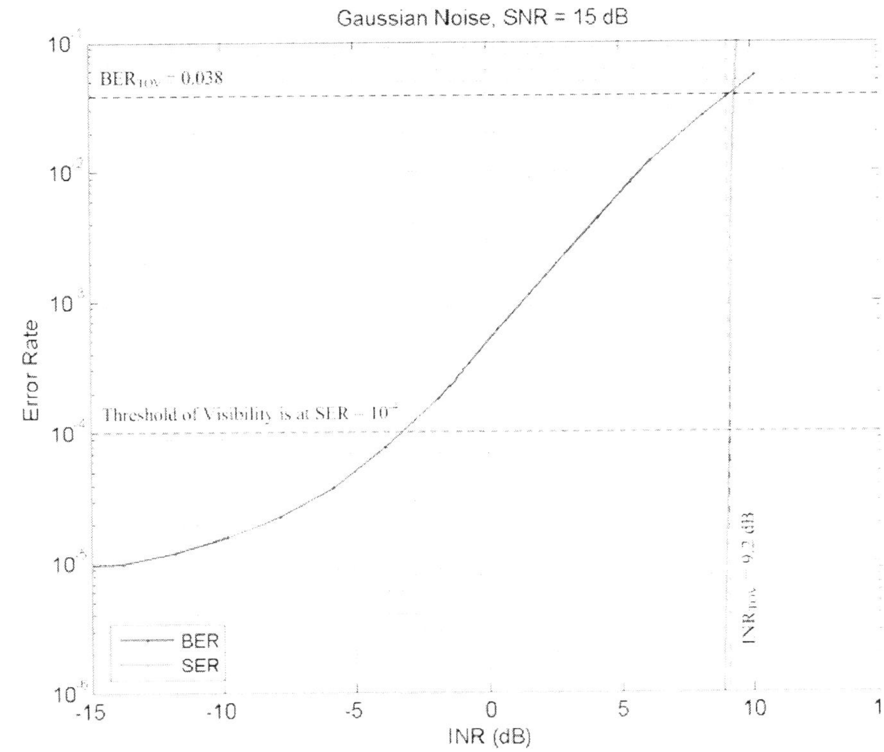

Figure 9. Illustration of threshold-of-visibility metrics.

Figures 10 – 15 plot INR_{TOV} and BER_{TOV} as a function of $1/\zeta$ in dB, where $\zeta = \{1.00, 0.50, 0.25, 0.125, 0.0625\}$ correspond to $10\log(1/\zeta) = \{0.00, 3.01, 6.02, 9.03, 12.04\}$ dB. The horizontal dashed reference line in each INR_{TOV} plot corresponds to DTV susceptibility that is strictly dependent on average power of the interference signal. As an example, figure 16 illustrates three gated-noise signals with different ζ but the same average power. If susceptibility were strictly dependent on average power, then the victim receiver would be equally susceptible to each of these signals.

For fixed *SNR* and τ_{on}, INR_{TOV} decreased (relative to the continuous-noise case) with decreasing ζ. Likewise, for fixed *SNR* and ζ, INR_{TOV} decreased with increasing τ_{on}. Hence, $INR_{TOV}(\tau_{on} = 10$ ns, $\zeta \sim 0.25)$ lay close to the horizontal reference line, while

$INR_{TOV}(\tau_{on} = 10{,}000$ ns, $\zeta = 0.0625)$ deviated the furthest. A notable anomaly to this trend occurred when $INR_{TOV}(\tau_{on} = 100$ ns, $\zeta = 0.0625)$ was greater than $INR_{TOV}(\tau_{on} = 100$ ns, $\zeta = 0.125)$.

These observations demonstrate that DTV susceptibility was dependent on temporal parameters of the interfering signal, i.e., τ_{on} and τ_{off}. If τ_{off} was less than or comparable to the reciprocal bandwidth of the victim receiver RRC filter, i.e., approximately $1/B_{DTV} = 51.3$ ns, then INR_{TOV} clustered near the horizontal reference line and DTV susceptibility was strictly dependent on average power of the interference signal (independent of τ_{on} and ζ). However, if τ_{off} was significantly greater than the reciprocal bandwidth of the victim receiver, then DTV susceptibility was clearly dependent on τ_{on} and ζ.

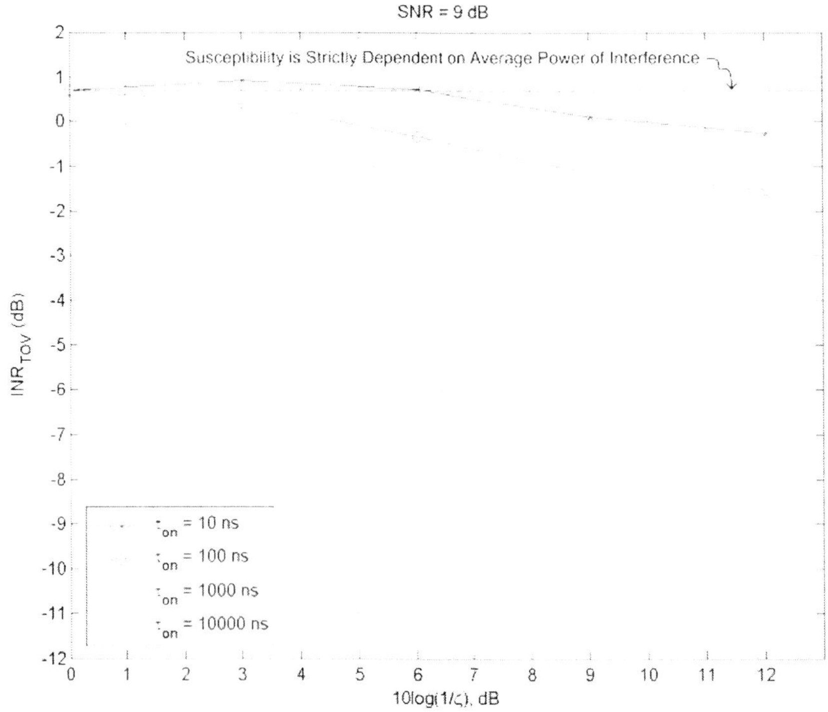

Figure 10. INR_{TOV} versus $10\log(1/\zeta)$ for a DTV receiver operating at SNR = 9 dB and exposed to GN interference.

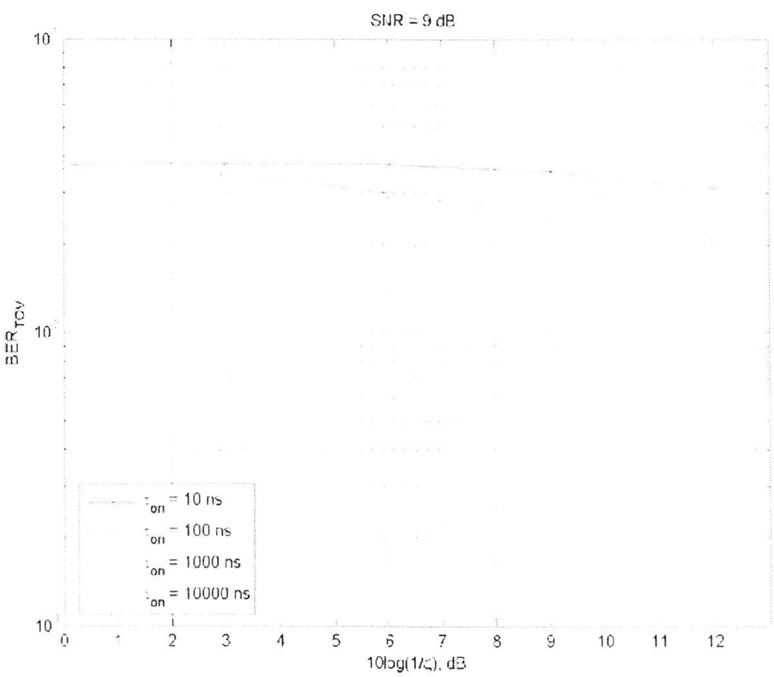

Figure 11. BER_{TOV} versus $10\log(1/\zeta)$ for a DTV receiver operating at SNR = 9 dB and exposed to GN interference.

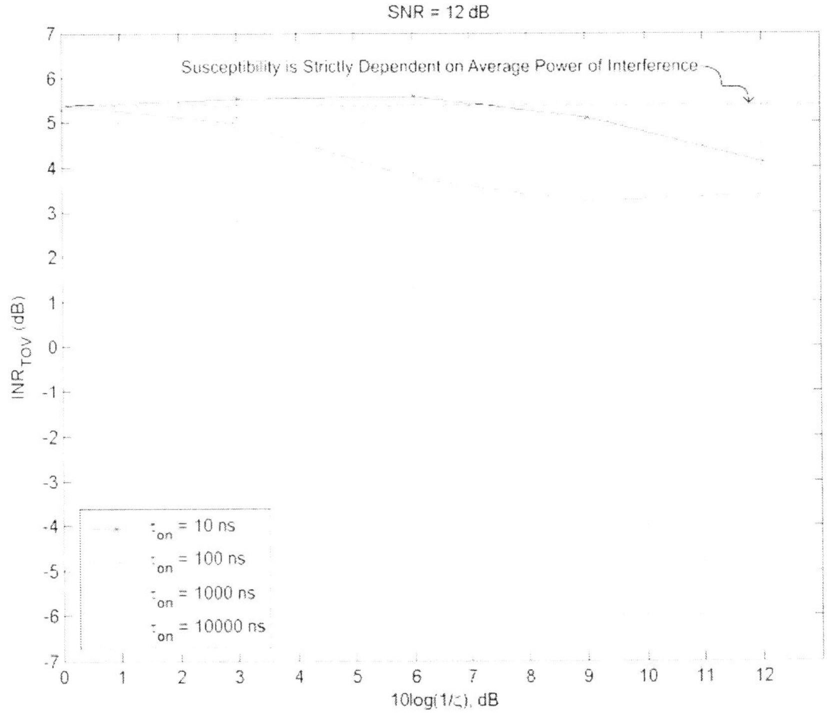

Figure 12. INR_{TOV} versus 10log(1/ζ) for a DTV receiver operating at SNR = 12 dB and exposed to GN interference.

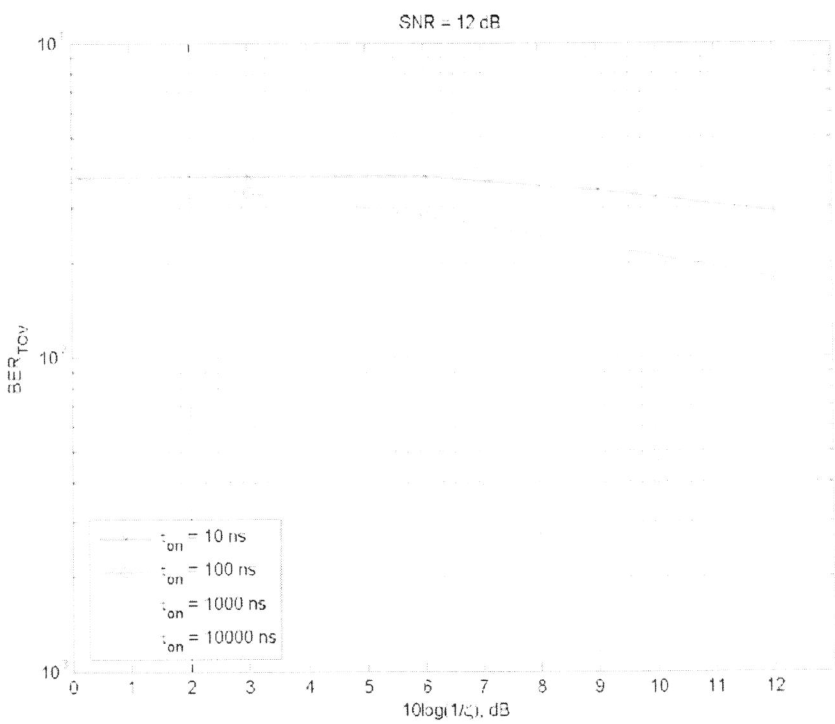

Figure 13. BER_{TOV} versus 1 0log(1 /ζ) for a DTV receiver operating at SNR = 12 dB and exposed to GN interference.

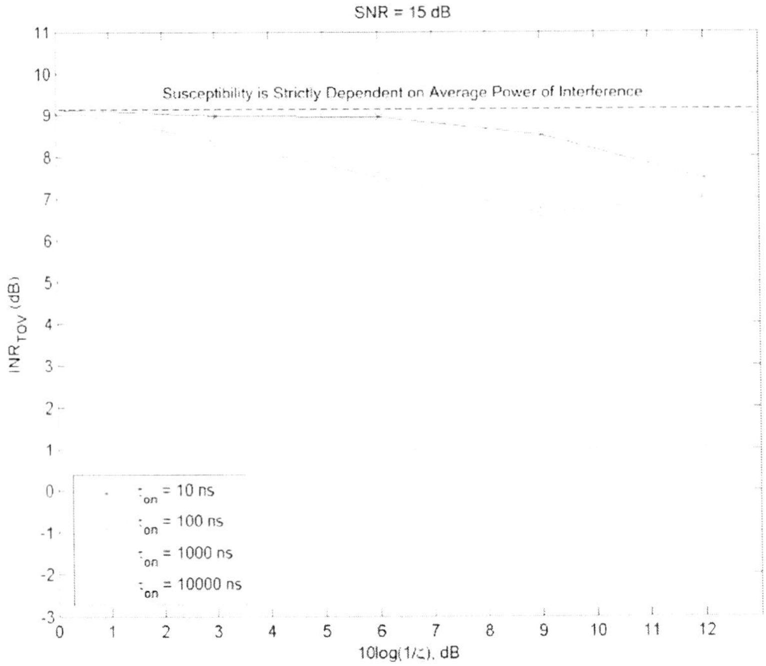

Figure 14. INR_{TOV} versus 10log(1/ζ) for a DTV receiver operating at SNR = 15 dB and exposed to GN interference.

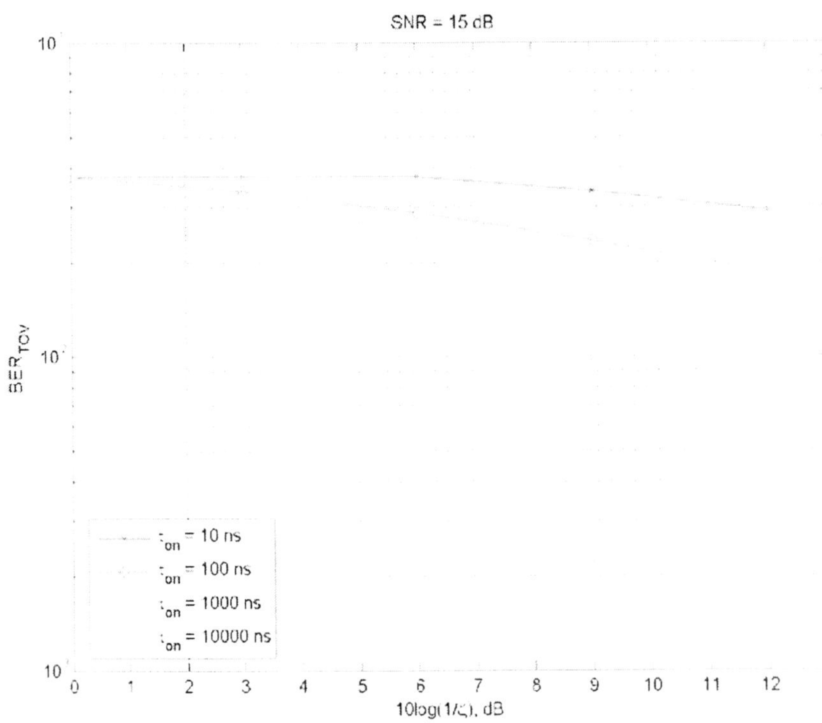

Figure 15. BER_{TOV} versus 1 0log(1 /ζ) for a DTV receiver operating at SNR = 15 dB and exposed to GN interference.

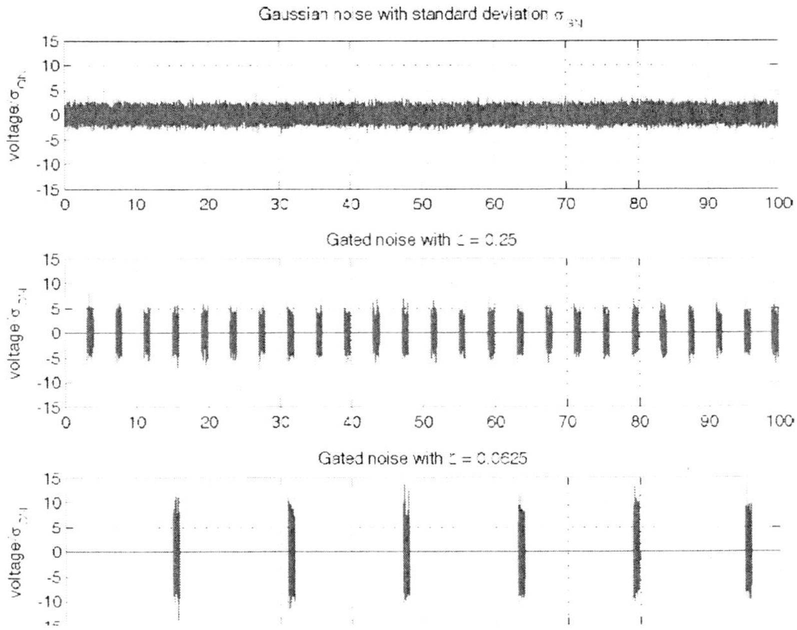

Figure 16. Simulated gated-noise signals with the same average power.

3. CHARACTERIZATION OF GN SIGNALS

Signal characterization of the complex-baseband (CBB) gated-noise signals was accomplished via temporal, amplitude, and spectral analyses. Appendix C provides a theoretical evaluation of the corresponding amplitude and spectral characteristics of gated-noise signals in an ultra-wide transmission bandwidth. To evaluate the effects of band-limiting, the GN signals were measured with the vector signal analyzer (VSA) under optimal measurement conditions at the RF output of the vector signal generator (VSG) and under TOV operational conditions at the IF. Appendix D provides a description of the measurement setup, RF measurement results in various bandwidths, i.e., 36, 10, 1, and 0.1 MHz, and IF measurement results at the three different *SNR*s. Unless otherwise noted, analyses in this section use RF measurements sampled at fs = 46.08 MSps (corresponding to VSA span = 36 MHz) and band-limited to $_{BDTV}$ with the victim receiver RRC filter applied in post-measurement processing.

3.1. Temporal Analysis

Temporal characteristics of GN signals were evaluated with crossing statistics, which provide probabilities of the time a signal crosses an amplitude threshold and stays either above or below that threshold. Historically, this type of analysis has been used to study pulse statistics of noise [8], hence the terms pulse duration and pulse interval depicted in figure 17. In this analysis of gated noise, their utility has been extended to the temporal statistics of noise bursts; accordingly, the terminology was changed to burst duration *(BD)* and burst interval *(BI)* as illustrated in figure 18. For our purposes, the 10th percentile of the cumulative distribution function *(CDF)* of BD and BI using a threshold equal to -70 dBm was optimal for quantifying the effects of band-limiting on temporal characteristics of the GN signals.

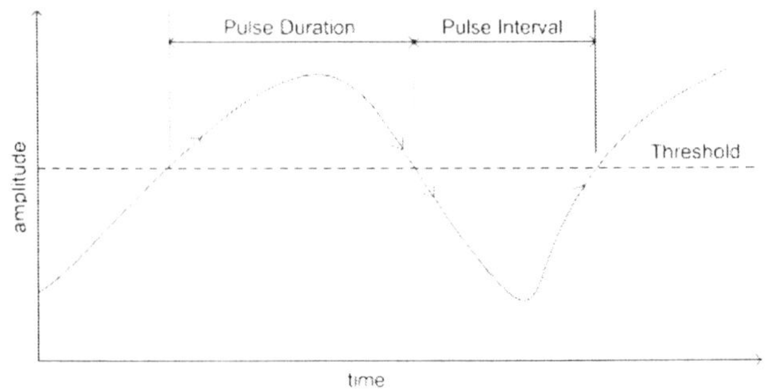

Figure 17. Illustration of pulse duration and pulse interval.

Table 3 summarizes the effects of band-limiting the GN signals to $_{BDTV.}$ Note that BD and BI are given only when gating was resolved and the CDFs provided meaningful estimates of the band-limited on- and off-times, respectively. The band-limited fractional on-time, ζ_{DTV}, was computed as $BD/(BD + BI)$. As expected, band-limiting lengthened on-times, shortened off-times, and increased fractional on-times. Band-limited metrics approached the gating parameters only when τ_{on} and τ_{of} were significantly larger than $1/B_{DTV.}$

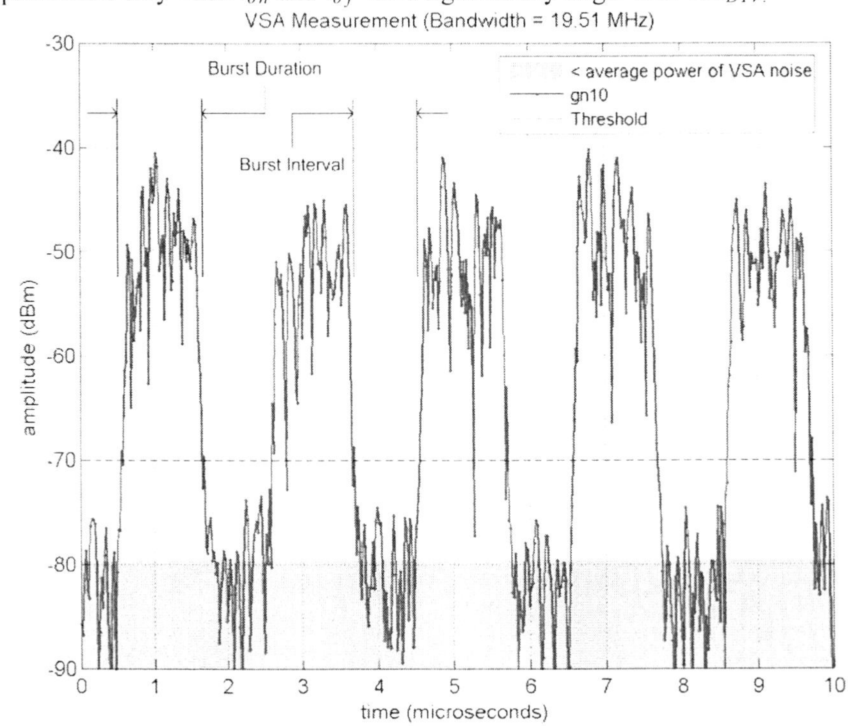

Figure 18. Illustration of burst duration and burst interval.

Table 3. Temporal Characteristics of GN Signals Band-Limited to B_{DTV}

	Gating Parameters			Band-Limited Metrics		
GN	τ_{on} (μs)	τ_{of} (μs)	ζ	BD (μs)	BI (μs)	ζ_{DTV}
01	~	0.00	1.0000			1.00
02	0.01	0.01	0.5000			1.00
03	0.01	0.03	0.2500			1.00
04	0.01	0.07	0.1250			
05	0.01	0.15	0.0625			
06	0.10	0.10	0.5000			
07	0.10	0.30	0.2500	0.2	0.2	0.50
08	0.10	0.70	0.1250	0.2	0.6	0.25
09	0.10	1.50	0.0625	0.2	1.4	0.13
10	1.00	1.00	0.5000	1.1	0.9	0.55
11	1.00	3.00	0.2500	1.1	2.9	0.28
12	1.00	7.00	0.1250	1.1	6.9	0.14
13	1.00	15.00	0.0625	1.1	14.9	0.07
14	10.00	10.00	0.5000	10.1	9.9	0.51
15	10.00	30.00	0.2500	10.1	29.9	0.253
16	10.00	70.00	0.1250	10.1	69.9	0.126
17	10.00	150.00	0.0625	10.1	149.9	0.063

3.2. Amplitude Analysis

Random amplitudes of the gated-noise signals were characterized with the amplitude probability distribution, which describes the probability that signal amplitude exceeds a certain value. In this chapter, *APD*s are plotted on Rayleigh graphs where Rayleigh-distributed amplitudes of complex-Gaussian noise appear as a negatively-sloped, straight line. The sorting method, documented in Appendix D of Part 1, was used to estimate and plot *APD*s of 2^{17} = 131,072 samples that are free of correlation introduced by the measurement system and post-measurement band-limiting.

Figures 19 – 22 provide composite *APD* plots of the measured GN signals. Each plot represents a specific τ_{on}, and each curve within a plot represents a different ζ. Dashed lines represent the *APD* of the continuous-noise signal, GN-0 1. The upper bound of the shaded region is the average power of the VSA noise, approximately -79.7 dBm in B_{DTV}.

Band-limiting had profound effects on the *APD*s of the GN signals. If τ_{of} exceeded $1/B_{DTV}$ a step-like *APD* occurred, where higher amplitudes primarily corresponded to elongated on-times and lower amplitudes to the shortened off-times. This was most pronounced for GN-17 (τ_{on} = 10,000 ns, ζ = 0.0625). If τ_{of} was less than or comparable to $1/B_{DTV}$ a negatively-sloped, straight *APD* occurred, corresponding to Rayleigh-distributed amplitudes due to elongated noise bursts overlapping each other. This was most pronounced for GN-02 (τ_{on} = 10 ns, ζ = 0.50).

Peak-to-average ratio is a metric derived from the *APD* as illustrated in figure 23 for GN-10. When interpreting *P/A* results, it is important to understand the bandwidth that peak and average powers were measured in, the duration or number of samples, and the statistical definition of peak in terms of percentage of samples. In this study, peak and average powers were measured in the same bandwidth to compute a single *P/A*. Also, peak power is defined as the amplitude exceeded 0.01% of the time, which corresponds to 13 of 131,072 samples. Figure C-2 shows how *P/A* changes with various peak definitions.

Table 4 summarizes measured *P/A* and *APD* characteristics for the band-limited GN signals. Also included are ultra-wide bandwidth limits to *P/A* derived in Section C. 1. *APD* characteristics are labeled RAYL for Rayleigh-distributed amplitudes or RG for resolved gating, i.e., step-like *APD* whose amplitude falls below VSA average noise power at least 10% of the time.

Band-limiting affected *P/A*. If τ_{of} far exceeded 1 $/B_{DTV}$, then *P/A* approached the ultra-wide transmission bandwidth limit. If τ_{of} was less than $1/B_{DTV}$, then *P/A* approached the 9.6-dB value for Rayleigh-distributed amplitudes. When τ_{of} was between these extremes, *P/A* transitioned between the limits.

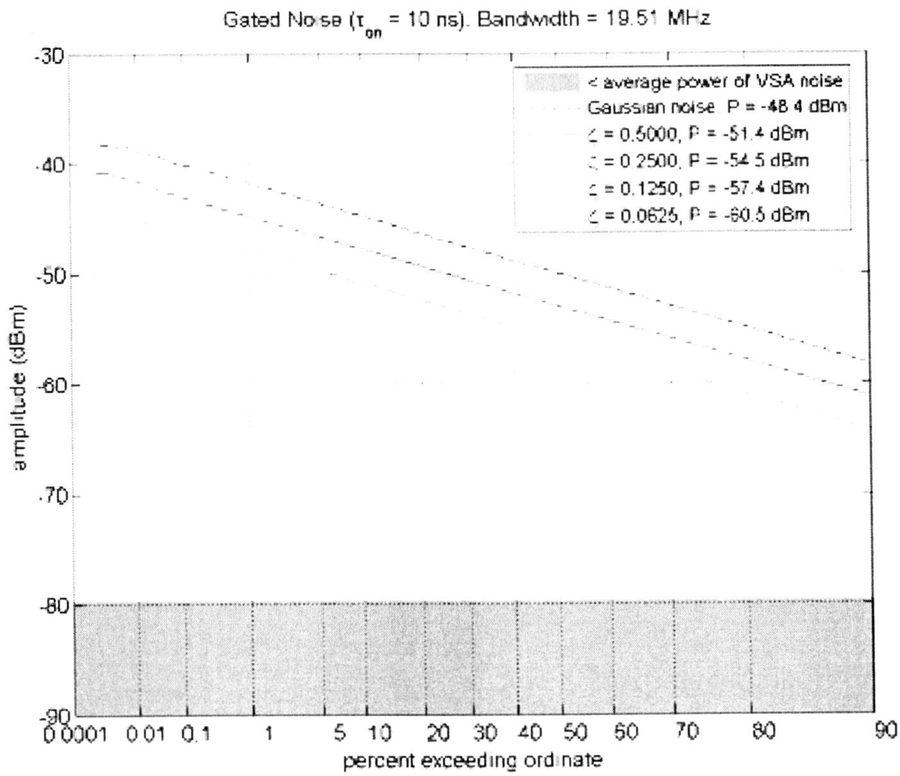

Figure 19. APDs of GN signals with τ_{on} = 10 ns band-limited to B_{DTV}.

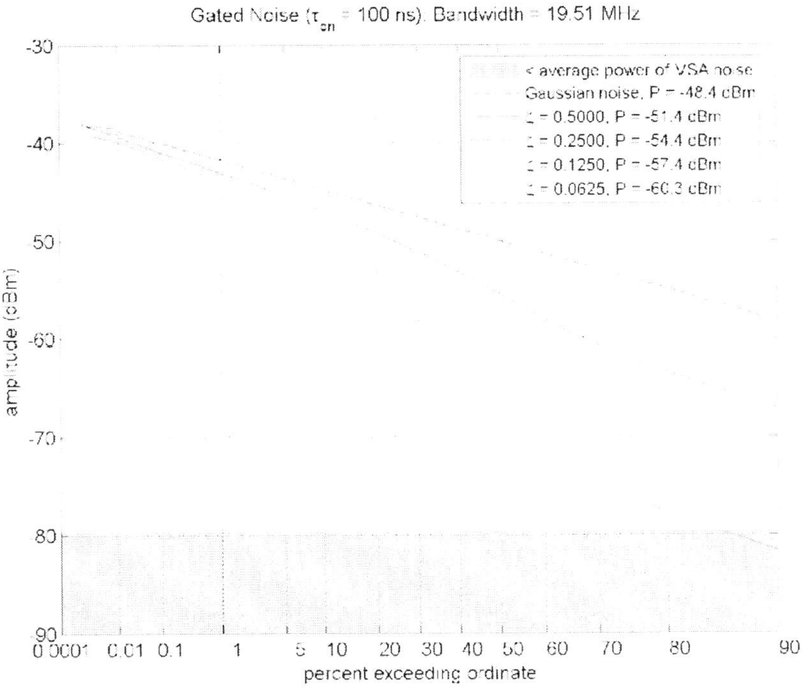

Figure 20. $APDs$ of GN signals with τ_{on} = 100 ns band-limited to B_{DTV}.

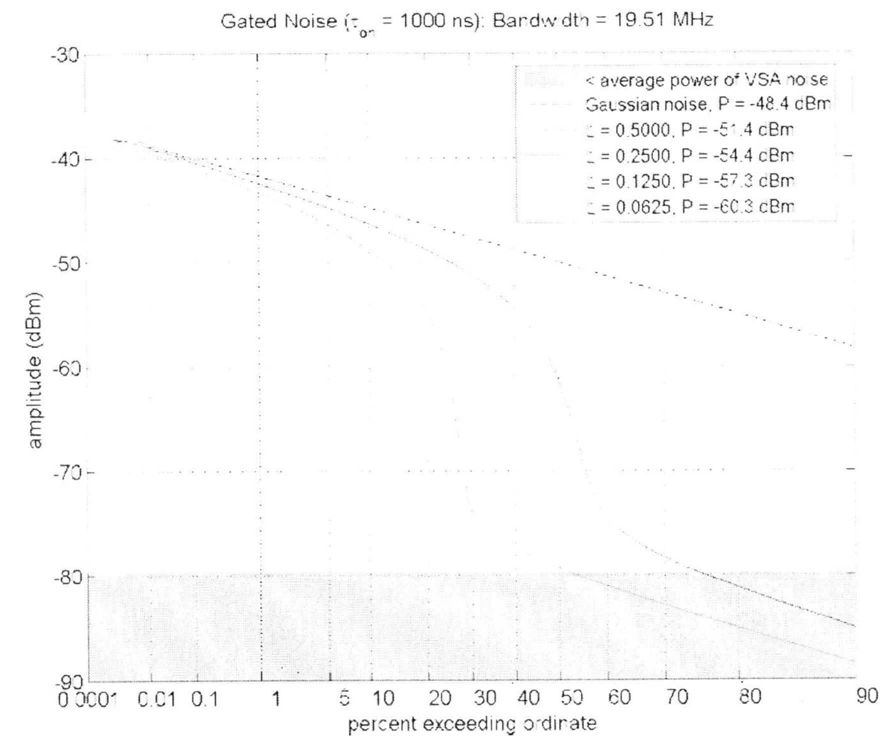

Figure 21. $APDs$ of GN signals with τ_{on} = 1,000 ns band-limited to B_{DTV}.

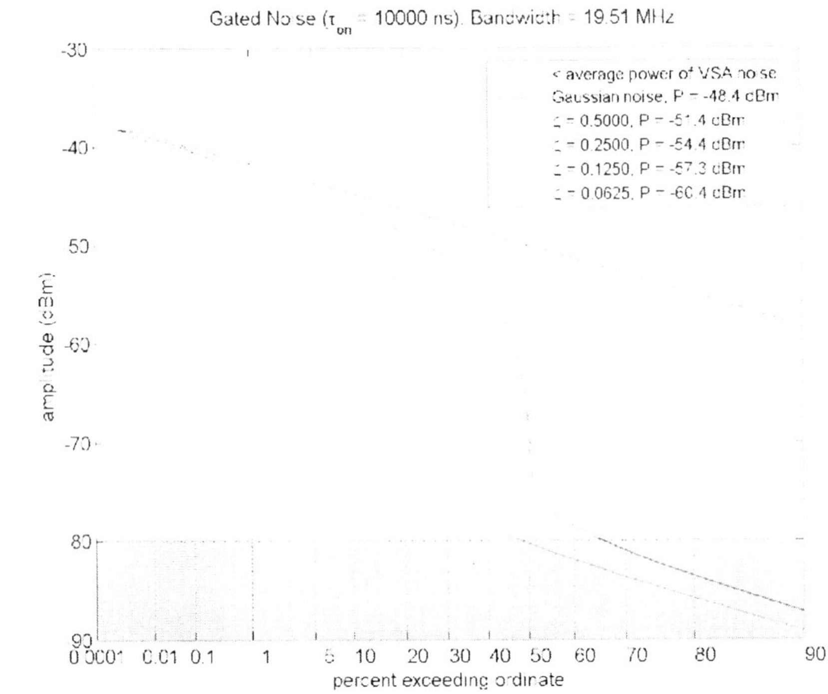

Figure 22. *APD*s of GN signals with τ_{on} = 10,000 ns band-limited to B_{DTV}.

Table 4. Amplitude Characteristics of GN Signals Band-Limited to B_{DTV}

GN	Ultra-wide Bandwidth Limits		Band-Limited Metrics	
	P/A (dB)	*APD*	*P/A* (dB)	*APD*
01	9.6	RAYL	9.6	RAYL
02	12.3	RG	9.9	RAYL
03	14.9	RG	9.7	RAYL
04	17.5	RG	11.5	
05	20.1	RG	14.2	
06	12.3	RG	12.2	
07	14.9	RG	14.6	RG
08	17.5	RG	17.3	RG
09	20.1	RG	19.6	RG
10	12.3	RG	12.4	RG
11	14.9	RG	15.0	RG
12	17.5	RG	17.6	RG
13	20.1	RG	20.3	RG
14	12.3	RG	12.3	RG
15	14.9	RG	15.1	RG
16	17.5	RG	17.8	RG
17	20.1	RG	20.1	RG

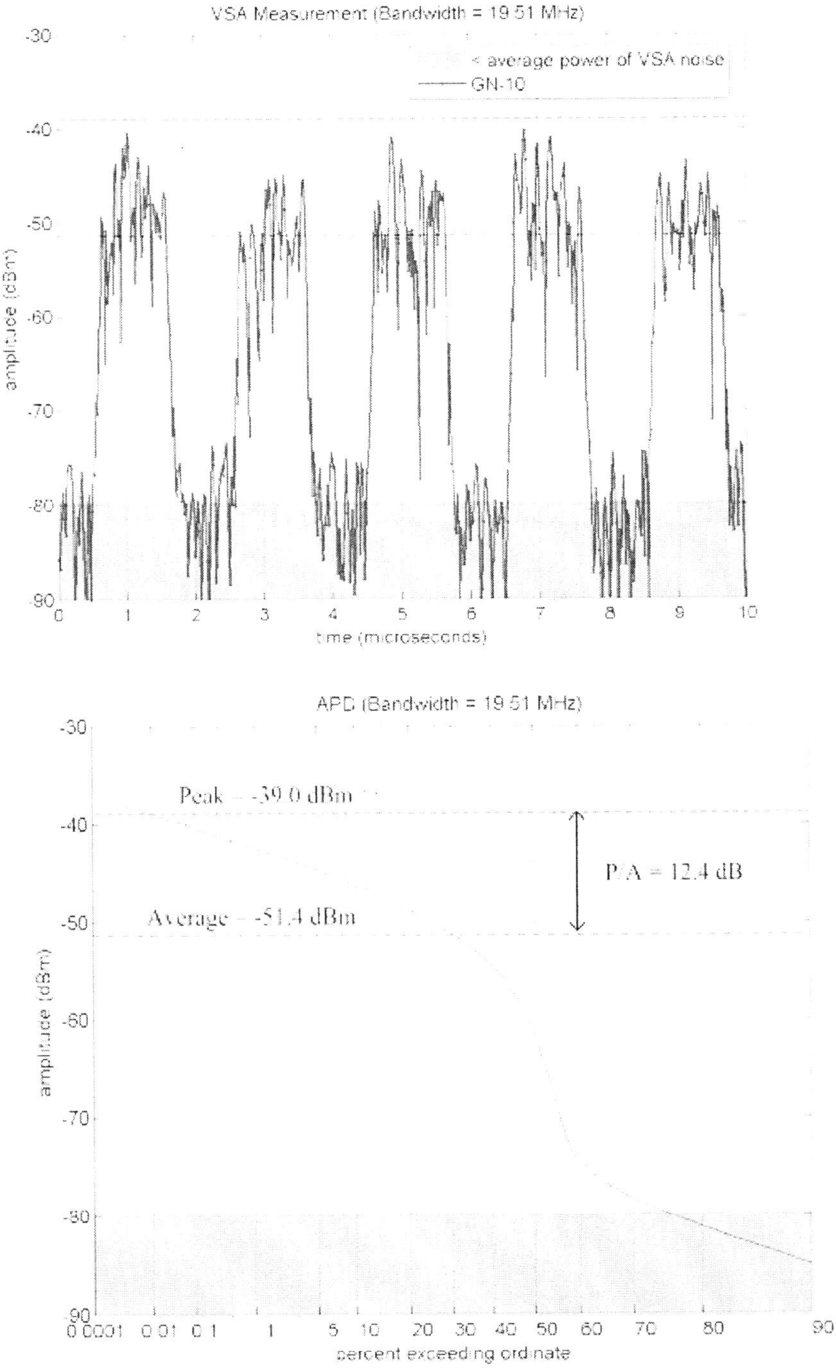

Figure 23. Illustration of P/A for GN-10 band-limited to B_{DTV}.

3.3. Spectral Analysis

Spectral characteristics of the gated-noise signals were characterized with the power spectral density, which represents the average power present per unit bandwidth as a function of frequency. Section C.2 provides a theoretical analysis of the *PSD* for gated Gaussian noise. In short, it can be considered to be constant over the bandwidth of interest.

Figures 24 – 27 provide *PSD*s of the GN signals over the entire VSA bandwidth, i.e., the victim receiver RRC filter was not applied to the data in post-measurement processing. *PSD*s were calculated by averaging squared spectrum magnitudes of rectangular-windowed, non-overlapping blocks of signal data. The periodic nature of the GN signals required that the block size be an integer multiple of the gating period. To have the same Δf for all *PSD*s, the block size must also be a common integer multiple of all gating periods. For the GN signals, these requirements were satisfied with a block size of 160 μs, which corresponds to Δf = 6.2 kHz. This allowed 500 blocks to be averaged, given the 4 million samples available in the acquired data sets.

The figures show that the power density of the gated signals decreased in proportion to ζ. In addition, the spectral spike at 0 Hz was due to VSG local-oscillator feed-through. The influence of the VSG local-oscillator feed-through to test results was discussed in Appendix C of Part 1.

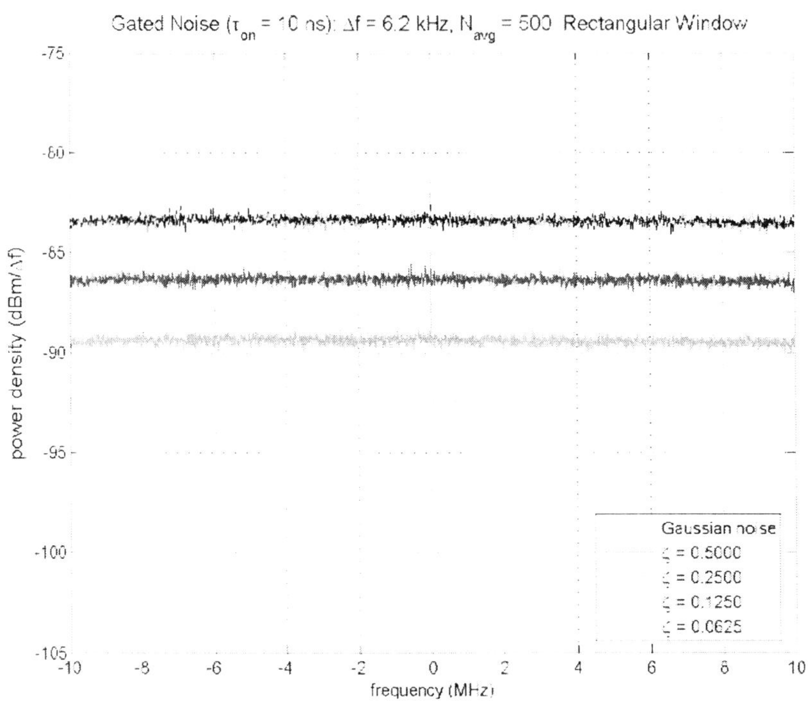

Figure 24. *PSD*s of GN signals with τ_{on} = 10 ns.

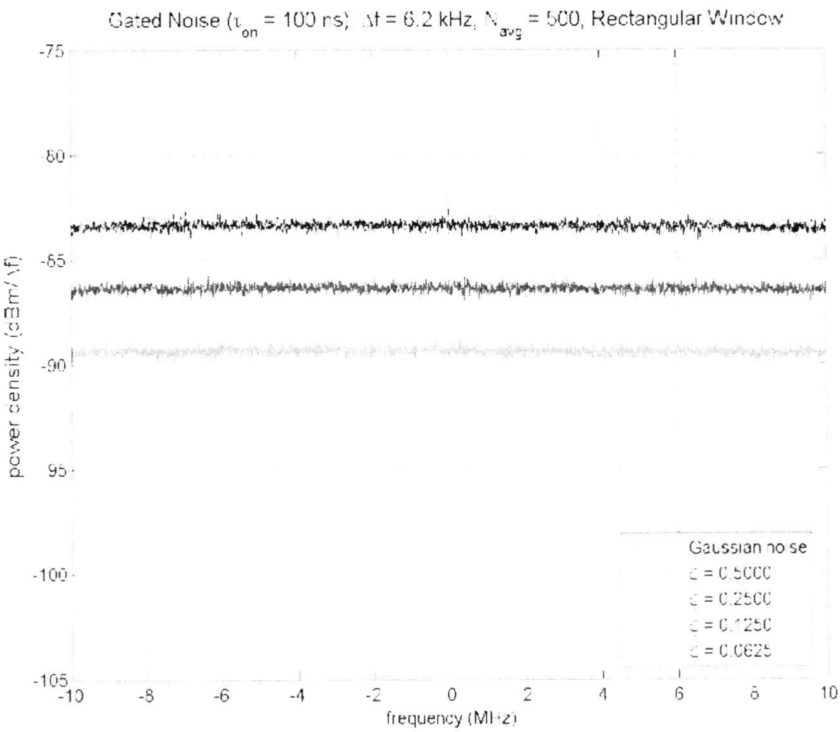

Figure 25. PSDs of GN signals with τ_{on} = 100 ns.

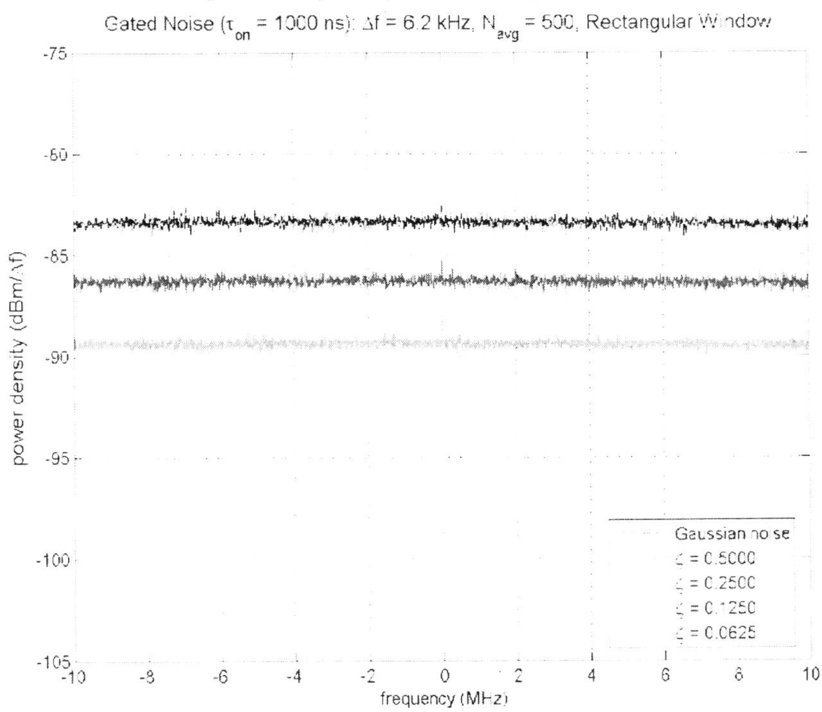

Figure 26. PSDs of GN signals with τ_{on} = 1,000 ns.

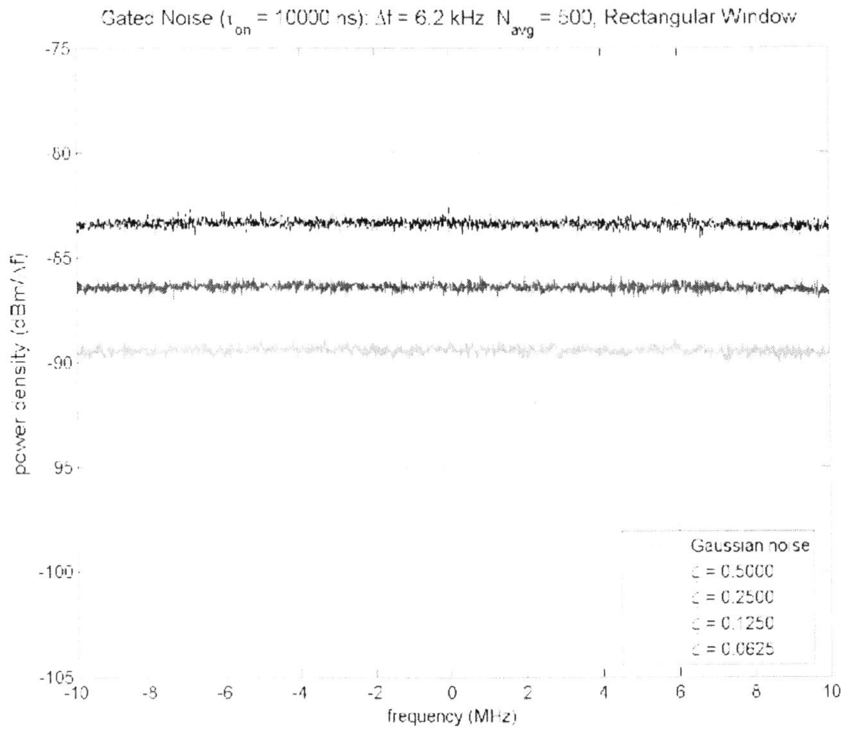

Figure 27. *PSDs* of GN signals with τ_{on} = 10,000 ns.

4. CONCLUSION

The objective of this experiment was to measure and evaluate the susceptibility of a C-band satellite DTV receiver to gated-noise interference. Results in this chapter have been measured with a reliable test system and repeatable procedures. Validation against theory is demonstrated in Appendix B. Results are expected to be fundamental in understanding interference caused by DS-UWB and MB-OFDM signals. In these concluding remarks, findings are summarized and their significance explained.

Temporal analyses showed that band-limiting the GN interference signals elongated on-times and shortened off-times, which had profound effects on corresponding *APDs*. When τ_{of} was less than or comparable to the receiver reciprocal bandwidth, the elongated on-times overlapped, producing Rayleigh amplitude statistics. When τ_{of} was several times greater than the reciprocal bandwidth, the elongated on-times remained distinct, producing gated-noise amplitude statistics.

These temporal and amplitude observations help explain the susceptibility results. When τ_{of} was less than or comparable to the reciprocal bandwidth, DTV susceptibility results were strictly dependent on average power of the band-limited interference. This is because Rayleigh-distributed amplitudes correspond to zero-mean Gaussian noise, which is completely characterized by its variance or average power. Moreover, the DTV receiver was least susceptible to signals that resembled continuous noise after band-limiting. Conversely, when τof was several times greater than the reciprocal bandwidth, gating was resolved and susceptibility results were clearly dependent on the gating parameters τon and ζ. The DTV receiver was more susceptible to the gated-noise signals that maintained some semblance of a gated signal after band-limiting. More specifically, DTV susceptibility increased with increasing τon and decreasing ζ.

Interestingly, peak-to-average ratios of the band-limited GN signals were also dependent on τon and ζ. In fact, as demonstrated in figures 28 – 30, there was some correlation between measured DTV susceptibility and P/A, *i.e.*, DTV was more susceptible to GN interference with higher P/A. However, these figures also show distinct separation between τon curves, which indicates that P/A alone cannot serve as a sole predictor of DTV interference susceptibility.

In summary, DTV susceptibility to gated-noise interference cannot be predicted by interference power characteristics alone. Indeed, it is essential to consider the temporal parameters of the interfering signal as well as the bandwidth of the victim receiver. Additionally, FEC performance trends, characterized by $_{BERTOV}$, behaved similar to DTV susceptibility trends, leading to the conclusion that DTV susceptibility is determined largely by FEC performance.

Caution is advised when extending these results to small ζ. For example, it is entirely likely that low susceptibility could be achieved for each τon provided ζ is sufficiently small, which may create conditions more favorable to FEC performance or simply provide longer off-times where errors are unlikely. Determination of these conditions is beyond the scope of this chapter.

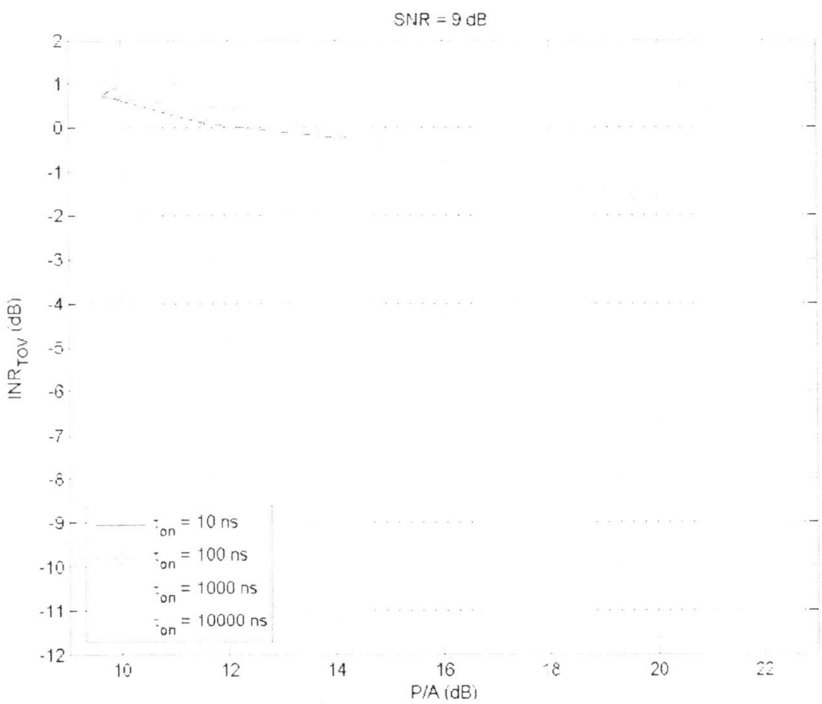

Figure 28. INR_{TOV} versus P/A for a DTV receiver operating at SNR = 9 dB and exposed to GN interference.

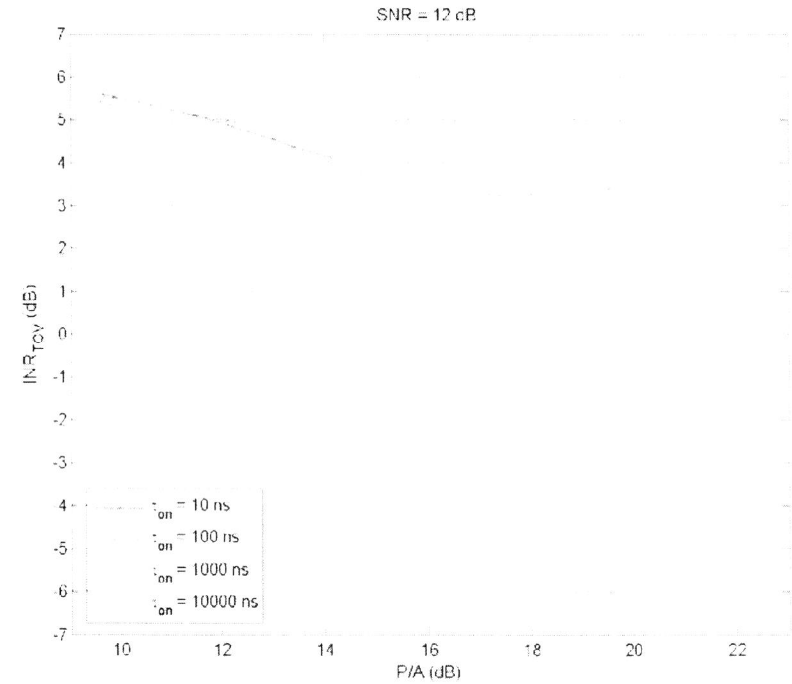

Figure 29. INR_{TOV} versus P/A for a DTV receiver operating at SNR = 12 dB and exposed to GN interference.

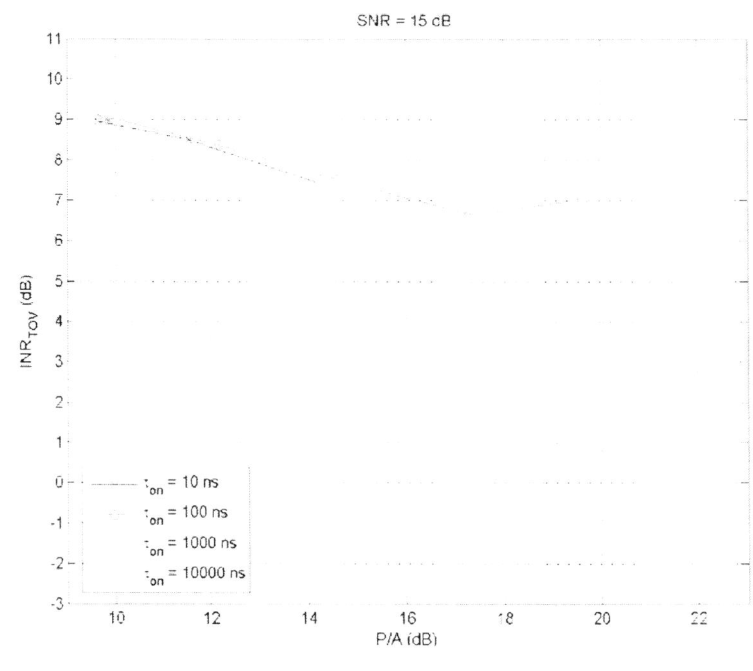

Figure 30. INR_{TOV} versus P/A for a DTV receiver operating at SNR = 15 dB and exposed to GN interference.

5. ACKNOWLEDGMENTS

The authors recognize John McCorkle of Freescale, Inc. for his leadership, insight, and insistence on the highest quality. We learned a great deal from working with John in all relevant technical areas.

This investigation required the services of a number of ITS engineers with a broad range of expertise, some of whom are not included in the list of authors. The authors recognize Brent Bedford for his expertise in test and measurement equipment and techniques; Randy Hoffman, Richard Statz, and Steve Engelking for developing much of the low level software routines needed to automate the experiment via software control; Stephen Wolf for his expertise in video quality assessment; Frank Sanders for sharing knowledge learned in previous interference tests on C-band earth station receivers [9]; Dr. Roger Dalke for his knowledge of random processes and experimental uncertainty analyses; Andrew Rogers and Michael Mitchell for their attention to detail in hardware production and signal characterization measurements; and Dr. William Kissick for his support and leadership.

6. REFERENCES

[1] "First Report and Order in the matter of ET Docket 98-153 (Revision of Part 15 of the Commission's Rules Regarding Ultra-Wideband Transmission Systems)," *FCC 02-48*, adopted Feb. 14, 2002, released Apr. 22, 2002.

[2] W.A. Kissick, Ed., "The temporal and spectral characteristics of ultrawideband signals," *NTIA Report*. 01-383, Jan. 2001.

[3] J.R. Hoffman, M.G. Cotton, R.J. Achatz, R.N. Statz, and R.A. Dalke, "Measurements to determine potential interference to GPS receivers from ultrawideband transmission *systems," NTIA Report*. 01-384, Feb. 2001.

[4] J.R. Hoffman, M.G. Cotton, R.J. Achatz, and R.N. Statz, "Addendum to NTIA Report 01-384: Measurements to determine potential interference to GPS receivers from ultrawideband transmission systems," *NTIA Report*. 01-389, Sep. 2001.

[5] J.R. Hoffman, E.J. Haakinson, and Y. Lo, "Measurements to determine potential interference to public safety radio receivers from ultrawideband transmission systems," *NTIA Report*. 03-396, Apr. 2003.

[6] M. Cotton, R. Achatz, J. Wepman, B. Bedford, "Ultrawideband interference potential: Part 1 – Procedures to characterize ultrawideband emissions and characterize interference susceptibility of C-band satellite digital television receivers," *NTIA Report*. TR-05-41 9, Feb. 2005.

[7] J.C. Whitaker, "Digital television transmission systems," Ch. 13 in *Standard Handbook of Video and Television Engineering*, J.C. Whitaker, Ed., New York, NY: McGraw-Hill, 2000.

[8] L.R. Espeland and A.D. Spaulding, "Amplitude and time statistics for atmospheric radio noise," *ESSA Technical Memorandum*, ERLTM-ITS. 253, Sep. 1970.

[9] F. Sanders, "Measurements of pulsed co-channel interference in a 4-GHz digital earth station receiver," *NTIA Report*. 02-3 93, May 2002.

[10] ETS 300 421: "Digital broadcasting systems for television, sound, and data services: Framing structure, channel coding, and modulation for 11/12 GHz satellite services," *European Telecommunications Standards Institute,* Dec. 1994.

[11] ETSI TR 101 290: "Digital Video Broadcasting (DVB); Measurement guidelines for DVB systems," *European Telecommunications Standards Institute,* v1.2.1, 2001.

APPENDIX A: EXTENSION OF EXPERIMENT TO GN(MB) SIGNALS

This appendix provides susceptibility and characterization measurements for gated-noise signals that emulate MB-OFDM, i.e., GN(MB) signals. Recall that MB-OFDM is an ultrawideband (UWB) signal that hops between 14 bands according to a specified time-frequency code; it is specified by the number of bands *(b)* it hops between and the number of consecutive dwells *(d)* the signal stays in a particular band. To a relatively narrowband victim receiver, MB-OFDM appears as gated interference. Table A-1 provides the MB-OFDM parameters, i.e., b and d, and corresponding gating parameters, i.e., on-time (τ_{on}), off-time (τ_{of}), and fractional on-time (ζ). Gating parameters are illustrated in figure A-1 for GN(MB)-03.

Table A-1. GN(MB) Gatin^g Parameters

GN	MB-OFDM Parameters		Gating	Parameters	
(MB)	b	d	τon (μs)	τof (μs)	ζ
01	1	1	0.24	0.07	0.7758
02	3	1	0.24	0.70	0.25 86
03	3	2	0.55	1.32	0.2586
04	7	1	0.24	1.95	0.1108
05	7	2	0.55	3.82	0.1108
06	7	6	1.80	11.32	0.1108
07	13	1	0.24	3.82	0.0597
08	13	2	0.55	7.57	0.0597
09	13	12	3.68	45.07	0.0597

From the perspective of a victim receiver, MB-OFDM has two gating processes associated with: (1) hopping and (2) insertion of zero-prefix and guard-intervals in each MB-OFDM symbol. Each MB-OFDM symbol is comprised of a zero-prefix, data block, and guard interval; the MB-OFDM symbol period is the sum of its constituents, i.e., $T_{OFDM} = T_{ZP} + T_{DATA} + T_{GI}$, where $T_{ZP} = 60.6$ ns, $T_{DATA} = 242.4$ ns, and $T_{GI} = 9.5$ ns. Fractional on-time is calculated as $\zeta = T_{DATA} / bT_{OFDM}$. τon is approximated by including the zero-prefix and guard interval between consecutive dwells and excluding those adjacent to hopping transitions,

$$\tau_{on} \approx dT_{OFDM} - T_{ZP} - T_{GI} \quad .$$

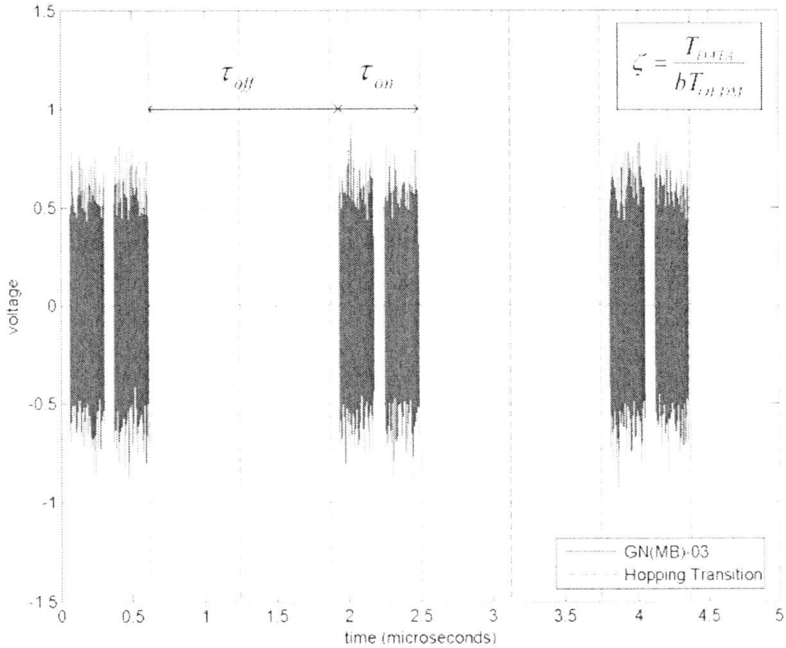

Figure A-1. Simulated gated-noise signal, GN(MB)-03.

Accordingly, τ_{of} is approximated as

$$\tau_{off} \approx (b-1)dT_{OFDM} + T_{ZP} + T_{GI} \quad .$$

A.1. DTV Susceptibility to GN(MB) Interference

Figures A-2 – A-7 provide measured segment error rate *(SER)* and pre-Viterbi bit error rate *(BER)* versus interference-to-noise ratio *(INR)*. These composite graphs for GN(MB) signals correspond to figures 3 – 8 for GN signals in the main text.

Figures A-8 – A-13 plot $_{INRTOV}$ and $_{BERTOV}$ versus $1/\zeta$ in dB, where $\zeta = \{0.78, 0.26, 0.11, 0.06\}$ correspond to $10\log(1/\zeta) = \{1.10, 5.87, 9.55, 12.24\}$ dB. These composite graphs for GN(MB) signals correspond to figures 10 – 15 for GN signals. Table A-2 provides INR_{TOV} and BER_{TOV} for GN(MB) signals as table 2 does for GN signals.

There is no one-to-one correspondence for gating parameters between the two sets. For example, τ_{on} § {0.24, 0.55, 1.80, 3.68} μs for GN(MB) are different than $\tau_{on} = \{0.01, 0.10, 1.00, 10.00\}$ μs for GN and $10\log(1/\zeta) = \{1.10, 5.87, 9.55, 12.24\}$ dB for GN(MB) are different than $10\log(1/\zeta) = \{0.00, 3.01, 6.02, 9.03, 12.04\}$ dB for GN. In spite of these discrepancies, GN(MB) results lie where expected in the region bounded by 0.10- to 10.00-μs τ_{on} curves.

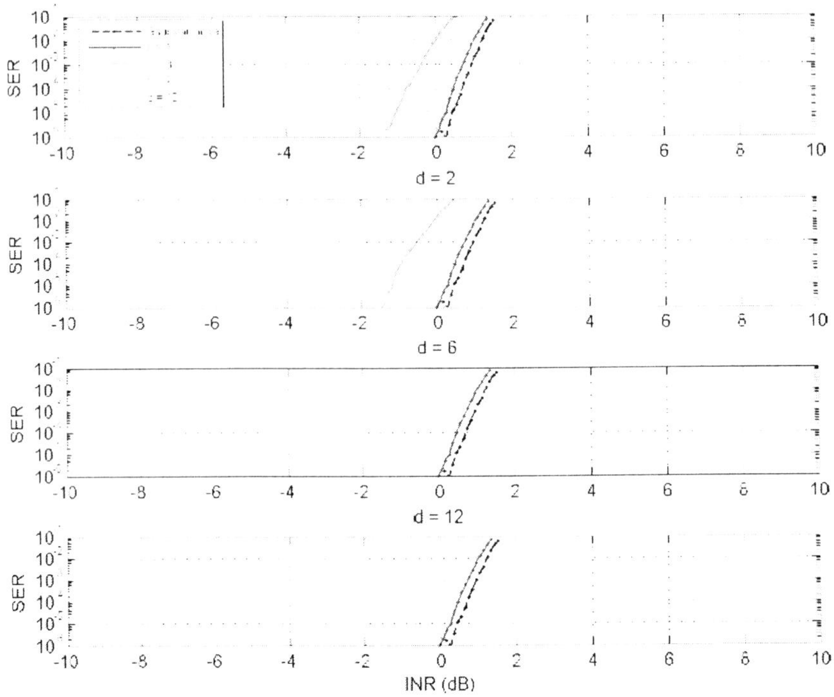

Figure A-2. SER versus INR for a DTV receiver operating at SNR = 9 dB and exposed to GN(MB) interference.

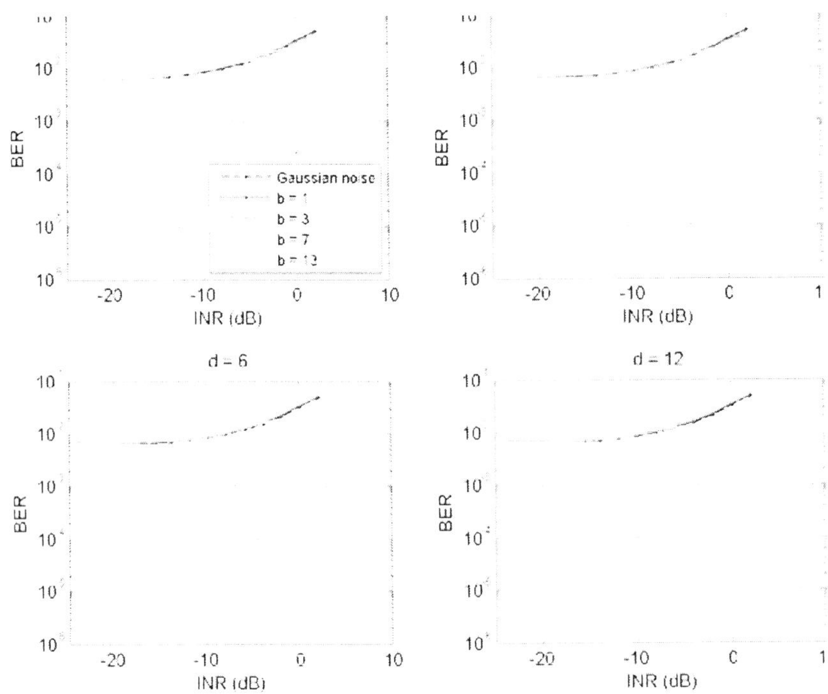

Figure A-3. BER versus INR for a DTV receiver operating at SNR = 9 dB and exposed to GN(MB) interference.

Table A-2. Measured DTV Susceptibility and FEC Performance for GN(MB) Interference

GN (MB)	Gating Parameters			$SNR = 9$	dB	$SNR = 12$	dB	$SNR = 15$	dB
	τ_{on} (µs)	τ_{off} (µs)	ζ	INR_{TOV} (dB)	BER_{TOV}	INR_{TOV} (dB)	BER_{TOV}	INR_{TOV} (dB)	BER_{TOV}
01	0.24	0.07	0.7758	0.5	0.037	5.2	0.036	8.5	0.036
02	0.24	0.70	0.2586	-0.7	0.030	3.7	0.027	7.0	0.028
03	0.55	1.32	0.2586	-0.9	0.028	3.5	0.025	6.5	0.025
04	0.24	1.95	0.1108	-1.9	0.022	2.3	0.018	5.8	0.019
05	0.55	3.82	0.1108	-3.2	0.018	0.8	0.013	4.2	0.013
06	1.80	11.32	0.1108	-5.0	0.015	-1.2	0.009	2.3	0.009
07	0.24	3.82	0.0597	-3.3	0.016	0.7	0.012	4.3	0.011
08	0.55	7.57	0.0597	-5.1	0.014	-1.1	0.009	2.3	0.008
09	3.68	45.07	0.0597	-7.4	0.011	-3.3	0.005	-0.3	0.005

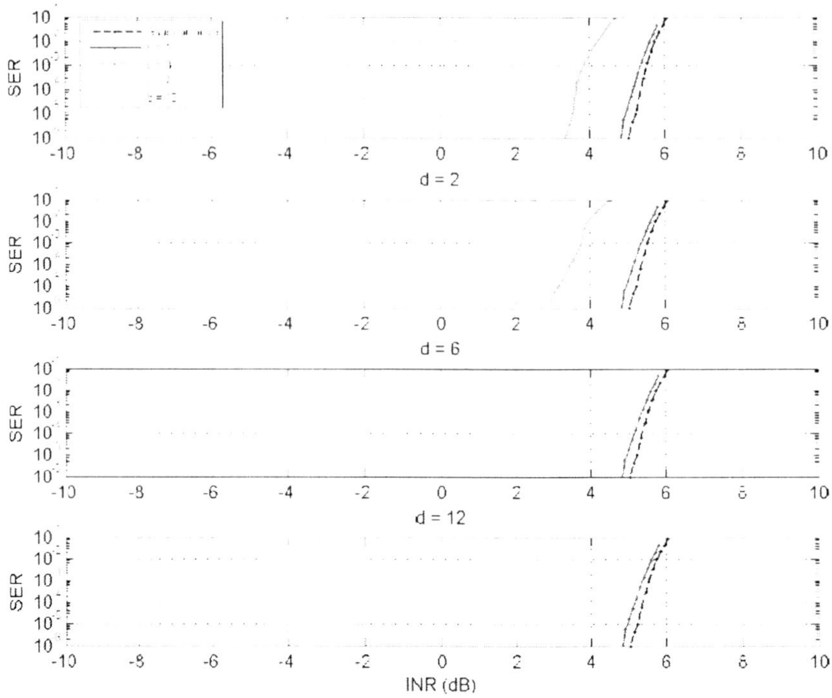

Figure A-4. SER versus INR for a DTV receiver operating at SNR = 12 dB and exposed to GN(MB) interference.

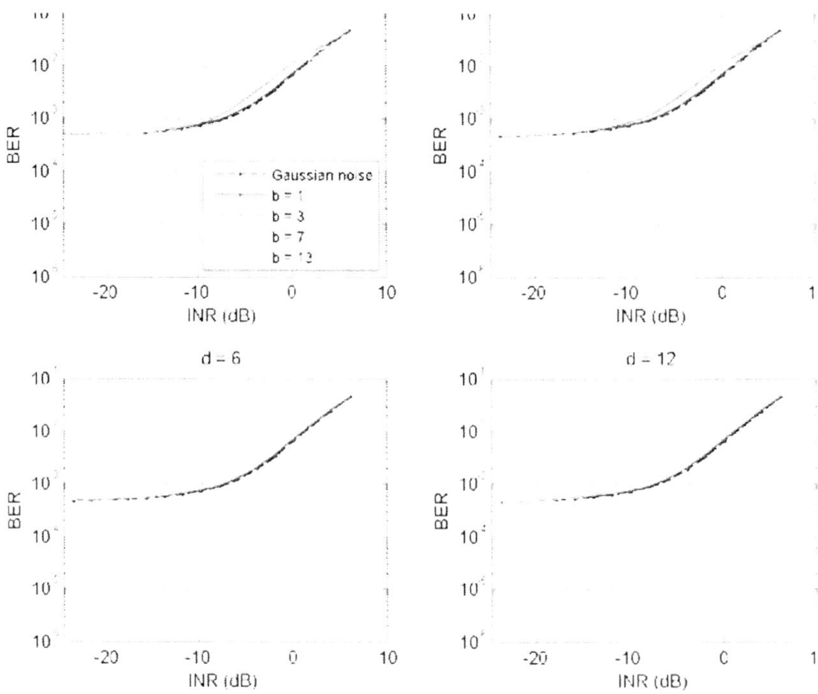

Figure A-5. BER versus INR for a DTV receiver operating at SNR = 12 dB and exposed to GN(MB) interference.

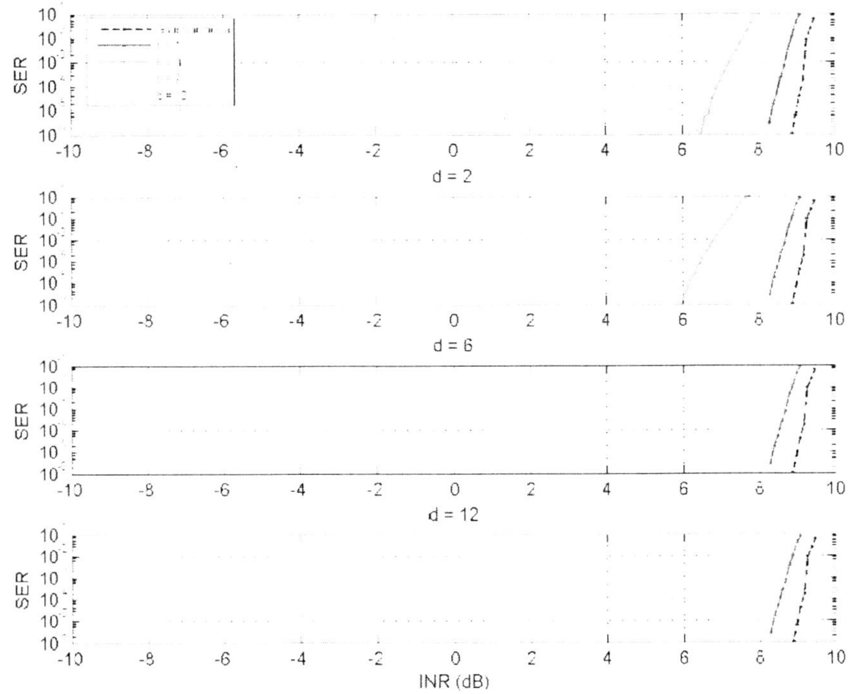

Figure A-6. SER versus INR for a DTV receiver operating at SNR = 15 dB and exposed to GN(MB) interference.

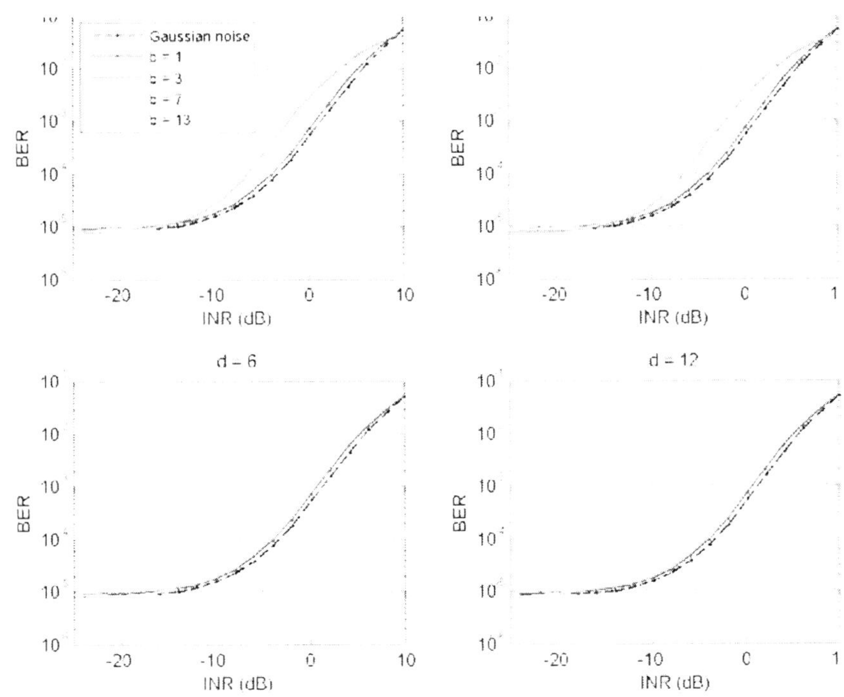

Figure A-7. BER versus INR for a DTV receiver operating at SNR = 15 dB and exposed to GN(MB) interference.

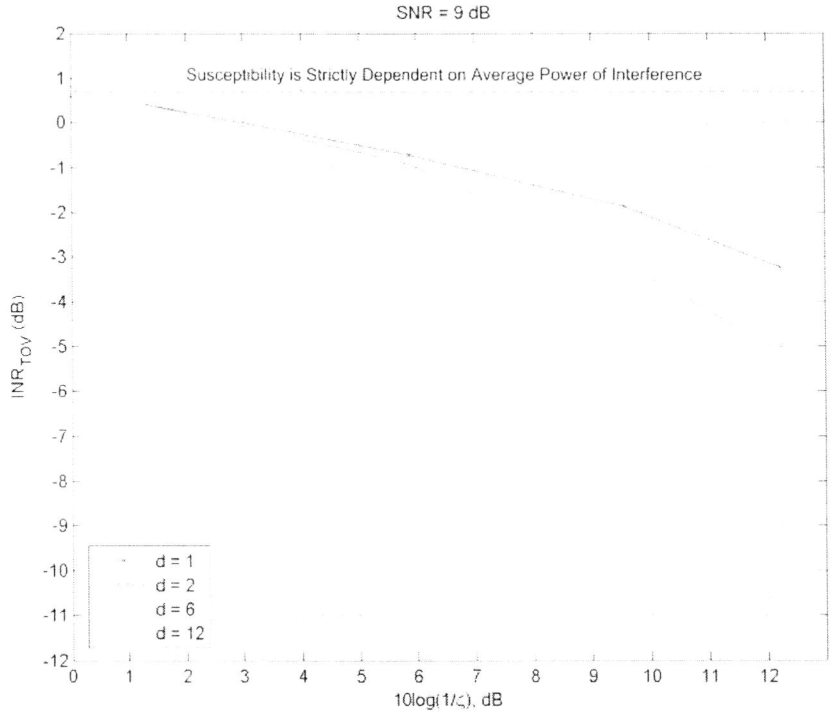

Figure A-8. INR_{TOV} versus 10log(1/ζ) for a DTV receiver operating at SNR = 9 dB and exposed to GN(MB) interference.

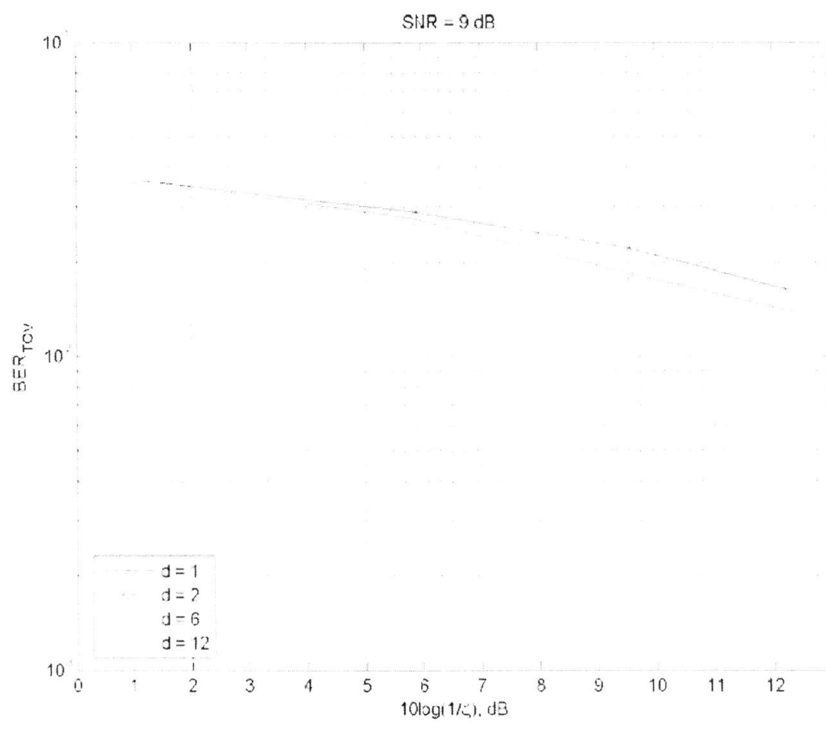

Figure A-9. BER_{TOV} versus 10log(1/ζ) for a DTV receiver operating at SNR = 9 dB and exposed to GN(MB) interference.

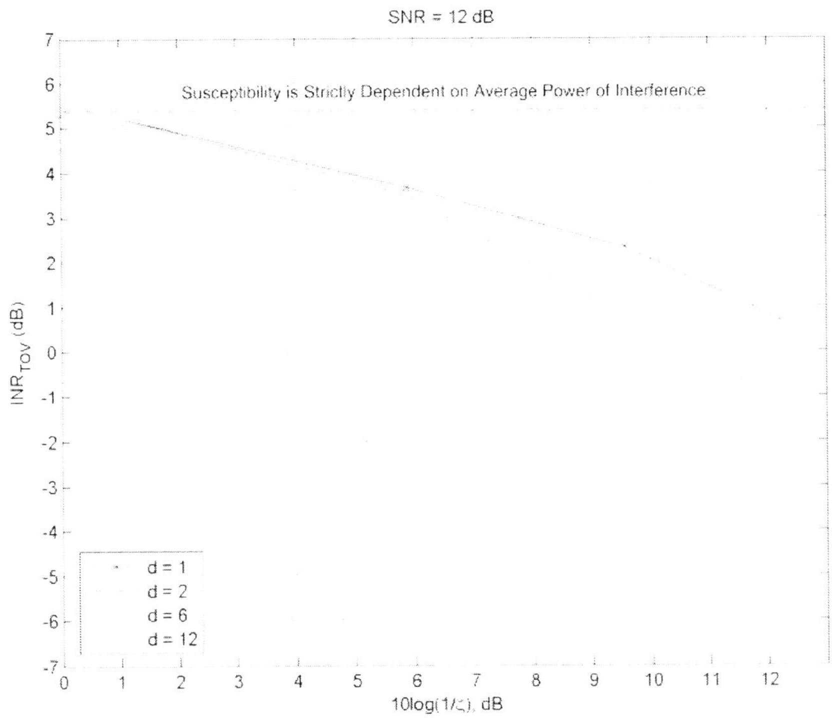

Figure A-10. INR_{TOV} versus 10log(1/ζ) for a DTV receiver operating at SNR = 12 dB and exposed to GN(MB) interference.

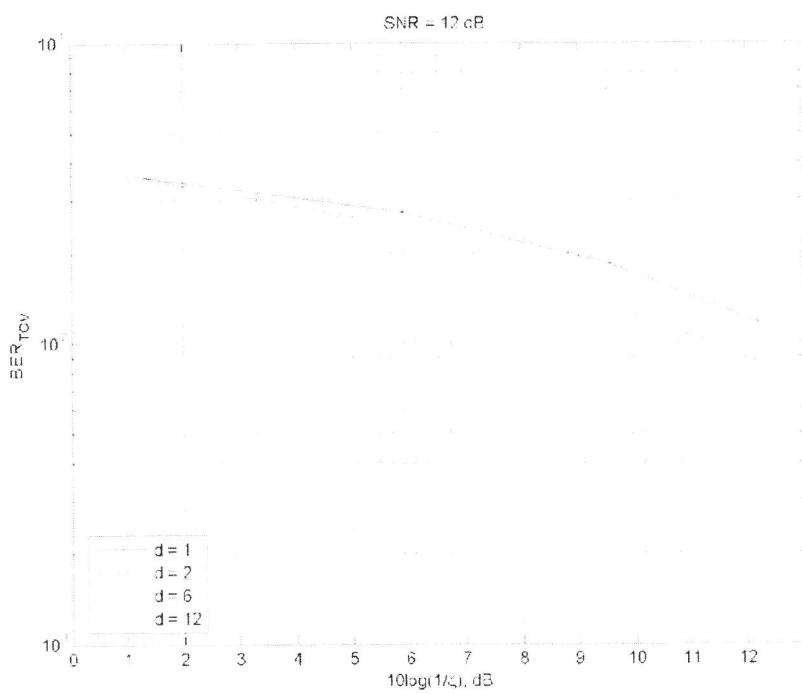

Figure A-11. BER_{TOV} versus 1 0log(1 /ζ) for a DTV receiver operating at SNR = 12 dB and exposed to GN(MB) interference.

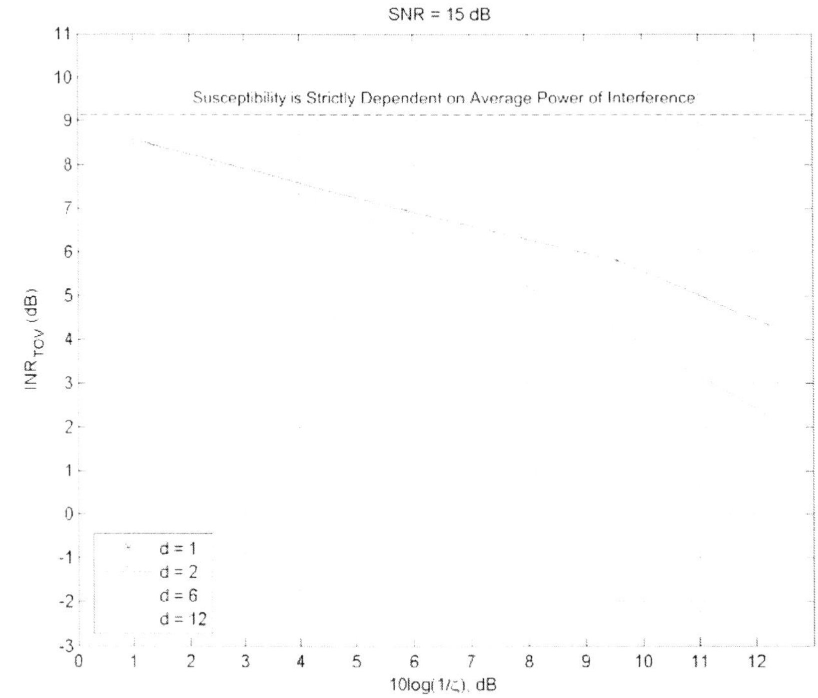

Figure A-12. INR_{TOV} versus 10log(1/ζ) for a DTV receiver operating at SNR = 15 dB and exposed to GN(MB) interference.

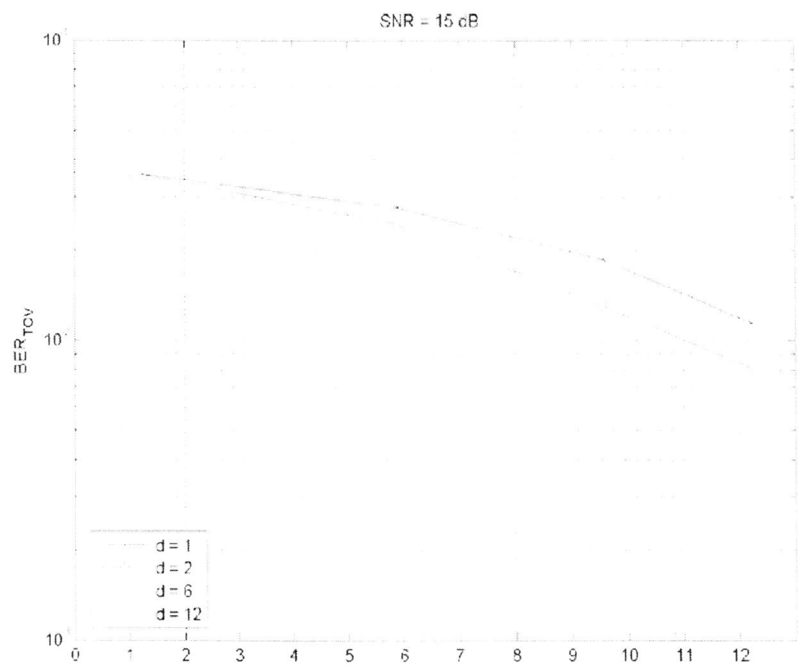

Figure A- 13. BER_{TOV} versus 1 0log(1 /ζ) for a DTV receiver operating at SNR = 15 dB and exposed to GN(MB) interference.

A.2. Characterization of GN(MB) Signals

Table A-3 provides band-limited temporal metrics compared to the approximate gating parameters of the GN(MB) signals. Table A-3 provides temporal characteristics of GN(MB) signals as table 3 does for GN signals.

Table A-3. Temporal Characteristics of GN(MB) Signals Band-Limited to B_{DTV}

GN	Gating Parameters			Band-Limited Metrics		
(MB)	τon (μs)	τof (μs)	ζ	BD (μs)	BI (μs)	ζDTV
01	0.24	0.07	0.7758			
02	0.24	0.70	0.2586	0.4	0.6	0.40
03	0.55	1.32	0.2586	0.7	1.2	0.37
04	0.24	1.95	0.1108	0.4	1.8	0.18
05	0.55	3.82	0.1108	0.7	3.7	0.16
06	1.80	11.32	0.1108	1.9	11.2	0.15
07	0.24	3.82	0.0597	0.4	3.7	0.10
08	0.55	7.57	0.0597	0.7	7.4	0.09
09	3.68	45.07	0.0597	3.8	44.9	0.08

Figures A-14 and A-15 provide amplitude probability distributions (APDs) of single- and multi-dwell GN(MB) signals, respectively. Table A-4 provides band-limited amplitude metrics compared to the corresponding ultra-wide bandwidth limits. Table A-4 provides amplitude characteristics of GN(MB) signals as table 4 does for GN signals.

Table A-4. Amplitude Characteristics of GN(MB) Signals Band-Limited to B_{DTV}

GN	Ultra-wide Bandwidth Limits		Band-Limited Metrics	
	P/A (dB)	APD	P/A (dB)	APD
01	10.6	RG	10.5	
02	14.8	RG	14.7	RG
03	14.8	RG	14.6	RG
04	18.0	RG	18.0	RG
05	18.0	RG	17.9	RG
06	18.0	RG	17.8	RG
07	20.3	RG	20.2	RG
08	20.3	RG	20.1	RG
09	20.3	RG	19.9	RG

Figures A-16 – A-17 provide power spectral densities (PSDs) for the GN(MB) signals. PSDs were calculated as described in Section 3.3. Block length was chosen as 170.6 μs (Δf = 5.9 kHz), which allowed 500 averages. The PSD of GN(MB)-09 (b = 13, d = 12) was excluded because it would have forced the number of averages to be less than 500. In figure A-17, the PSD of GN(MB)-05 (b = 7, d = 2) is directly beneath GN(MB)-06 (b = 7, d = 6).

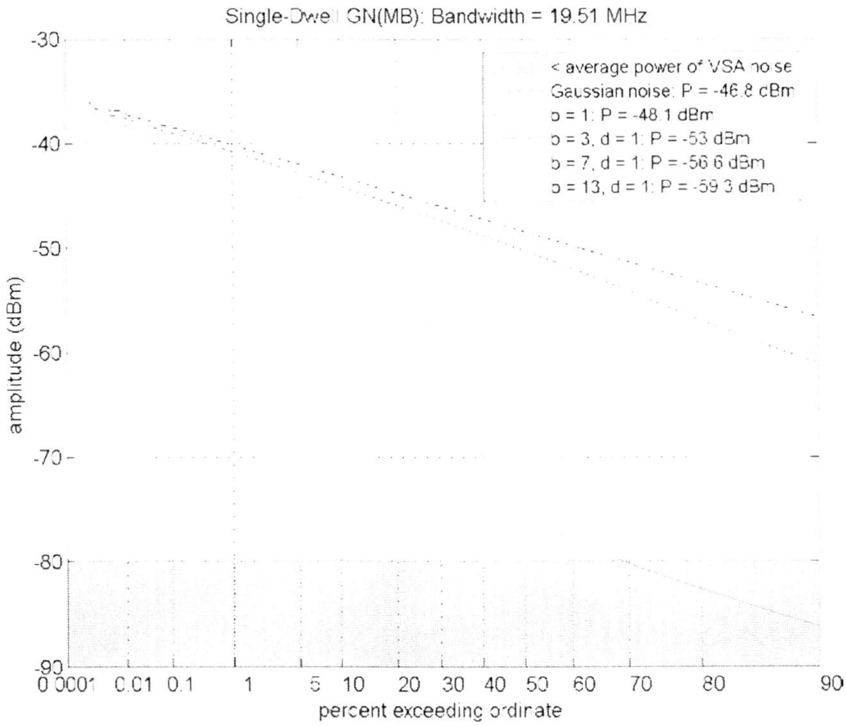

Figure A-14. $APDs$ of single-dwell GN(MB) signals band-limited to B_{DTV}.

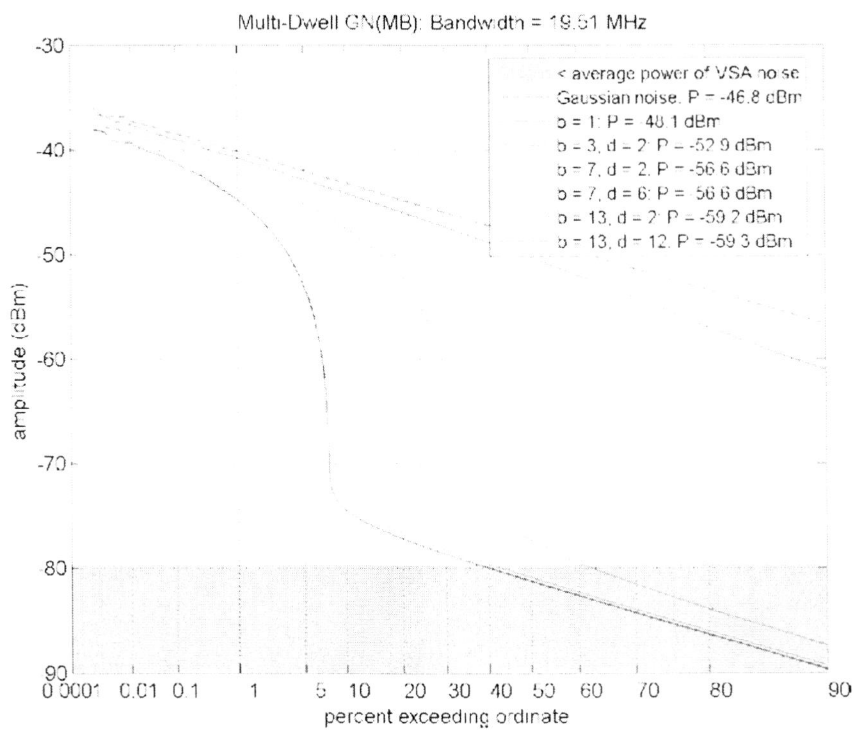

Figure A-15. $APDs$ of multi-dwell GN(MB) signals band-limited to B_{DTV}.

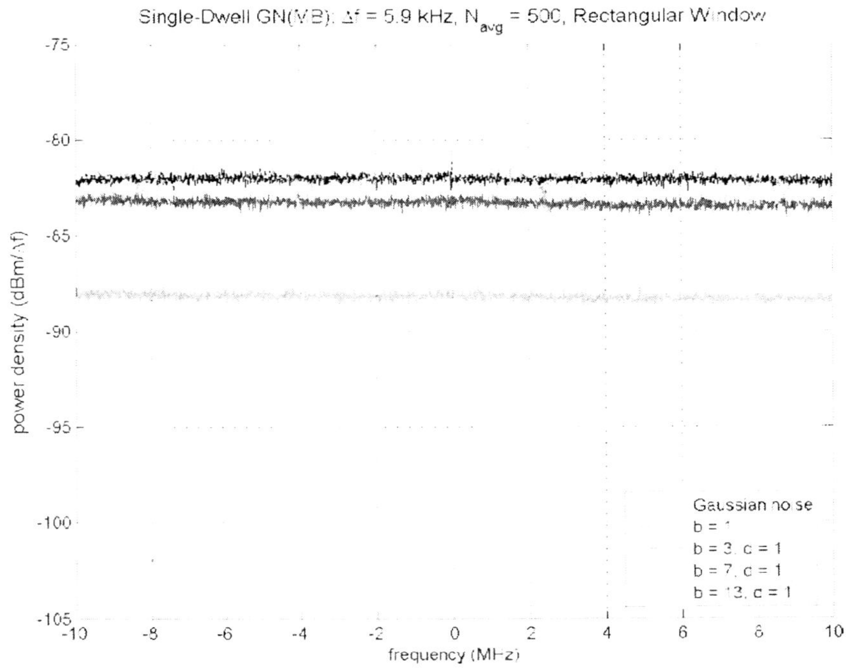

Figure A-16. PSDs of single-dwell GN(MB) signals.

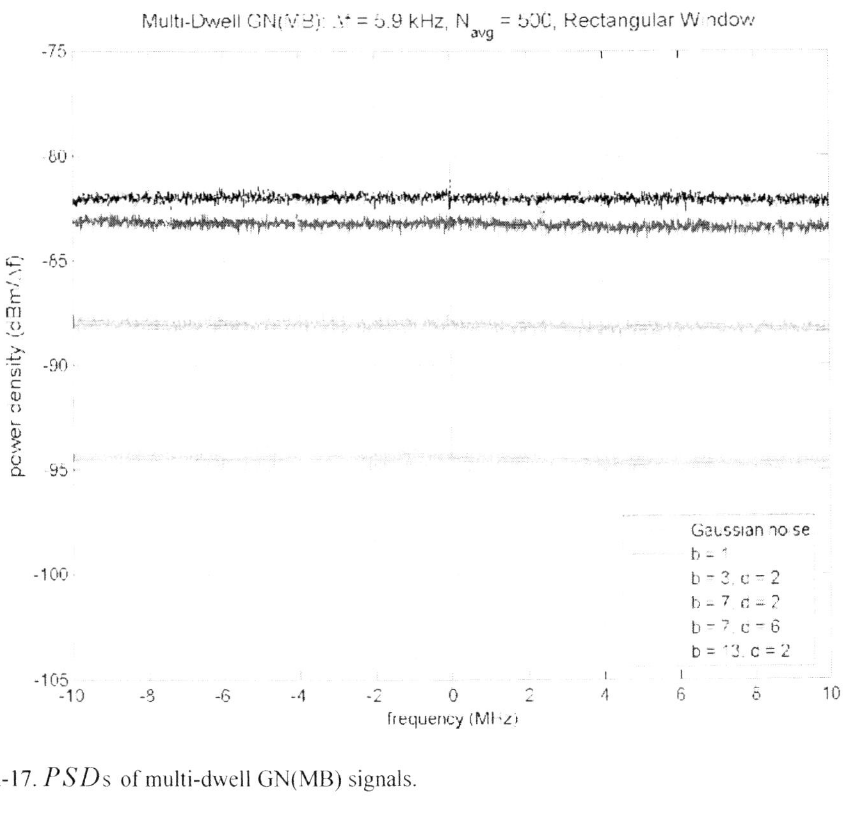

Figure A-17. PSDs of multi-dwell GN(MB) signals.

APPENDIX B: VERIFICATION OF SUSCEPTIBILITY RESULTS

Susceptibility of C-band satellite digital television (DTV) to gated-noise interference was measured with the test system and procedures described in Part 1 [6]. These tests measured effects on DTV signal quality due to increasing levels of interference. DTV signal quality metrics collected during the tests include modulation error ratio (MER), pre-Viterbi bit error rate (BER), and post-Reed-Solomon segment error rate (SER).

From the point of view of interference susceptibility tests, it is essential that signal power measurements and DTV signal quality metrics be precisely defined. This appendix gives an overview of the test system, provides precise measurement definitions for signal-to-noise ratio (SNR), interference-to-noise ratio (INR), MER, BER, and SER, and provides some degree of validation via comparison of theoretical and measured DTV signal quality metrics for Gaussian noise degradation.

B.1. Test System Overview

The test system was designed to inject carefully characterized interference into an operating satellite DTV receiver and measure susceptibility of the receiver with precisely defined DTV signal quality metrics. Figure B-1 is a block diagram of the test system hardware. In this figure BPF, LPF, and LNB correspond to bandpass filter, lowpass filter, and low-noise block downconverter.

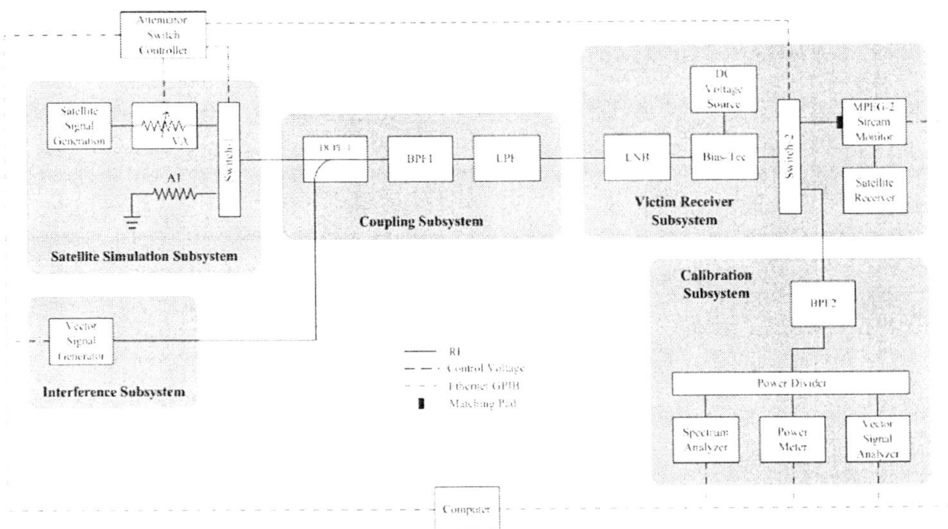

Figure B-1. Block diagram of DTV interference susceptibility test system.

The test system:

- simulated interference signals in software, so the widest range of UWB signal parameters were available for testing;
- generated interference signals with a vector signal generator (VSG), so *INR* was precisely controlled;
- characterized interference signals with a vector signal analyzer (VSA), which provided both amplitude and phase information for post-measurement analyses;
- simulated the satellite signal, so that scene content and *SNR* were controlled;
- conductively coupled signals to the satellite receiver, so uncertainties introduced by radio channel interference, attenuation, and distortion were eliminated;
- evaluated performance degradation objectively and quantitatively with DTV signal quality metrics from an MPEG-2 stream monitor;
- was completely characterized, so that its effects on the test were known;
- was under software control to eliminate procedural errors.

For this test, the operational scenario is summarized in table B-1. Compression, modulation, and error correction schemes comply with digital video broadcast (DVB) recommendations described in *ETS 300 421 [10]*.

B.2. Signal Calibration

At the beginning of each test, the system was calibrated to ensure that all hardware was operating properly and *SNR* and *INR* were correct. Figure B-2 provides a simplified model of the four calibration measurements: (m_1) VSA noise, (m_2) combined system noise and VSA noise, (m_3) combined DTV signal, system noise, and VSA noise, and (m_4) combined interference, system noise, and VSA noise. For these measurements, DTV signal power was set to a level corresponding to the appropriate *SNR*, the interfering signal power was set to a level that produced an accurate VSA measurement, and a 36-MHz VSA span was used. The satellite signal path (SSP) and interference signal path (ISP) were characterized in Appendix F of Part 1.

Table B-1. Operational Scenario for DTV Interference Susceptibility Tests

Transponder Center Frequency	3820 MHz
Modulation	QPSK with root-raised cosine filter ($\alpha = 0.35$)
Compression	MPEG-2
Data Rate	26.970353 Mbps
Symbol Rate	$RS = 19.510468$ Mbaud
Segment Rate	$Rseg = 17{,}932$ segments/second
Reed-Solomon Error Correction	$RRS = 188/204$
Convolutional Error Correction	$Rconv = 3/4$, $K = 7$
Interleaving	Depth = 12 bytes
Signal-to-Noise Ratio	{8.5, 11.5, 14.5} dB

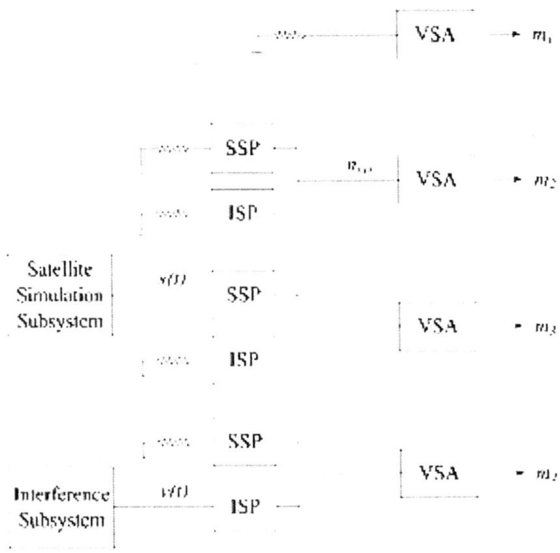

Figure B-2. Block diagrams for calibration measurements.

The following equations define these measurements

$$m_1(t) = n_{vsa}(t) + n_0(t)$$

$$m_2(t) = n_{sys}(t) - n_{vsa}(t)$$

$$m_3(t) = G_{SSP}\, s(t) + n_{sys}(t) + n_{vsa}(t)$$

$$m_4(t) = G_{ISP}\, v(t) - n_{sys}(t) - n_{vsa}(t)$$

where $n_{vsa}(t)$ is the internal noise of the VSA instrument, $n_0\ (t)$ is thermal noise, $n_{sys}(t)$ is the system noise, $s(t)$ is the satellite signal, $v(t)$ is the interference signal, and G_{SSP} and G_{ISP} are gains of SSP and ISP, respectively. The noise analysis was simplified by assuming the gain of the VSA to be one; this is justified because gain of the VSA is divided out in the power ratio calculations.

Measurement results were then band-limited in post-measurement processing by the victim receiver RRC filter

$$P_1 = \left\langle \left| m_1(t) * h_{rrc}(t) \right|^2 \right\rangle$$

where k is the measurement index, $h_{rrc}(t)$ is the impulse response of the victim receiver RRC filter, $\langle\,\rangle$ is the time-average operator, and is the convolutional operator. Power estimates were derived from approximately 2.8 milliseconds of data, i.e., $2^{17} = 131{,}072$ samples.

Power of the system noise (P_N), satellite signal (P_S), and interference signal (P_I) are calculated as

$$P_N \approx P_{[illegible]} - P_1$$

$$P_S = P_{[illegible]} - P_{[illegible]}$$

$$P_I = P_{[illegible]} - P_{[illegible]} \quad .$$

where P_N is a good estimate because $n_{sys}(t) >> n_0(t)$.

Finally, SNR and INR are computed as

$$SNR = \frac{P_S}{P_N}$$

$$INR = \frac{P_I}{P_N}$$

Example calculations are provided in the validation portion of this appendix (see table B-3).

B.3. Measured DTV Signal Quality

The MPEG-2 stream monitor was used to measure DTV signal quality. DTV signal quality metrics and their reference points within the receiver include *MER* and *BER* at the output of the quadrature demodulator, or equivalently at the input of the Viterbi decoder, and *SER* at the output of the Reed-Solomon (RS) decoder.

During the test, the following procedure was followed at each discrete interference level:

1. Interference was turned off and the receiver was allowed to acquire and demodulate the desired satellite signal in the absence of interference for 20 seconds.
2. Interference was applied for approximately 200 seconds.
3. *MER, BER,* cumulative transport error flag *(TEF),* and UTC time were sampled during the time of interference exposure at about 3 times per second and written to a data file.

DTV signal quality metrics were acquired with coarse and fine interference-level resolution. For coarse resolution, interference level was increased in 2-dB steps, beginning at a level that had negligible effect on DTV signal quality and ending where *SER* was in excess of 0.1. Fine-resolution acquisition began 2 dB below the level where the first *TEF* occurred during coarse-resolution acquisition and interference was incremented in 0.1 -dB steps until 0.1 *SER* was again achieved.

B.3.1. MER

Measured MER is defined as

$$MER = 10\log_{10}\left(\frac{\sum_{n=1}^{N}\left(I_n^2 + Q_n^2\right)}{\sum_{n=1}^{N}\left(\delta I_n^2 + \delta Q_n^2\right)}\right),$$

where I_n and Q_n are the ideal in- and quadrature-phase samples of the demodulated signal, δI and δQ are the error between actual and ideal samples, and N is the number of samples used for estimation. The MPEG-2 stream monitor utilized in this experiment derives *MER* from 512 samples collected at a rate of approximately 25 Hz.

B.3.2. BER

Measured pre-Viterbi bit error rate is defined as

$$BER = \frac{bit\ error\ count}{bit\ count}.$$

The MPEG-2 stream monitor estimates *BER* with the circuit illustrated in figure B-3. This circuit compares received bits to the estimated transmit bits. *BER* measurements use 2^{26} = 67,108,864 bits which provide good precision to approximately $7{\cdot}10^{-5}$ *BER* assuming 100 errors are needed.

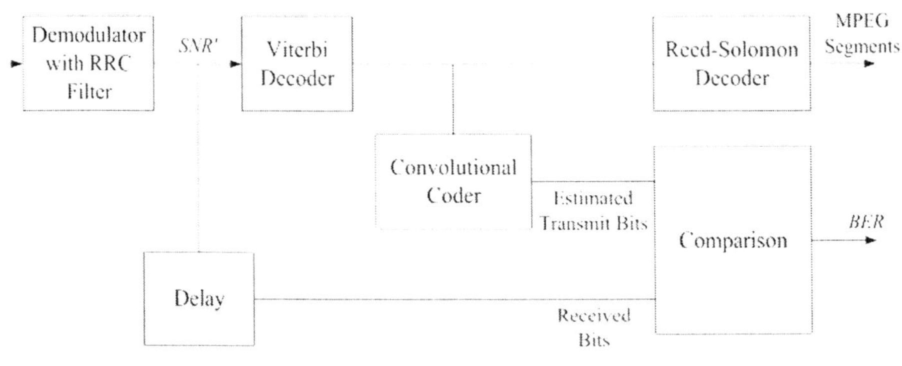

Figure B-3. *BER* estimation circuit used by MPEG-2 stream monitor.

Figure B-4 displays simulation results for the *BER* estimation circuit, where no interference was present. Good agreement was found for *SNR'* > 4.5 dB corresponding to *BER* < 0.05, where *SNR'* is referenced to the output of the demodulator. Poor agreement at lower *SNR'* was due to errors in the estimated transmit bits.

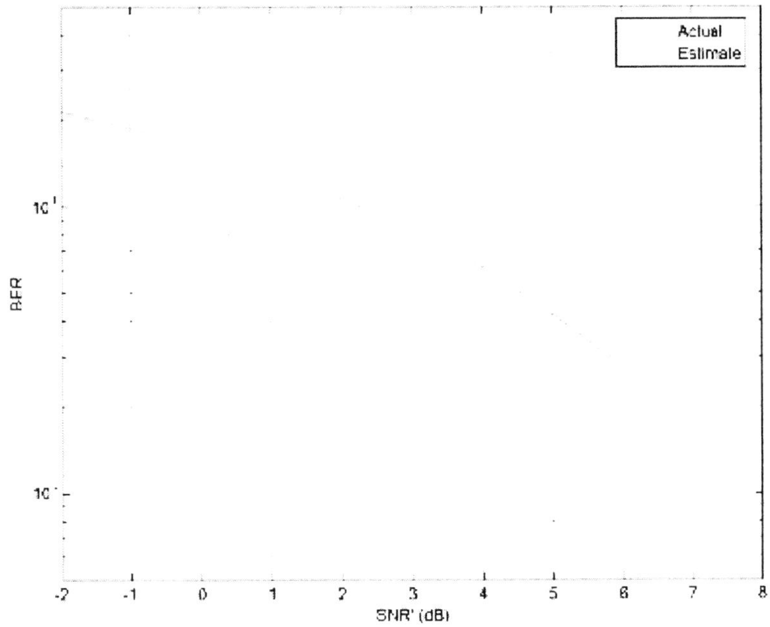

Figure B-4. Simulation of BER estimation circuit used by MPEG-2 stream monitor.

B.3.3. SER

Measured SER is defined by

$$SER = \frac{\text{segment error count}}{\text{segment count}} = \frac{TEF(t_2) - TEF(t_1)}{(t_2 - t_1)R_{seg}} .$$

where $TEF(t_k)$ is the cumulative number of segment errors at UTC time t_k and R_{seg} is the number of segments per second specified by the DTV operational scenario. Measured SER was computed from approximately 200 seconds of data. During this time period, approximately 3.6 million segments were transmitted. Each segment is considered to be an independent susceptibility trial.

Satellite and interference signals were not without systematic uncertainty. More specifically, sinusoidal variations of approximately 0.1 -dB peak-to-peak with periods of 20 – 50 minutes were measured in signal amplitudes with a power meter after the LNB. Subsequent SER fluctuations were approximately an order of magnitude, which is explained by the high sensitivity of SER to change in signal-to-noise ratio (see table B-2). This SER fluctuation is only evident during long measurement periods when interference is either absent or has negligible effect on SER. For the interference tests considered in this chapter, the 0.1 -dB increase in interference power every 220 seconds increased SER at a rate that rendered the SER systematic fluctuations indistinguishable.

B.4. Theoretical DTV Signal Quality for Gaussian Noise Degradation

This section provides theoretical expressions of *MER, BER,* and *SER* for Gaussian noise degradation. For a Gaussian noise radio environment, degradation is predictable from the signal-to-noise ratio at the output of the demodulator *(SNR ')*.

B.4.1. MER

For Gaussian noise degradation, theoretical *MER* is equivalent to *SNR '* in dB, that is

$$MER = 10\log(SNR') \quad .$$

B.4.2. BER

Theoretical probability of a bit error for QPSK, comparable to measured *BER,* is computed as

$$P_{QPSK} = Q\left(\sqrt{SNR'}\right) \quad .$$

where Q represents the Q-function defined as

$$Q(z) = \frac{1}{2\pi}\int_{z}^{\infty} \exp\left(\frac{-\xi^2}{2}\right) d\xi$$

B.4.3. SER

Probability of post-Viterbi bit error for QPSK is upper bounded by

$$P_{conv} \leq \frac{1}{K}\sum_{d=d_{free}}^{\infty} w(d)Q\left(\sqrt{d \cdot SNR'}\right) \quad .$$

where constraint length K is specified by the operational scenario, d_{free} is the free distance of the code, and $w(d)$ are the code weights.

Theoretical calculation of *SER* requires converting P_{conv} to P_{byte}, *i.e.*, the probability of a byte error. The conversion factor can be determined mathematically with the assumption of independent bit errors [11]; however, if more accuracy is desired the effects of correlated bit errors, caused by the Viterbi algorithm, can be simulated. Simulations of Gaussian noise degradation near the *SNR '* required for $PRS = 10^{-4}$ have shown a probability of a byte error approximately equal to three times the probability of a bit error, i.e., $P_{byte} \approx 3P_{conv}$.

For the RS scheme specified in table B-1, a segment error occurs when more than (204 – 188)/2 or 8 byte errors occur in a single segment. Hence, if post-Viterbi byte errors are independent, P_{byte} is small, and the number of bytes per segment is large, then the upper bound of the probability of a post RS segment error can be estimated by

$$P_{RS} = P(m > M) \leq 1 - \sum_{l=0}^{M} \frac{\lambda^{l}}{l!} e^{-\lambda} \quad .$$

where M =8 and $\lambda = 204 P_{byte}$ for the RS(204, 188) code.

Figure B-5 shows P_{QPSK}, P_{conv}, and P_{RS} as a function of SNR'. Table B-2 provides P_{QPSK} and P_{RS} at various SNR' near the threshold of visibility, i.e., $P_{RS} = 10^{-4}$. Notice that P_{RS} is sensitive to changes in SNR', varying as much as an order of magnitude for each 0. 1-dB increment in SNR'.

In contrast, P_{QPSK} is relatively insensitive. Such differences between P_{QPSK} and P_{RS} are attributed to the forward error correction (FEC).

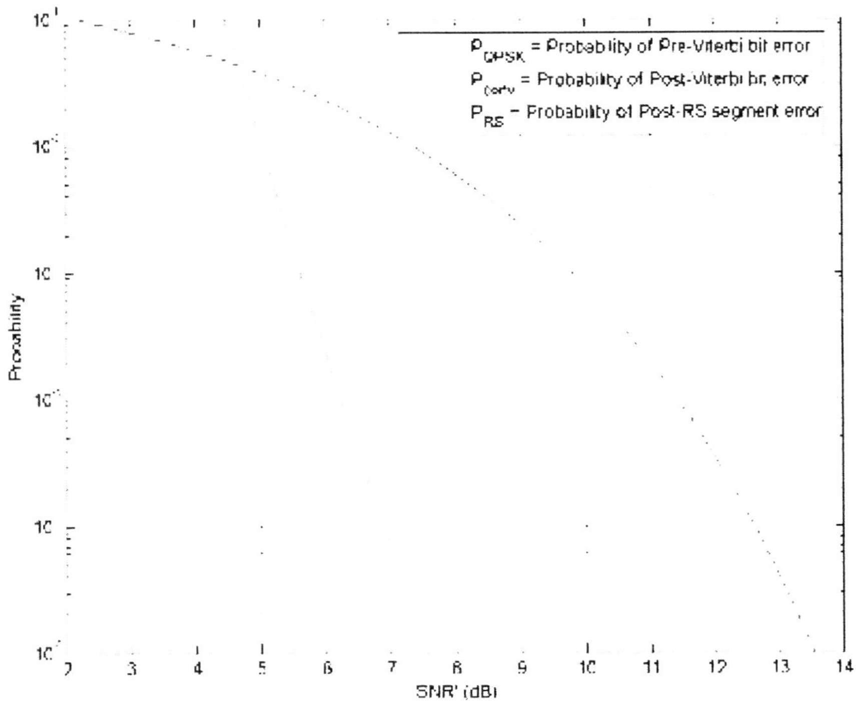

Figure B-5. Theoretical DTV receiver performance versus SNR'.

Table B-2. Probability of Pre-Viterbi Bit Error and Post-RS Segment Error Near TOV

SNR' (dB)	P_{QPSK}	P_{RS}
5.0	0.038	0.27
5.1	0.036	0.046
5.2	0.034	0.0045
5.3	0.033	0.00029
5.4	0.031	0.000014

B.5. Validation of Measured DTV Quality Metrics for Gaussian Noise Degradation

In this section, some degree of validation is accomplished by comparing theoretical predictions to measured DTV quality metrics for Gaussian noise degradation. Two scenarios are evaluated: (1) *BER* versus *SNR* in the absence of interference, and (2) *BER* and *SER* versus *INR* for a DTV receiver exposed to continuous noise interference. It is important to understand that results in this subsection relate to Gaussian noise degradation and cannot be generalized without further research.

B.5.1. BER as a Function of SNR in the Absence of Interference

The test system was set up as shown in figure B-1, except the interference signal path was terminated. *SNR, BER,* and *SER* were measured (as described in Sections B.2 and B.3) at various satellite signal powers. Figure B-6 shows measured *BER* and theoretical P_{QPSK} versus *SNR*.

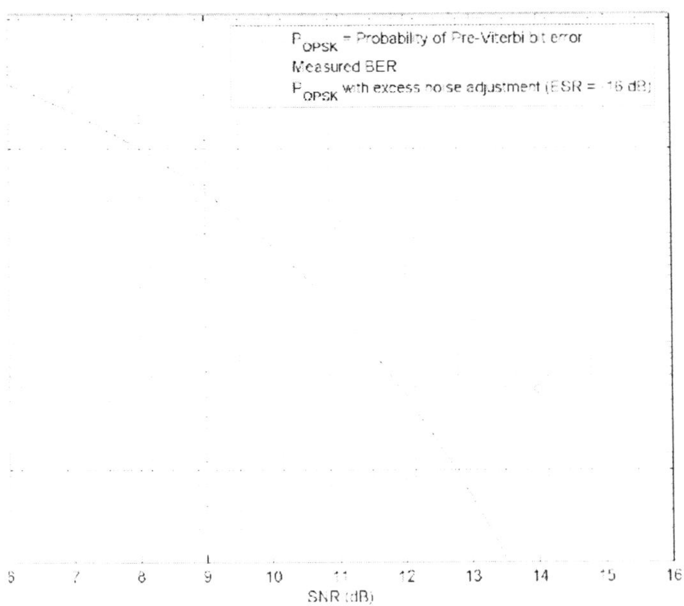

Figure B-6. Theoretical versus measured bit error rate performance of a DTV receiver.

Presumably, the primary difference between theory and measurement was due to distortions introduced by imperfect demodulation that effectively reduced the signal-to-noise ratio. Good agreement with measured *BER* was achieved when an excess noise term, proportional to the average power of the satellite signal, was used to model demodulator imperfections. More specifically, at the output of the demodulator

$$SNR' = \frac{P_S}{P_N + P_\varepsilon} = \frac{SNR}{1 + ESR \cdot SNR} \quad . \tag{B-1}$$

where P_Σ is the average power of the excess noise and excess-noise-to-signal ratio was determined empirically as $ESR = P_\Sigma / P_S = -16$ dB. Using this model, $SNR = \{9, 12, 15\}$ dB measured during the DTV susceptibility test corresponds to $SNR' = \{8.2, 10.5, 12.5\}$ dB.

B.5.2. BER and SER as a Function of INR for Gaussian-Noise Interference

In this subsection, an "effective signal-to-noise ratio" concept is developed to validate the GN-0 1 measurements. This concept is based on the assumption that receiver noise, excess noise, and GN-01 interference are all Gaussian noise processes. Theoretical expressions provided in Section B.4 are applicable since the addition of these noise processes results in a Gaussian noise process with an average power equal to the sum of its constituents. Hence, in the case of GN-01, the effective signal-to-noise ratio at the output of the demodulator is expressed as

$$SNR'_{eff} = \frac{P_S}{P_N + P_I + P_\varepsilon} = \frac{SNR}{1 + INR + ESR \cdot SNR} \quad .$$

Figures B-7 and B-8 provide theoretical versus measured *BER* and *SER*, respectively, for a DTV receiver operating at the different *SNR*s and exposed to a range of Gaussian noise interference levels. Table B-3 provides theoretical and measured test results for Gaussian noise interference that caused degradation at the TOV. Calibration measurements with the VSA provided P_1, P_2, P_3, and P_4. These measurements were used to compute P_S, P_I, *SNR*, and *INR*. Interestingly, P_S / P_4, which corresponds to the ratio of signal power to the combined power of interference and system noise, was approximately 5.5 dB in all cases. In general, theoretical results using the SNR'_{eff} model demonstrated good agreement with measured data.

Table B-3. Signal Powers at the TOV for Gaussian Noise Interference

Calibration Measurements			Susceptibility Test Results					Theoretical Results
P_2 (dBm)	P_3 (dBm)	P_4 (dBm)	P_S (dBm)	P_I (dBm)	P_S/P_4 (dB)	SNR (dB)	INR_{TOV} (dB)	INR_{TOV} (dB)
-47.8	38.4	44.4	-38.9	-47.1	5.5	8.9	0.7	0.5
-47.7	-35.4	-41.2	-35.7	-42.3	5.5	12.0	5.4	5.1
-47.8	-32.6	-38.1	-32.7	-38.6	5.4	15.0	9.2	8.7

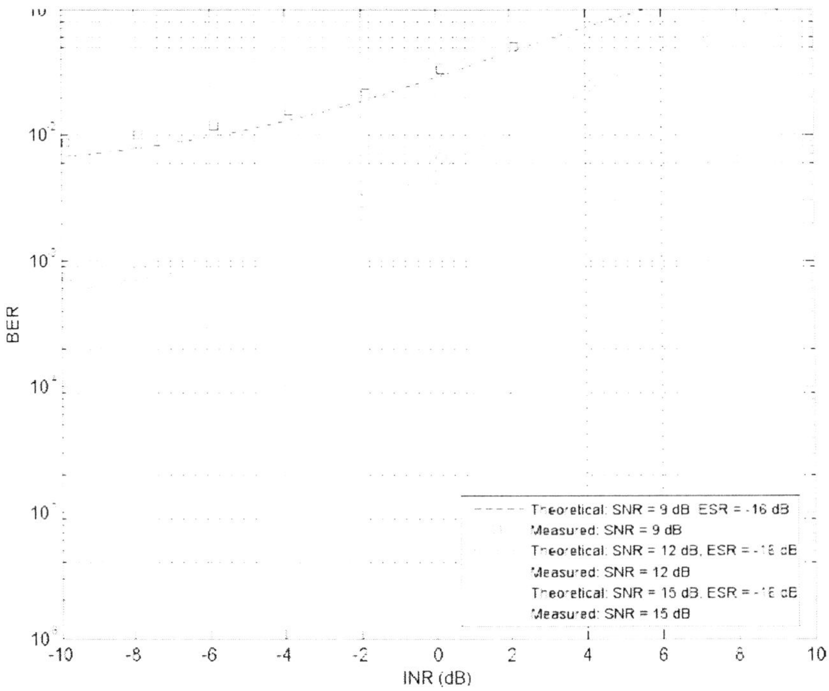

Figure B-7. Theoretical and measured BER for Gaussian noise interference.

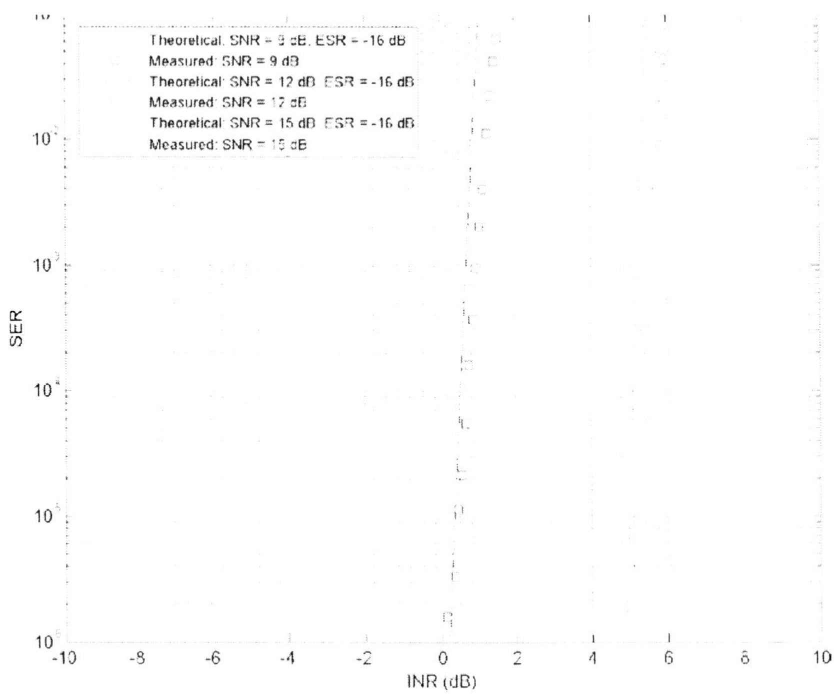

Figure B-8. Theoretical and measured SER for Gaussian noise interference.

APPENDIX C: THEORETICAL ANALYSIS OF GATED GAUSSIAN NOISE

This appendix provides theoretical analyses of gated Gaussian noise expressed mathematically as

$$u(t) = z(t)g(t;T,\tau_{on}) = [n(t) * h_{BPF}(t)]g(t;T,\tau_{on}) \quad .$$

where $z(t)$ is a Gaussian noise random process with an ultra-wide transmission bandwidth (B_{UWB}) much greater than the victim receiver bandwidth (B_{DTV}), $n(t)$ is a zero-mean white Gaussian noise random process, $h_{BPF}(t)$ is the impulse response of a band-pass filter that defines B_{UWB}, and * is the convolutional operator. The ideal gating function is described mathematically as

$$g(t;T,\tau_{on}) = \sum_{k=-\infty}^{\infty} R(t-kT;\tau_{on}) \quad , \text{ where } \quad R(t;\tau_{on}) = \begin{cases} 1 & (S_1 : \ 0 \le t < \tau_{on}) \\ 0 & (S_2 : \ \tau_{on} \le t < T) \end{cases}$$

and S_1 and S_2 refer to gated-on and gated-off states, respectively. The complex-baseband (CBB) representation, denoted by ˆ, of u(t) is modeled as

$$\hat{u}(t) = \hat{z}(t)g(t;T,\tau_{on}) \quad .$$

where $\hat{z}(t)$ is a complex Gaussian-distributed random variable with zero-mean. The variance of $\hat{z}(t)$ is $\sigma_t^2 = 2\sigma^2$ where σ^2 is the variance of the real and imaginary parts. The following amplitude probability distribution *(APD)* and power spectral density *(PSD)* expressions are derived in terms of $\hat{u}(t)$

C.1. Amplitude Probability Distribution

Computation of the *APD* for gated Gaussian noise begins with the total probability expression

$$\wp(A \le a) = \wp(A \le a \mid S_1)\wp(S_1) + \wp(A \le a \mid S_2)\wp(S_2) \quad . \tag{C-1}$$

where $\wp$ represents probability, $A = |\hat{u}(t)|$ is the amplitude random variable which takes on values of a ranging from 0 to ∞ , $\wp(S_k)$ represents the probability of being in state S_k , and $\wp(A \le a \mid S_k)$ represents the conditional probability of $A \le a$ given the state S_k.

Amplitudes are Rayleigh distributed in state S_k giving the probability density function *(PDF)*

$$q(a \mid S_1) = \frac{a}{\sigma^2}\exp\left(\frac{-a^2}{2\sigma^2}\right) \quad .$$

The corresponding cumulative distribution function *(CDF)* is

$$\wp(A \le a \mid S_1) = 1 - \exp\left(\frac{-a^2}{2\sigma^2}\right) \quad .$$

$\underline{a}$

Similarly, the amplitude is always 0 in state S_2 giving the *PDF*

$$q(a \mid S_2) = \delta(a) \quad .$$

where $\delta(a)$ is the delta function. The corresponding *CDF* is acquired by integrating the *PDF* over all amplitudes which yields the step function $U(a)$. Recall that $a \ge 0$, hence $U(a) = 1$ yielding

$$\wp(A \le a \mid S_2) = 1 \quad .$$

Probabilities of being in each state are expressed in terms of the fractional on-time (ζ)

$$\wp(S_1) = \zeta \quad ; \qquad \wp(S_2) = (1 - \zeta) \quad .$$

Substituting these expressions into the total probability expression (C-1) yields the *CDF*

$$\wp(A \le a) = 1 - \zeta \exp\left(\frac{-a^2}{2\sigma^2}\right) \quad .$$

The corresponding *APD* is obtained by subtracting the *CDF* from 1

$$\wp(A > a) = \zeta \exp\left(\frac{-a^2}{2\sigma^2}\right) \quad . \tag{C-2}$$

Figure C-1 provides *APDs* of normalized ($\sigma^2 = 1$) gated Gaussian noise with $\zeta = \{1, 0.5, 0.25, 0.125, 0.0625\}$ in an ultra-wide transmission bandwidth.

Peak-to-average ratio *(P/A)* is computed from the peak and root-mean-square (rms) amplitudes. Peak amplitude (a_{pk}) is defined by the percentile ρ. Replacing $\wp(A > a)$ with ρ and a with a_{pk} in equation (C-2) and solving for a_{pk} yields

$$a_{pk} = \sigma\sqrt{\ln(\zeta/\rho)} \quad .$$

This expression is only valid when $\zeta \geq \rho$.

Rms amplitude is found by substituting the probability densities given previously into the derivative of (C-1) with respect to a to obtain

$$q(a) = q(a \mid S_1)\wp(S_1) + q(a \mid S_2)\wp(S_2) = \zeta\frac{a}{\sigma^2}\exp\left(\frac{-a^2}{2\sigma^2}\right) + (1-\zeta)\delta(a) \quad .$$

and using this PDF to calculate rms amplitude

$$a_{rms} = \sqrt{\int_0^\infty a^2 q(a)da} = \sigma\sqrt{\zeta} \quad .$$

Finally, the expression for *P/A* is

$$P/A = \left(\frac{a_{pk}}{a_{rms}}\right)^2 = \frac{\ln(\zeta/\rho)}{\zeta} \quad .$$

Figure C-2 provides *P/A* as a function of ζ for gated noise in the ultra-wide transmission bandwidth; peak definitions considered in figure C-2 are $\rho = \{0.001, 0.01, 0.1, 1.0\}$ %.

C.2. Power Spectral Density

Derivation of the *PSD* for gated Gaussian noise begins with the autocorrelation function

$$R_{\hat{n}\hat{n}}(t,t+\tau) = E\{\hat{n}(t)\hat{n}^*(t+\tau)\} \quad .$$

where *E{ }* denotes expectation and * denotes conjugation. Pulling the deterministic gating function out of the expectation gives

The first complex-exponential term causes the autocorrelation to be time-varying. This variation is periodic, in that

$$R_{\hat{n}\hat{n}}(t,t-\tau) = g(t)g^*(t+\tau)R_{zz}(\tau) \quad .$$

and is removed by averaging over the gating period. The resulting time-averaged autocorrelation function is

$$\langle R_{\hat{n}\hat{n}}(\tau)\rangle = \frac{1}{T}\int_0^T R_{\hat{n}\hat{n}}(t,t+\tau)dt = \sum_{l=-\infty}^{\infty}\sum_{k=-\infty}^{\infty} c_l c_k^* \exp(-j2\pi k\tau/T)R_{zz}(\tau)\frac{1}{T}\int_0^T \exp(j2\pi(l-k)t/T)dt \quad .$$

which evaluates to

$$\langle R_{\hat{n}\hat{n}}(\tau)\rangle = \begin{cases} \sum_{k=-\infty}^{\infty} |c_k|^2 \exp(-j2\pi k\tau/T)R_{zz}(\tau) & (l = k) \\ 0 & (l \neq k) \end{cases} \quad .$$

The *PSD* is derived from the Fourier transform of the time-averaged autocorrelation function

$$\langle S_{\hat{n}\hat{n}}(f)\rangle = \int_{-\infty}^{\infty} \langle R_{\hat{n}\hat{n}}(\tau)\rangle \exp(-j2\pi f\tau)d\tau \quad .$$

Substitution yields the following expression for the *PSD* of gated Gaussian noise

$$\langle S_{\hat{n}\hat{n}}(f)\rangle = \sum_{k=-\infty}^{\infty} |c_k|^2 \delta(f + \frac{k}{T}) * S_{zz}(f) \quad .$$

Note that $S_{nn}(f)$ is constant, yielding

$$S_{zz}(f) = n_0 \left| H_{LPF}(f) \right|^2 .$$

where n_0 is the power density of the underlying white Gaussian process and $H_{LPF}(f)$ is the CBB representation of $H_{BPF}(f)$. This simplifies the *PSD* expression

$$\langle S_{ww}(f) \rangle = n_0 \sum_{k=-\infty}^{\infty} \left| c_k \right|^2 \delta(f - \frac{k}{T}) * \left| H_{LPF}(f) \right|^2 .$$

In this experiment we assume the transmission filter, $H_{LPF}(f)$, is constant over the bandwidth of the victim receiver. Hence, the *PSD* of gated noise is constant for all frequencies considered.

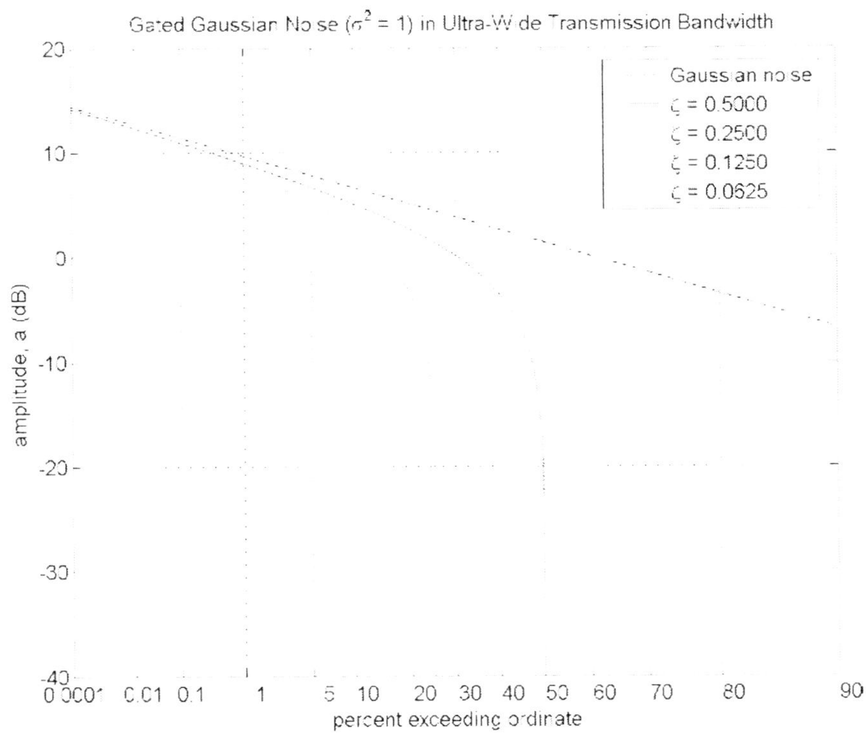

Figure C-1. Theoretical *APD*s of gated noise in B_{UWB}

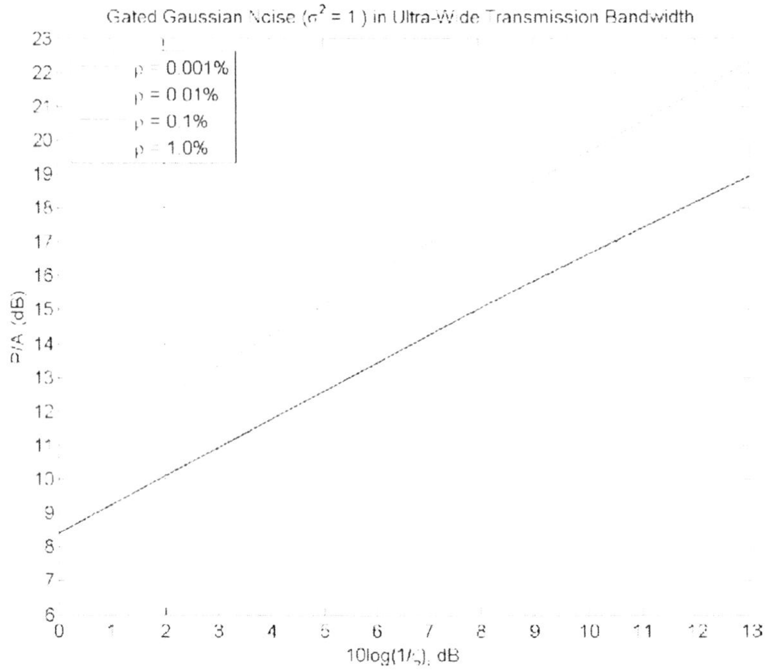

Figure C-2. Theoretical P/A of gated noise in B_{UWB}

APPENDIX D: CHARACTERIZATION MEASUREMENTS

This appendix provides a description of measurement setup and signal characterization measurements that complement those in Section 3.

D.1. Measurement Setup

Interference signals were measured under optimal conditions at the radio frequency (RF) output of the vector signal generator (VSG) as described in figure D- 1. The VSG was set up with the procedures described in Appendix B of Part 1 [6]. Measurements at RF were made with the vector signal analyzer (VSA) and digital oscilloscope (DO).

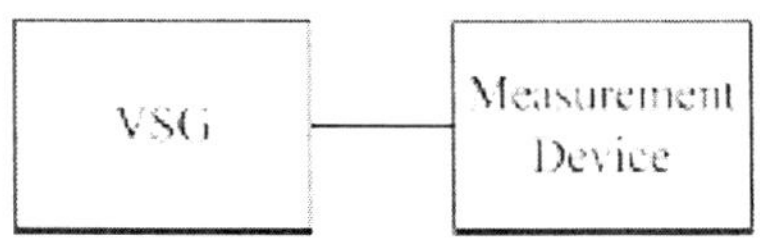

Figure D-1. Setup for RF signal characterization measurements.

The gated-noise signals, GN-02 – GN-17, were simulated by applying different gating signals to the continuous-noise signal, GN-0 1. This ensured that voltages of all signals were identical when gated on. Gated-noise signals, generated in hardware with the setup described by figure C-2 in Part 1, were also measured to verify corresponding VSG-generated signals.

Characteristics of the hardware-generated signals matched those of the VSG-generated signals and therefore are not included in this chapter.

For VSA characterization measurements at RF, the VSG output level was set to -28.19 dBm for GN-01 through GN-17 and -27.17 dBm for GN(MB)-01 through GN(MB)-09. At these levels, the VSG generated the "continuous" signals, GN-01 and GN(MB)-01, at an average power of approximately -45 dBm in a 36-MHz bandwidth at the output of the cable. This optimally utilized the dynamic range of the VSA as discussed in Appendix E of Part 1. VSA instrument settings are summarized in table D-1.

Table D-1. RF VSA Measurement Settings

Post-Processing Bandwidths (MHz)	Span (MHz)	Number of Samples	Center Frequency (MHz)	Input Range (dBm)
36, 19.51, 10	36	> 360,000		
1	4.5	> 450,000	3820	-30
0.1	0.5625	> 562,500		

DO measurements at the maximum sample rate, i.e., 20 GSps, provided real time-domain representations of RF signals. DO measurements were made with the VSG output level set to a higher value, i.e., 0 dBm, to compensate for low DO sensitivity at the maximum sample rate.

Interference signals were measured under TOV operational conditions at the IF output of the low-noise block downconverter (LNB) with the setup described in figure 20 of Part 1. The simplified block diagram of those components that affect the IF signal characterization measurements are shown in figure D-2. Bandpass filter BPF1 has a 1-dB bandwidth equal to 40 MHz. The VSG output level was set to INR_{TOV}. Measurements at IF were made with the VSA only. VSA settings for IF measurements were the same as those used for RF measurements described in table D- 1, except the center frequency was set to 1330 MHz.

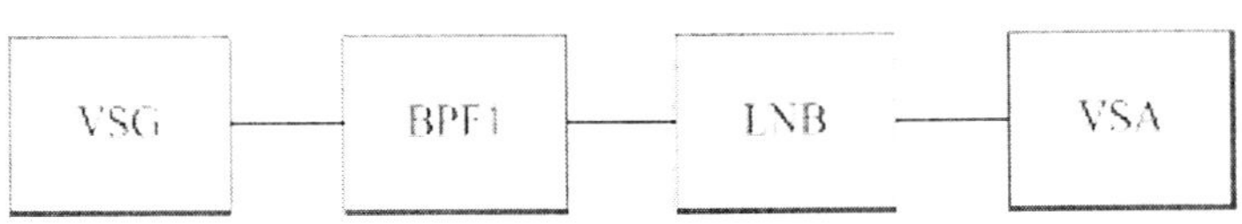

Figure D-2. Simplified setup for IF signal characterization measurements.

D.2. Results

This subsection provides amplitude statistics and time series of the gated-noise signals. Odd numbered plots in figures D-3 – D-54 provide RF measurement results organized as:

a. Real time series measured with DO and amplitude time series measured with the VSA and band-limited to B_{DTV} = 19.51 MHz.
b. *APD*s measured with a VSA and band-limited to 36, 10, 1, and 0.1 MHz as well as B_{DTV}.

Table D-2. RF P/A (dB) of GN and GN(MB) Signals Band-Limited to Different Bandwidths

Type	Index	B_{UWB}	36.0 MHz	B_{DTV}	10.0 MHz	1.0 MHz	0.1 MHz
GN	01	9.6	9.6	9.6	9.8	9.8	9.0
	02	12.3	9.7	9.9	9.5	9.7	9.6
	03	14.9	11.0	9.7	9.8	9.5	8.9
	04	17.5	13.1	11.5	9.4	9.5	10.1
	05	20.1	15.9	14.2	11.7	9.8	9.1
	06	12.3	11.8	12.2	11.2	9.5	9.3
	07	14.9	14.4	14.6	13.9	9.6	9.9
	08	17.5	16.8	17.3	16.5	9.7	9.4
	09	20.1	19.8	19.6	18.9	11.7	9.2
	10	12.3	12.5	12.4	12.5	11.1	10.1
	11	14.9	15.1	15.0	14.8	13.6	9.8
	12	17.5	17.9	17.6	17.6	16.3	9.4
	13	20.1	20.3	20.3	20.0	18.8	11.0
	14	12.3	12.6	12.3	12.2	12.3	10.3
	15	14.9	15.3	15.1	14.8	14.8	14.1
	16	17.5	17.4	17.8	17.9	17.6	15.7
	17	20.1	19.9	20.1	20.0	19.9	16.4
GN (MB)	01	10.6	10.5	10.5	10.1	9.6	9.7
	02	14.8	14.4	14.7	14.6	9.8	9.5
	03	14.8	14.7	14.6	14.5	11.6	9.5
	04	18.0	17.5	18.0	17.6	12.7	9.7
	05	18.0	17.6	17.9	17.5	14.7	9.5
	06	18.0	17.6	17.8	17.6	16.5	10.6
	07	20.3	20.1	20.2	19.7	15.0	9.4
	08	20.3	20.0	20.1	19.8	16.8	9.0
	09	20.3	20.1	19.9	19.5	19.0	15.0

Shaded cells highlight Rayleigh-amplitude statistics.

Even numbered plots in figures D-3 – D-54 provide IF measurement results organized as:

a. Amplitude time series measured with the VSA and band-limited to B_{DTV}.
b. *APD*s measured with the VSA and band-limited to B_{DTV}.

Table D-2 provides peak-to-average (P/A) values for the gated-noise signals measured at RF and band-limited to various bandwidths in post-measurement processing. The shaded cells highlight signals with Rayleigh-distributed amplitudes. Ultra-wide transmission bandwidth (B_{UWB}) limits for P/A were derived in Appendix C.

Table D-3 provides P/A values for gated-noise signals measured at IF under TOV operational conditions and band-limited to B_{DTV} in post-measurement processing.

Table D-3. IF *P/A* (dB) of GN and GN(MB) Signals at $INR_{TO\ V}$ and Band-Limited to B_{DTV}

Type	Index	$SNR \approx 9$ dB	$SNR \approx 12$ dB	$SNR \approx 15$ dB
GN	01	9.6	9.6	9.4
	02	9.8	9.8	9.9
	03	9.6	9.5	9.5
	04	10.4	11.1	11.1
	05	12.0	13.2	13.6
	06	10.8	11.1	11.7
	07	12.1	13.2	13.8
	08	14.1	15.2	16.4
	09	16.1	17.8	18.8
	10	11.2	11.4	11.8
	11	11.9	13.1	13.7
	12	13.0	14.6	15.8
	13	13.5	16.2	17.5
	14	10.7	11.5	12.0
	15	11.4	12.8	13.6
	16	12.0	13.8	15.3
	17	11.9	14.3	16.6
GN (MB)	01	10.2	10.2	10.3
	02	12.1	13.3	14.3
	03	12.0	13.4	13.9
	04	14.0	16.2	16.7
	05	13.5	15.4	16.4
	06	12.6	14.6	15.8
	07	15.0	17.8	18.7
	08	14.3	16.5	17.9
	09	12.7	15.2	17.4

* Shaded cells highlight Rayleigh-amplitude statistics.

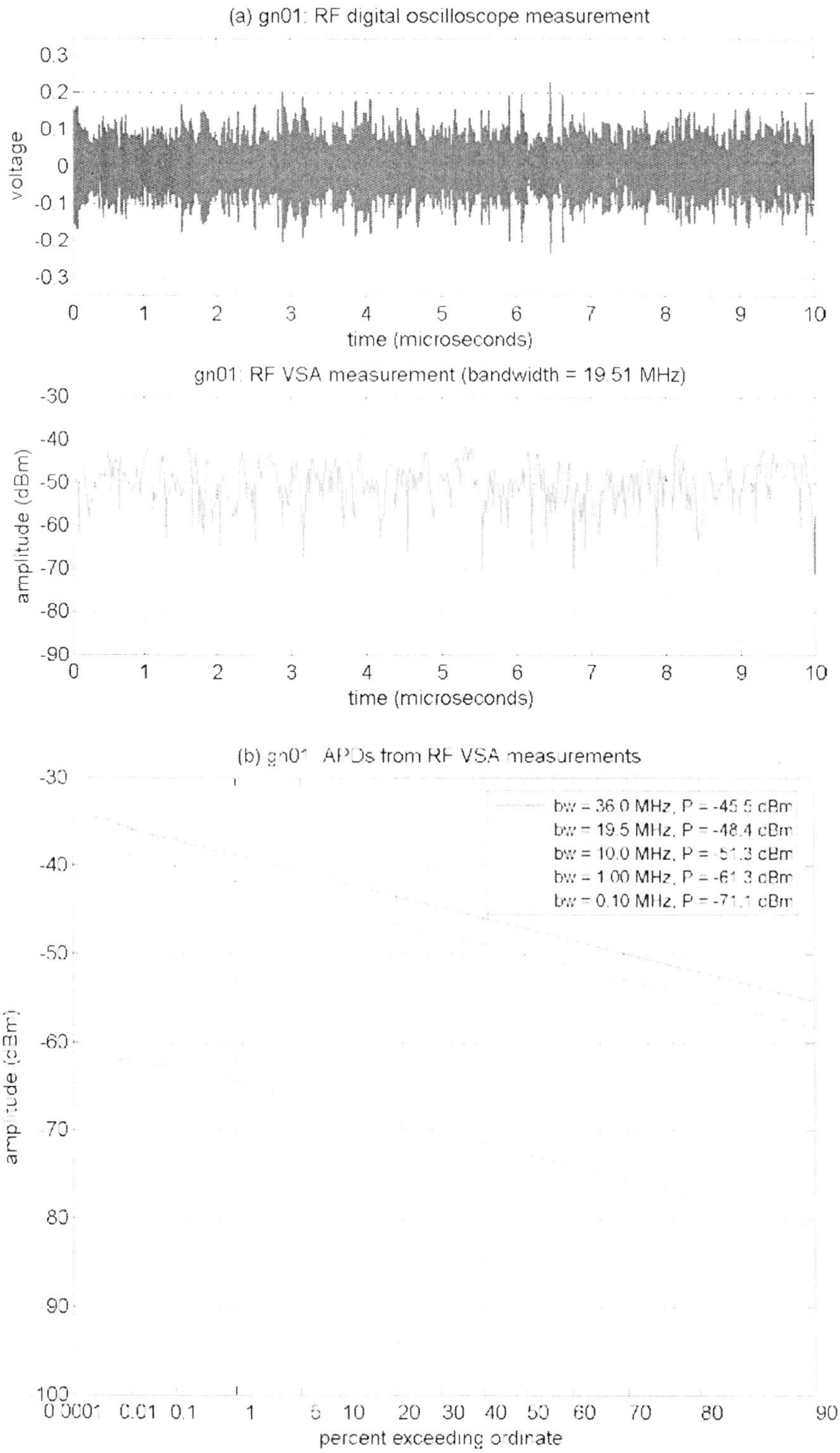

Figure D-3. RF measurements of GN-01, Gaussian noise.

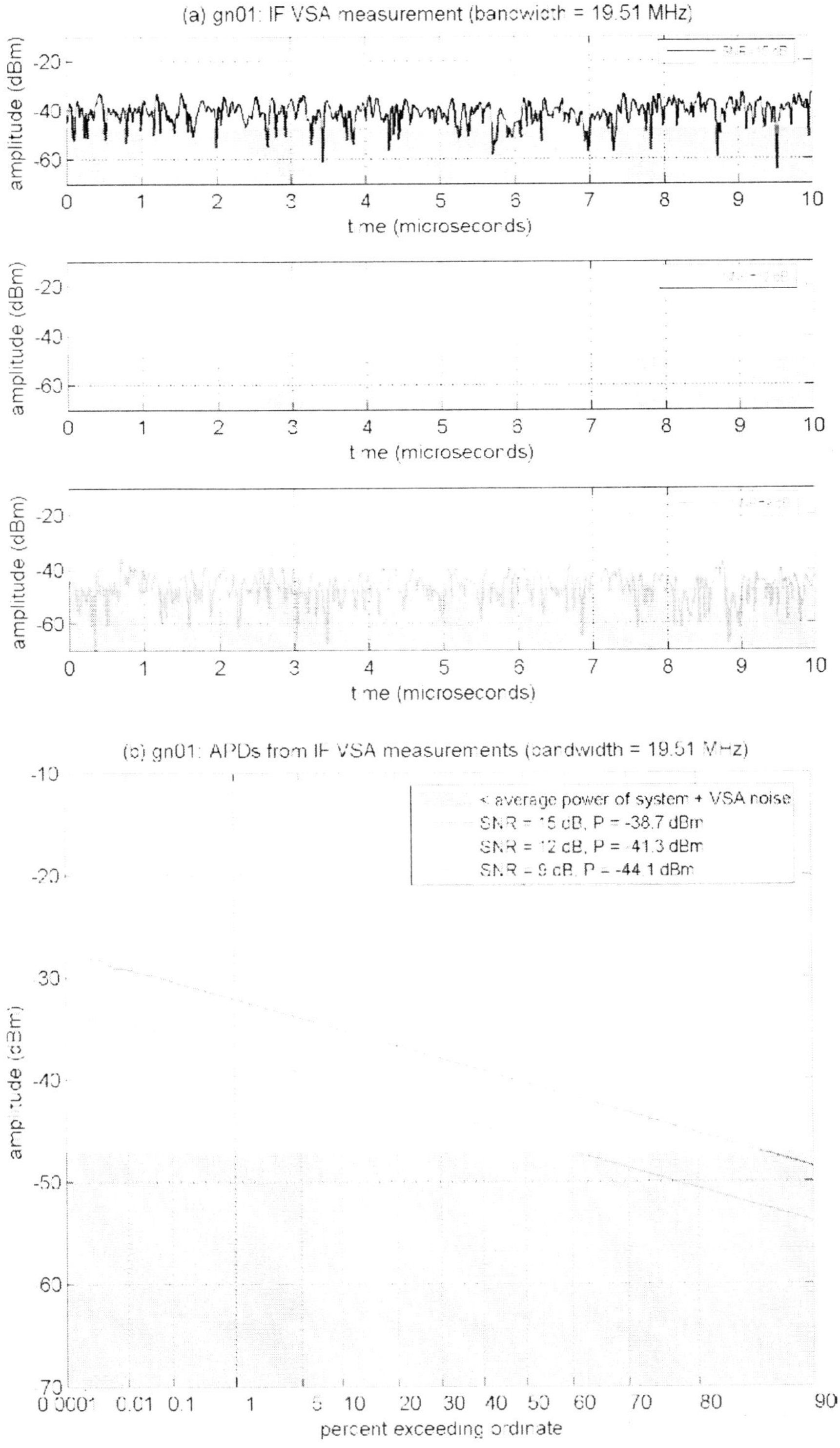

Figure D-4. IF measurements of GN-01 at INR_{TOV}.

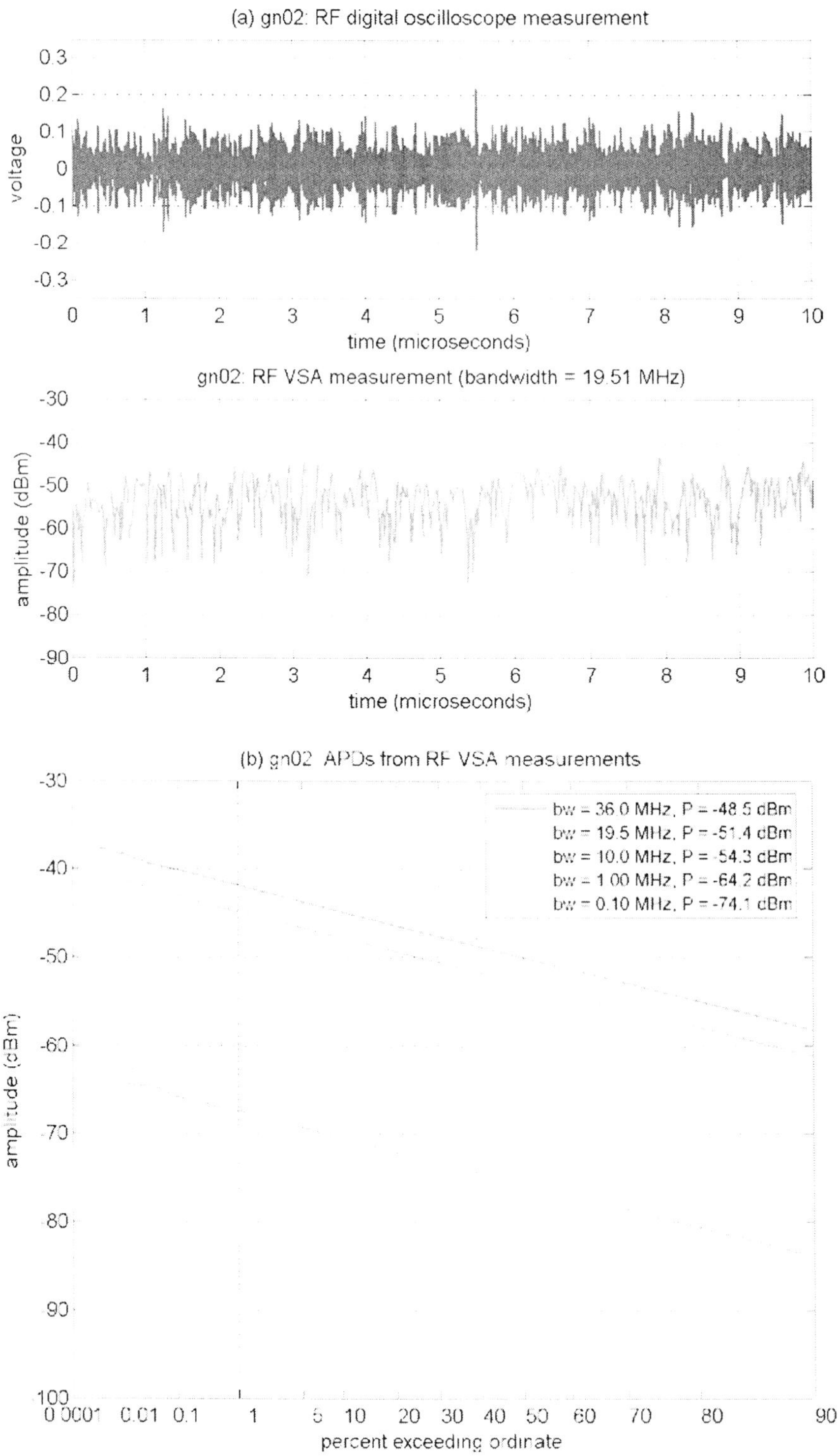

Figure D-5. RF measurements of GN-02 (τ_{on} = 10 ns, ζ = 0.5000).

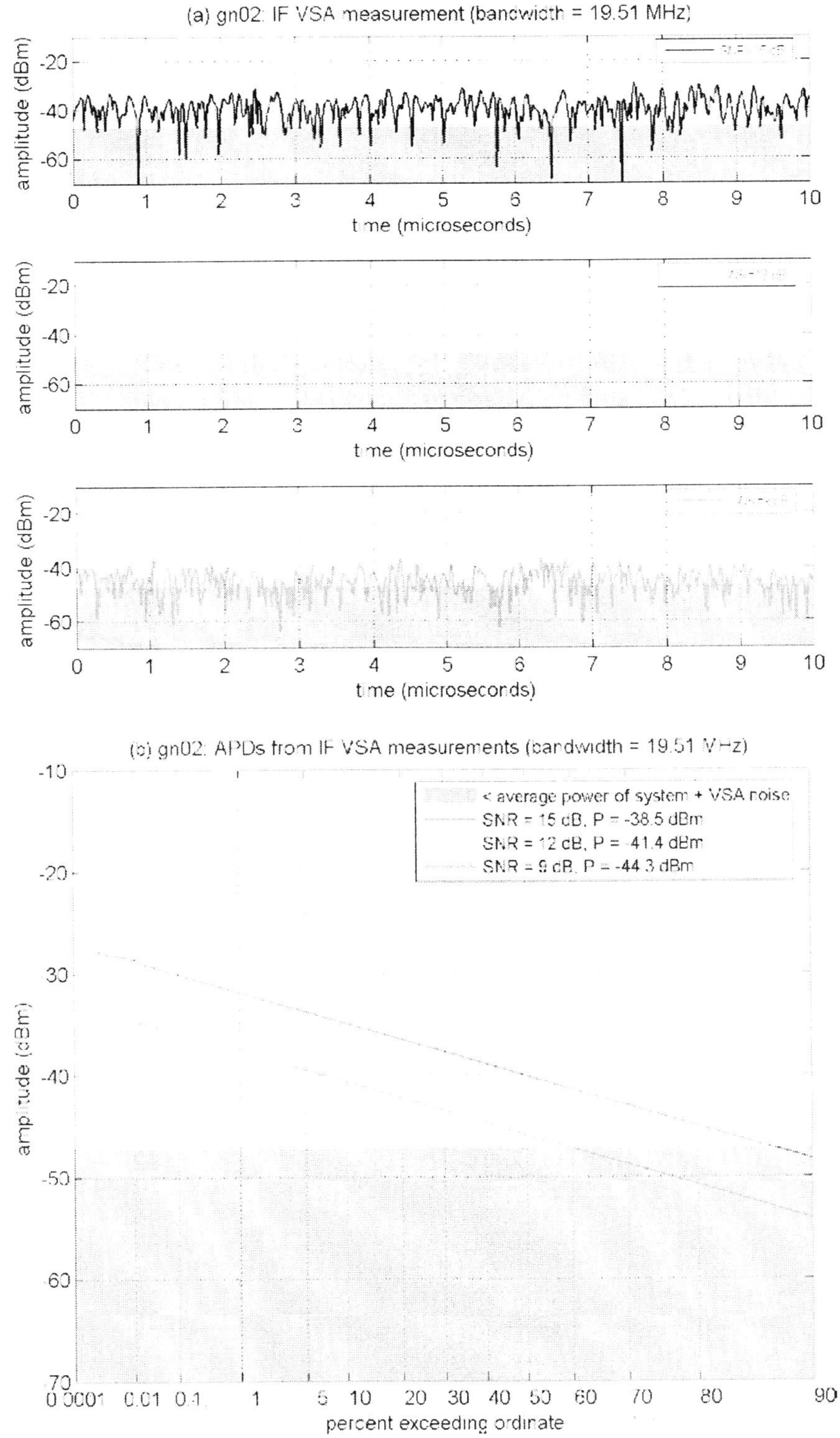

Figure D-6. IF measurements of GN-02 at INR_{TOV}.

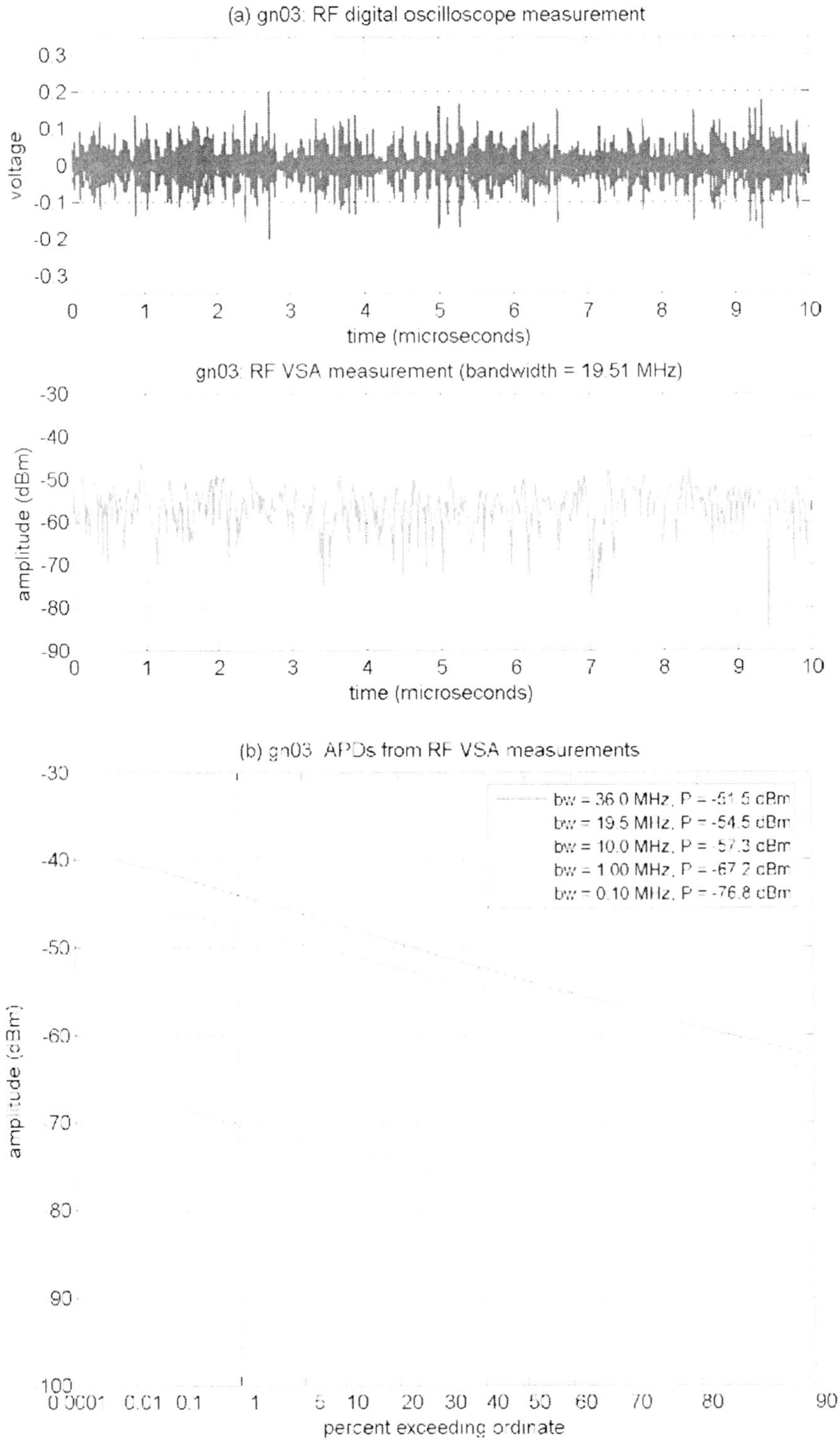

Figure D-7. RF measurements of GN-03 (τ_{on} = 10 ns, ζ = 0.2500).

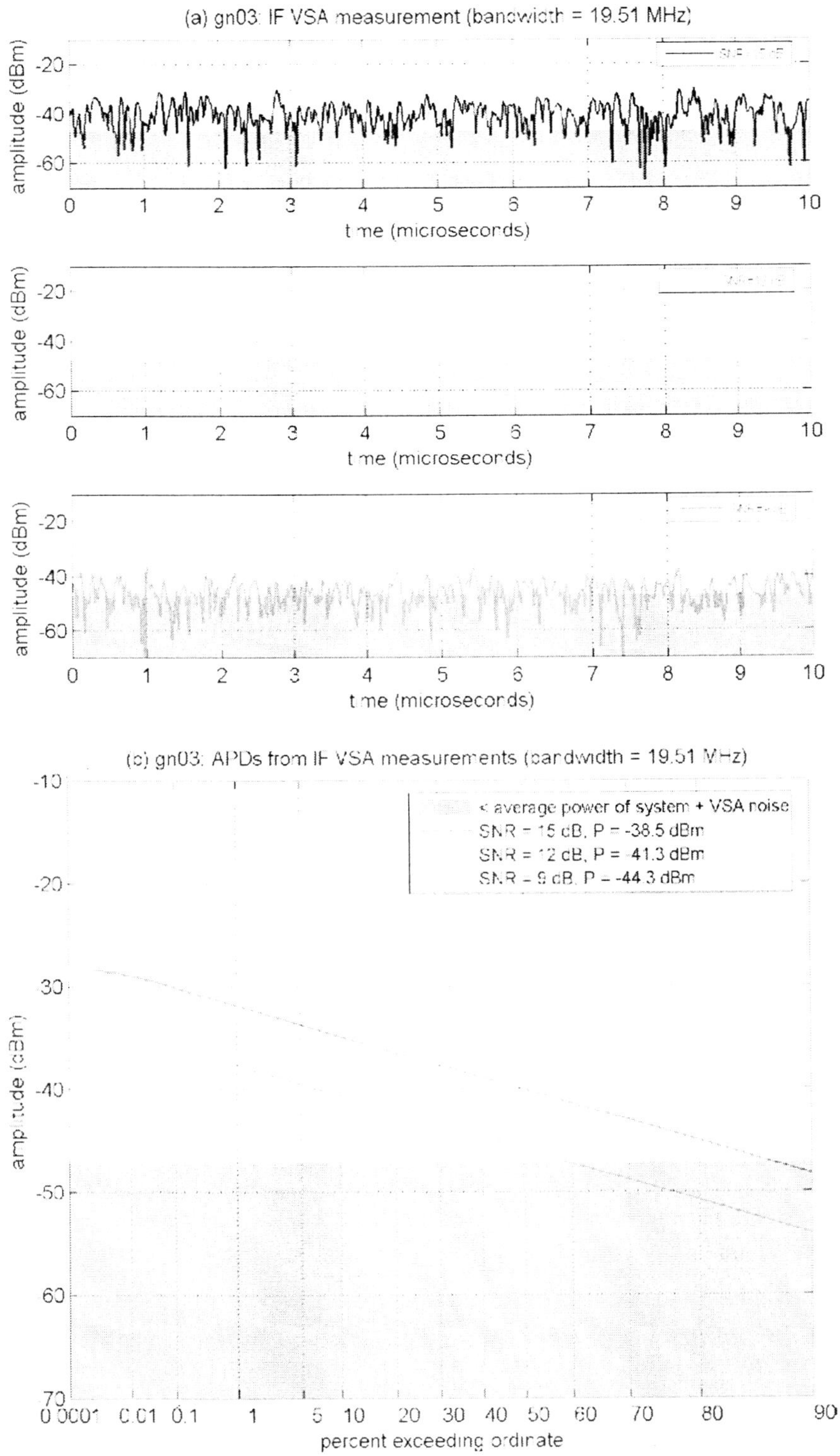

Figure D-8. IF measurements of GN-03 at INR_{TOV}.

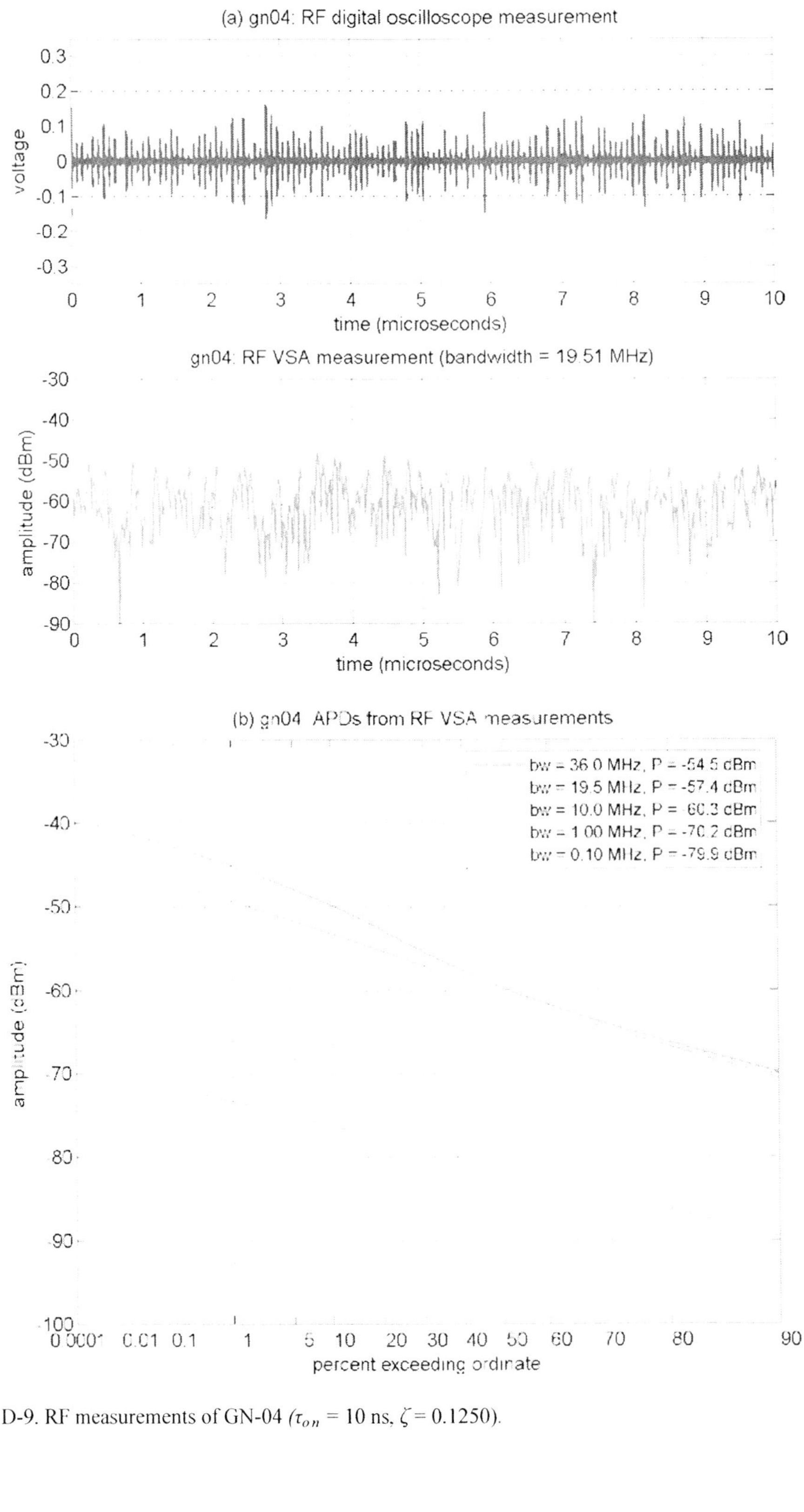

Figure D-9. RF measurements of GN-04 (τ_{on} = 10 ns, ζ = 0.1250).

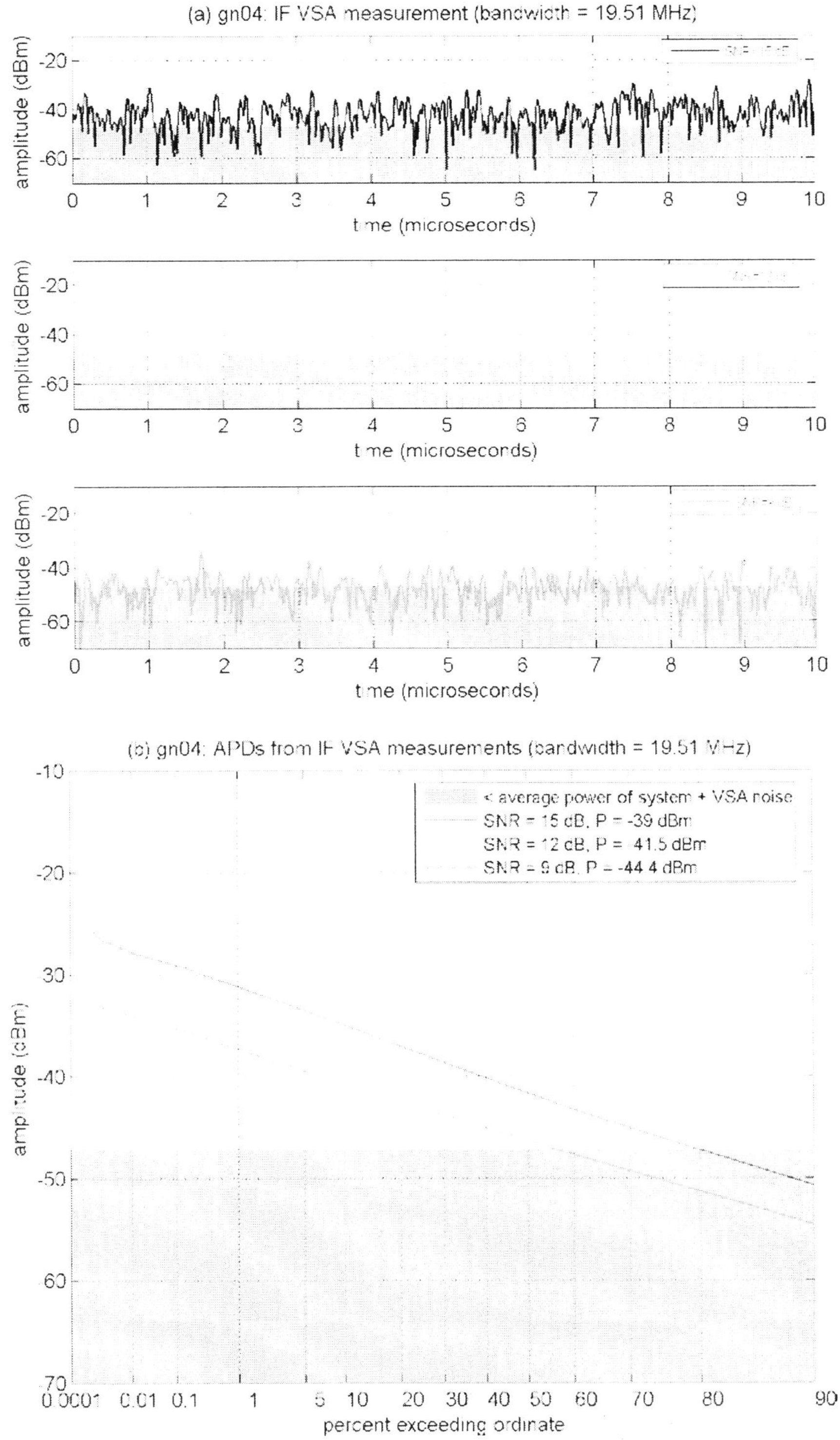

Figure D- 10. IF measurements of GN-04 at INR_{TOV}.

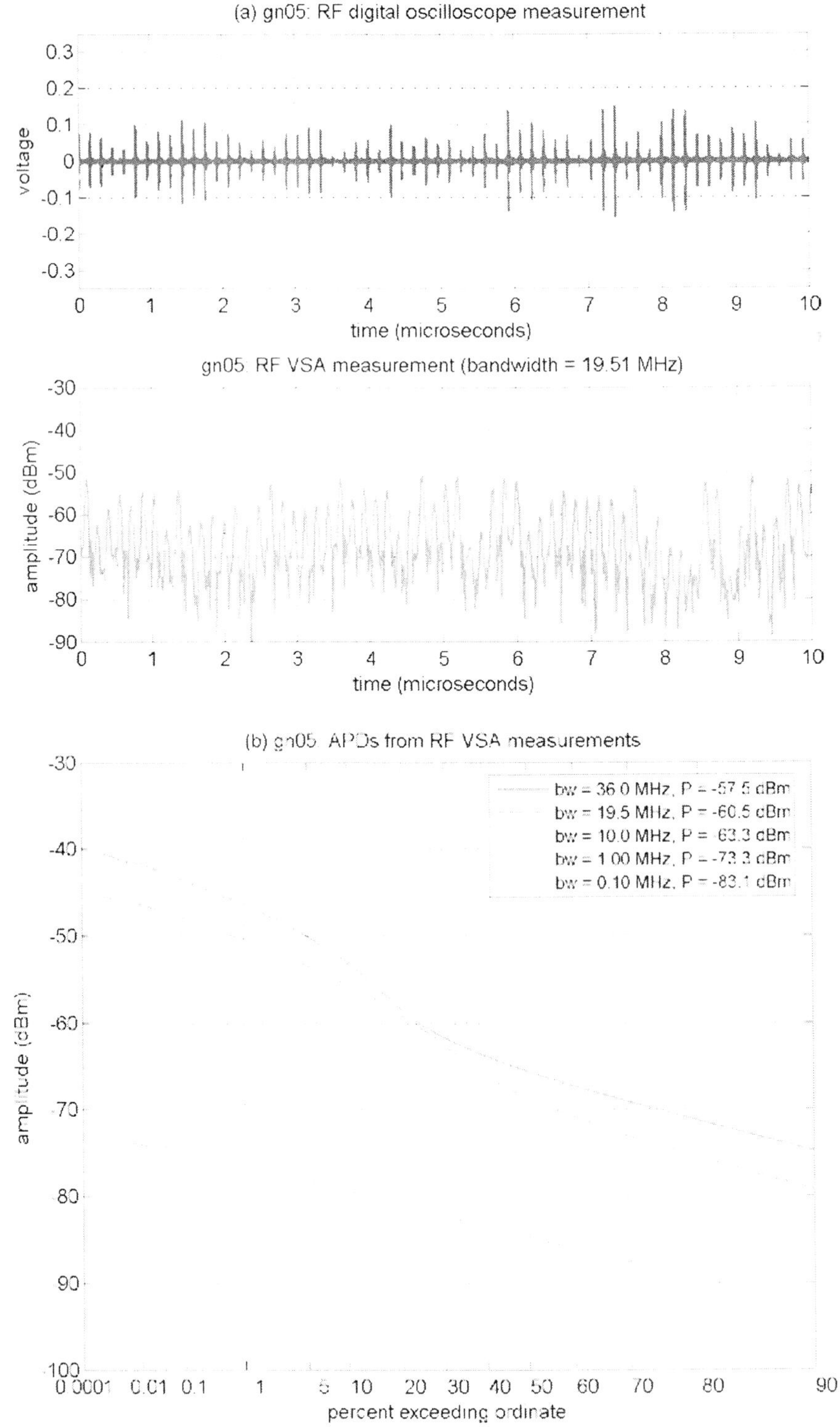

Figure D-11. RF measurements of GN-05 (τ_{on} = 10 ns, ζ = 0.0625).

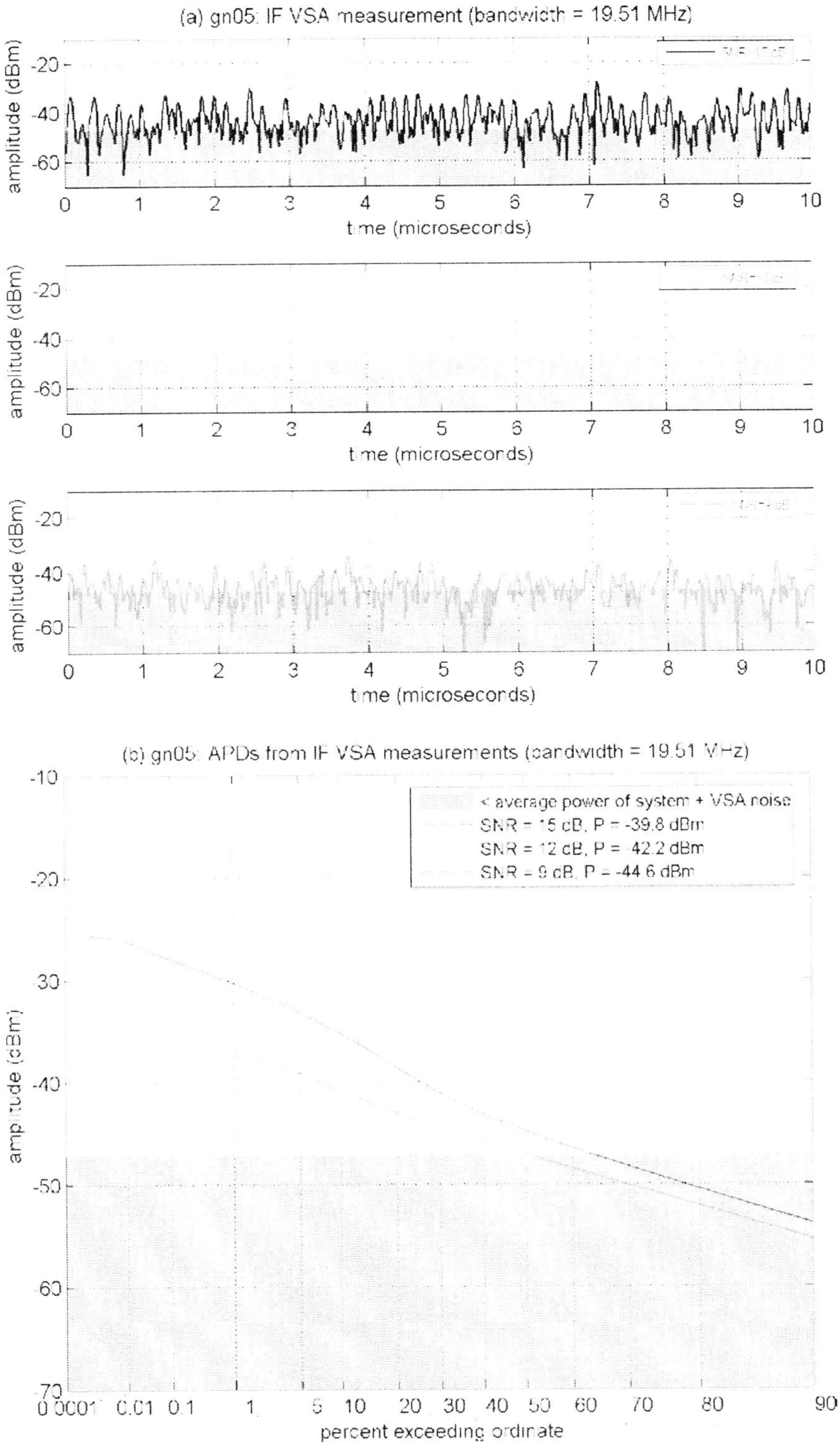

Figure D-12. IF measurements of GN-05 at INR_{TOV}.

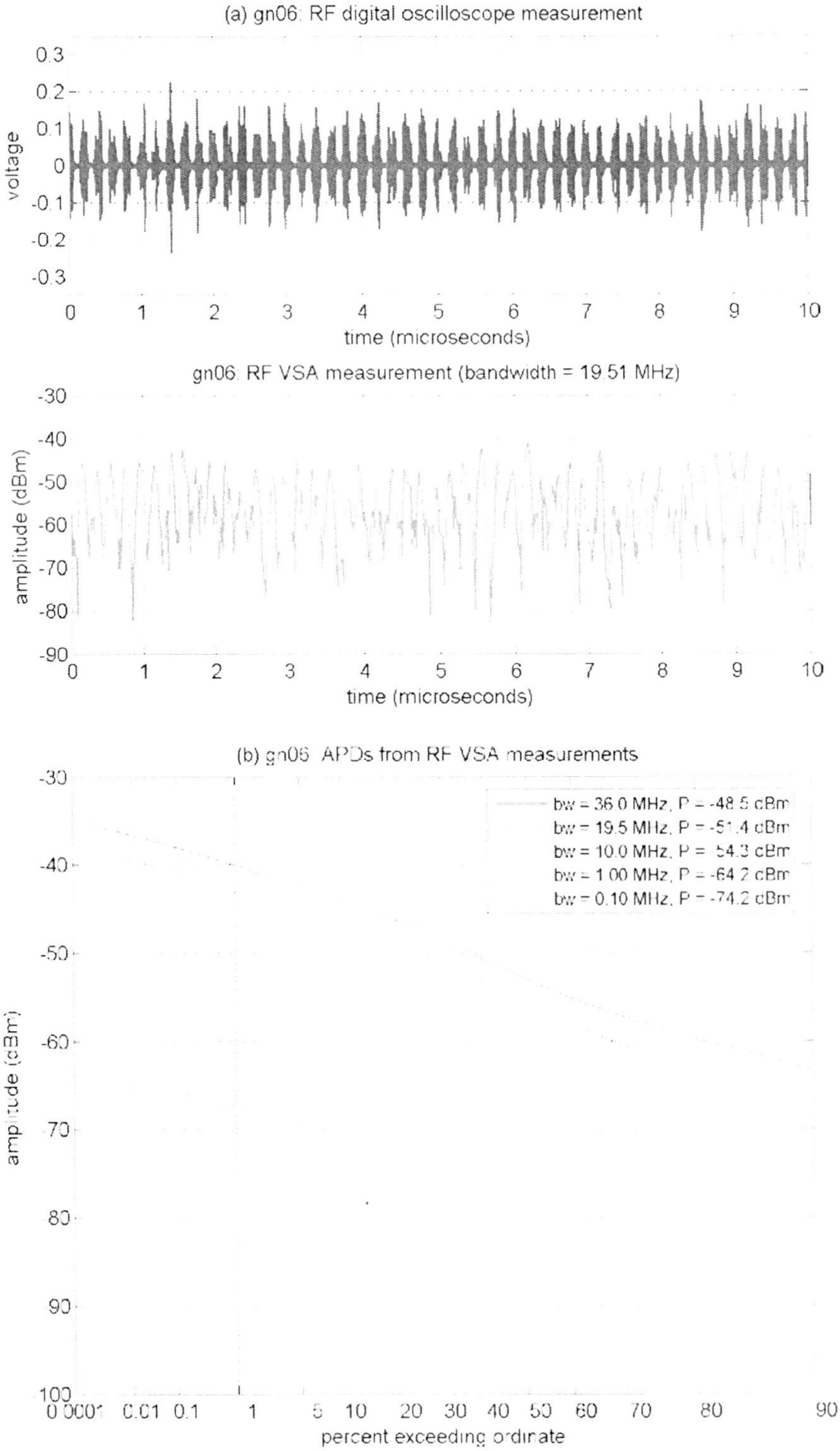

Figure D-13. RF measurements of GN-06 (τ_{on} = 100 ns, ζ = 0.5000).

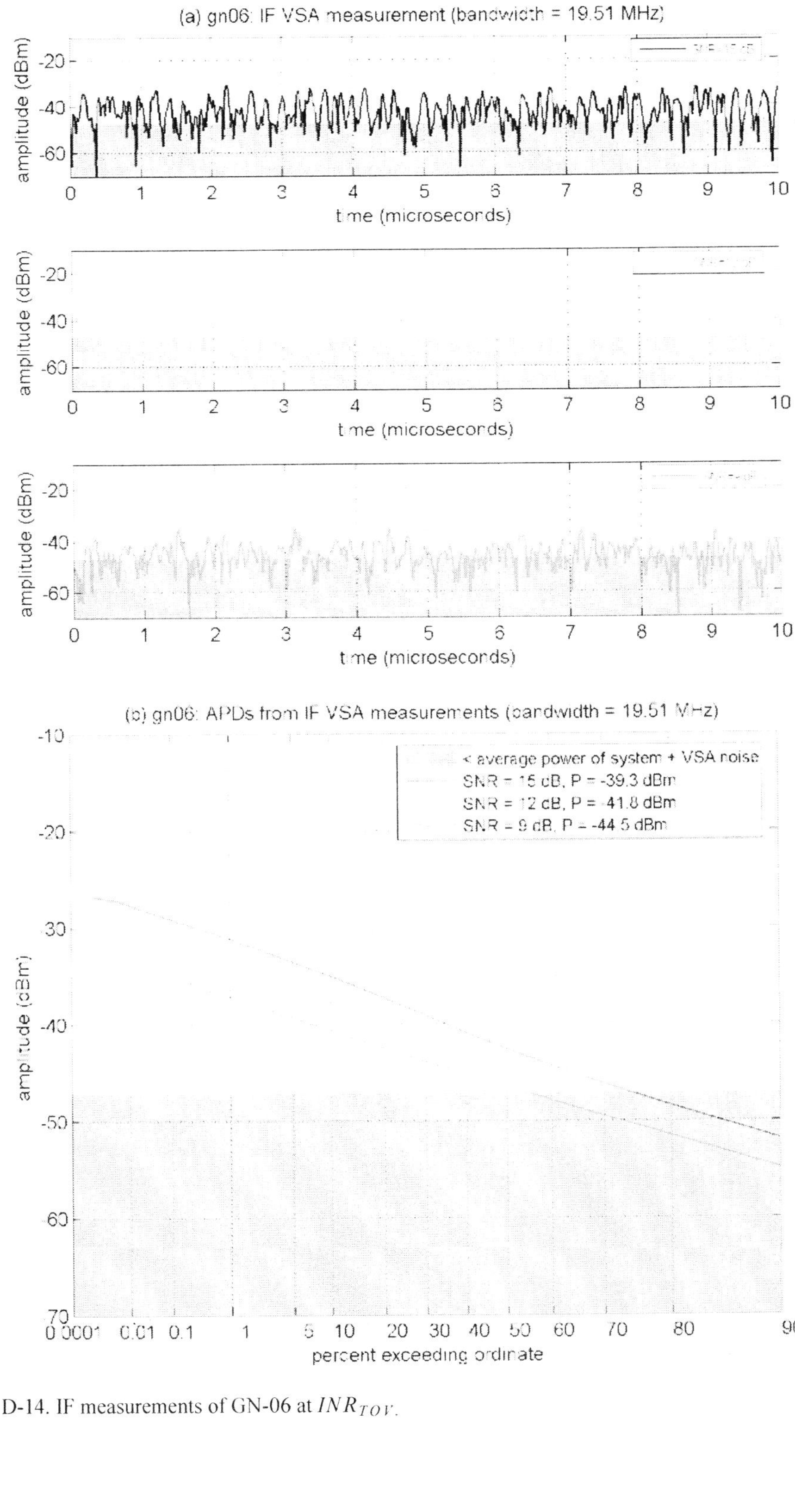

Figure D-14. IF measurements of GN-06 at INR_{TOV}.

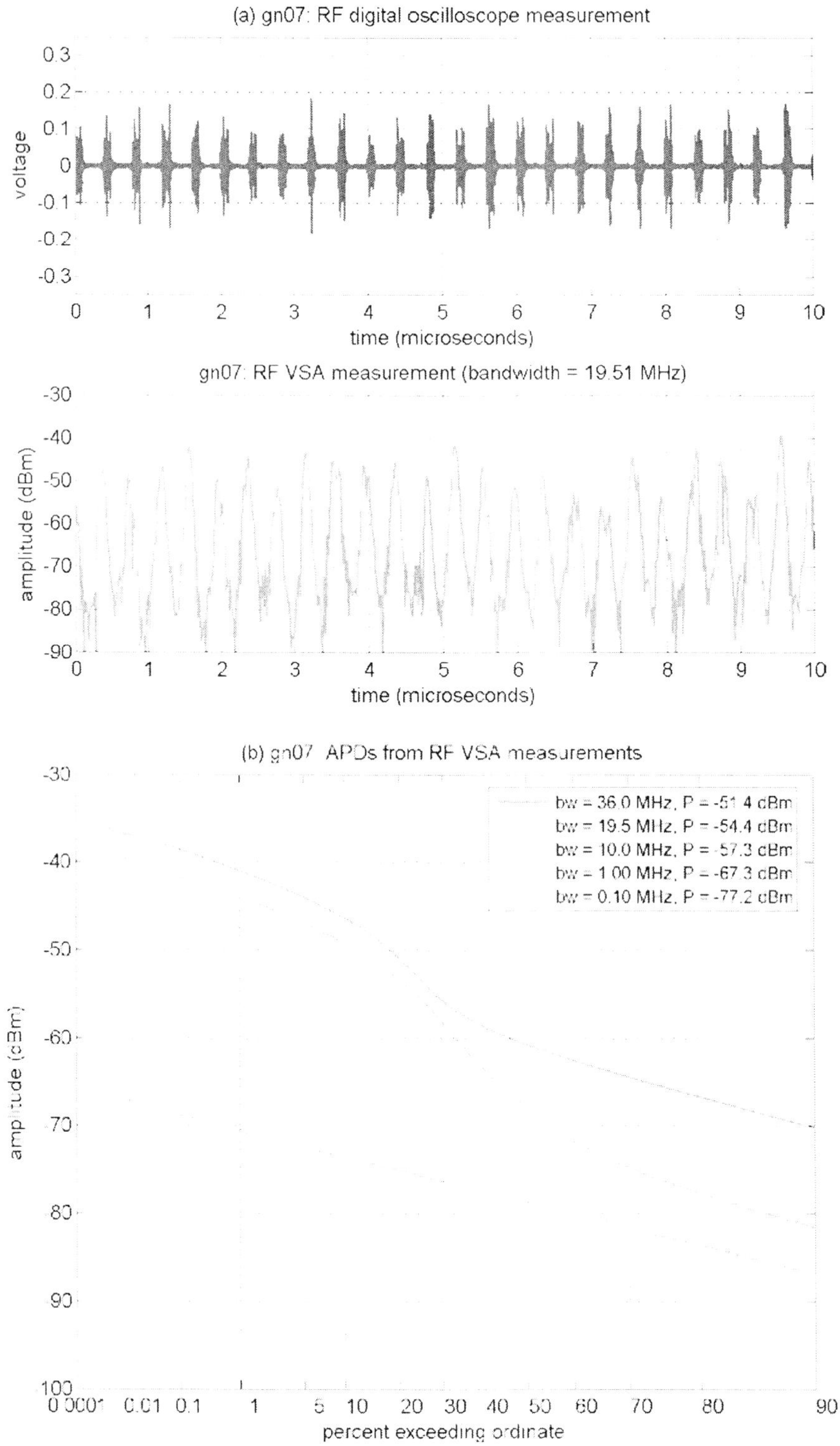

Figure D-15. RF measurements of GN-07 (τ_{on} = 100 ns, ζ = 0.2500).

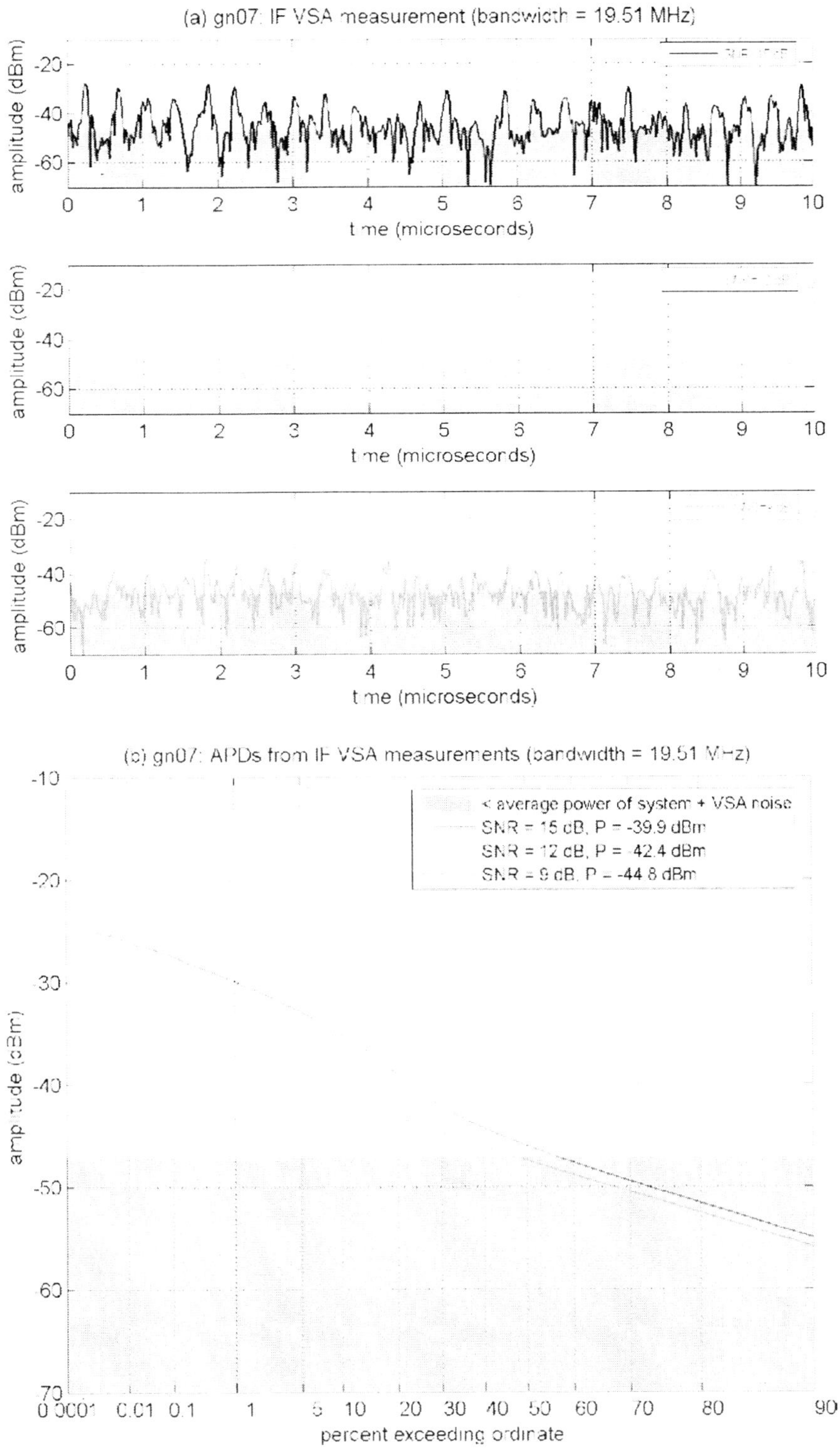

Figure D- 16. IF measurements of GN-07 at INR_{TOV}.

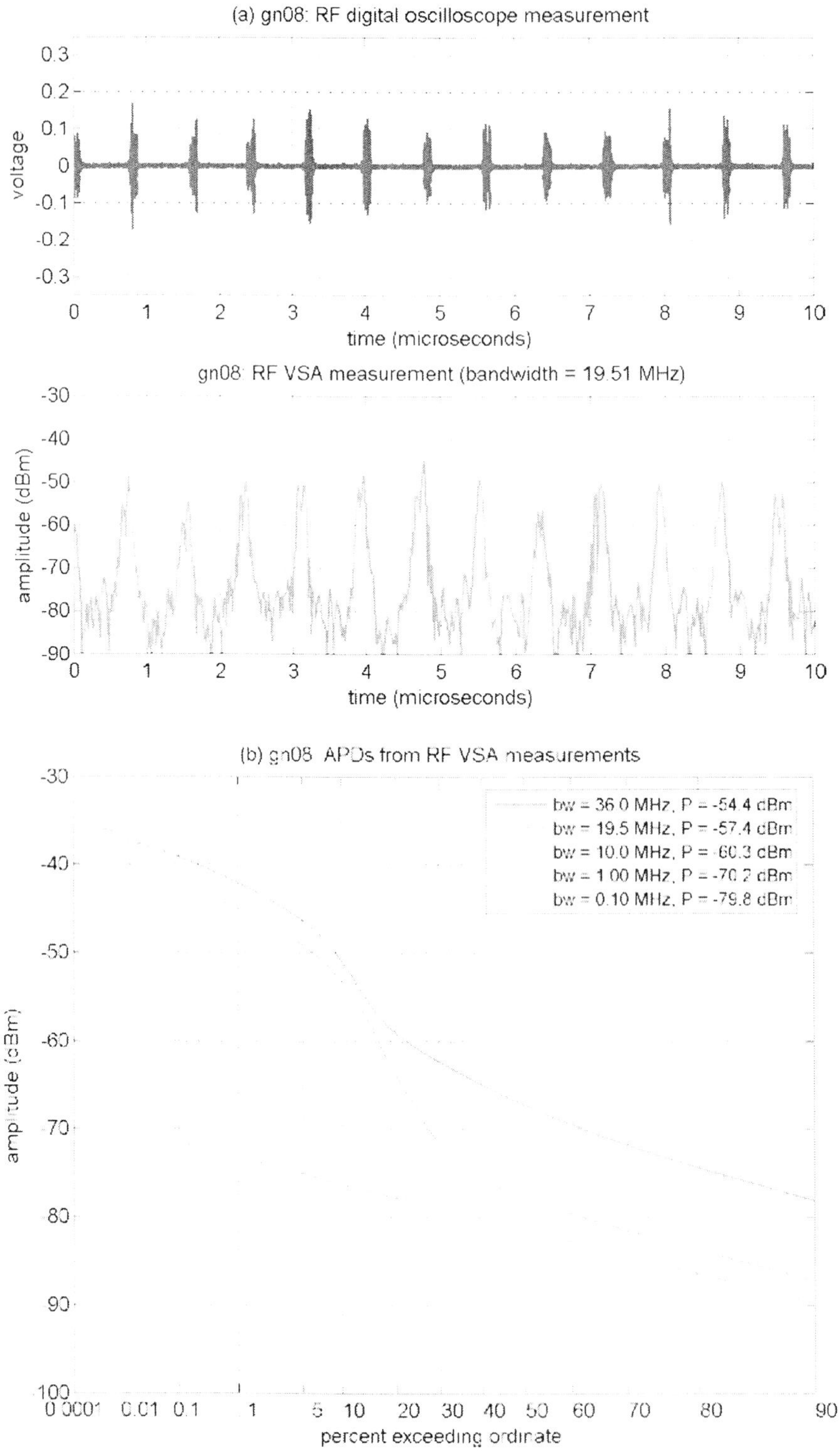

Figure D-17. RF measurements of GN-08 (τ_{on} = 100 ns, ζ = 0.1250).

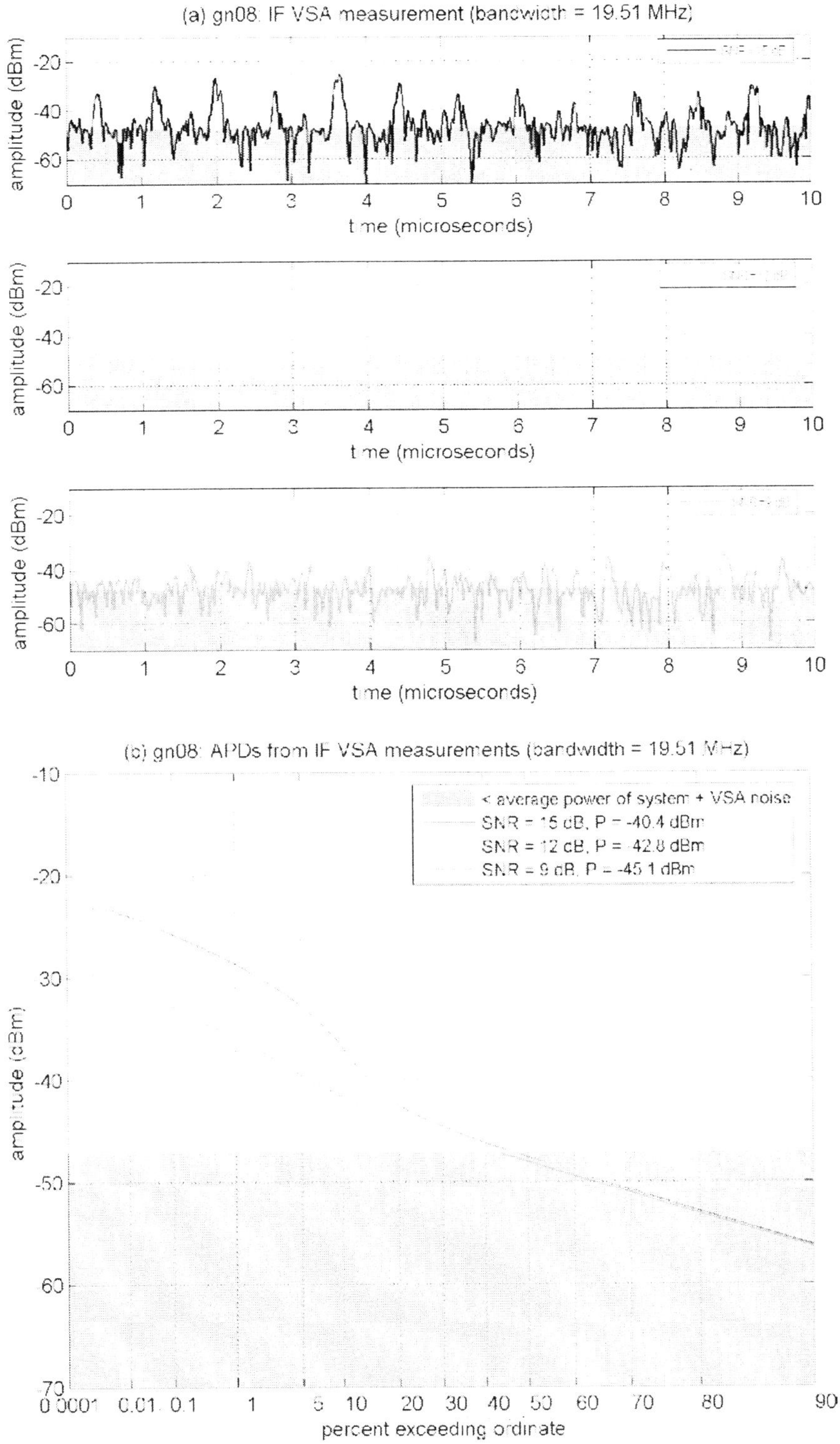

Figure D- 18. IF measurements of GN-08 at INR_{TOV}.

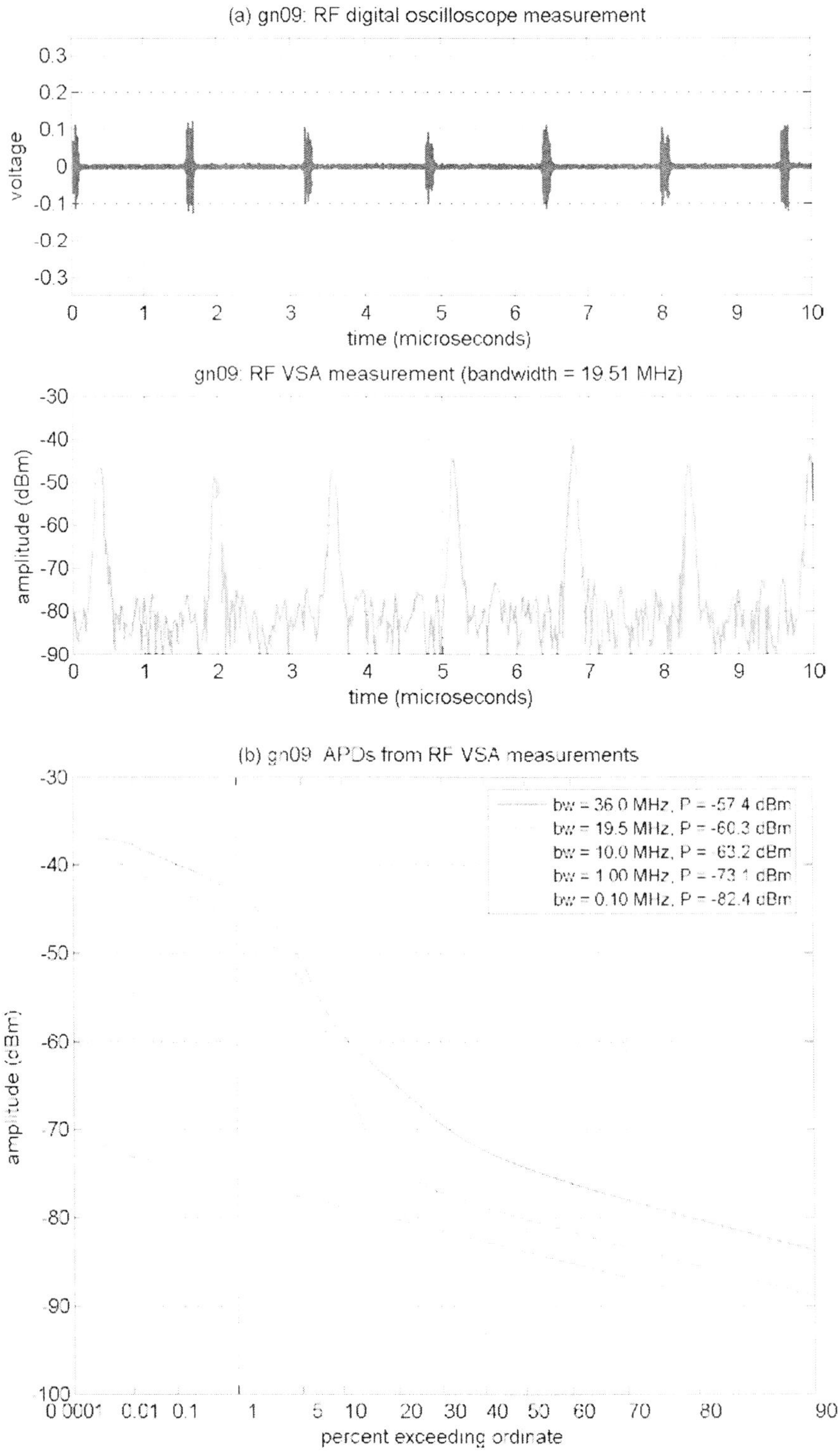

Figure D-19. RF measurements of GN-09 (τ_{on} = 100 ns, ζ = 0.0625).

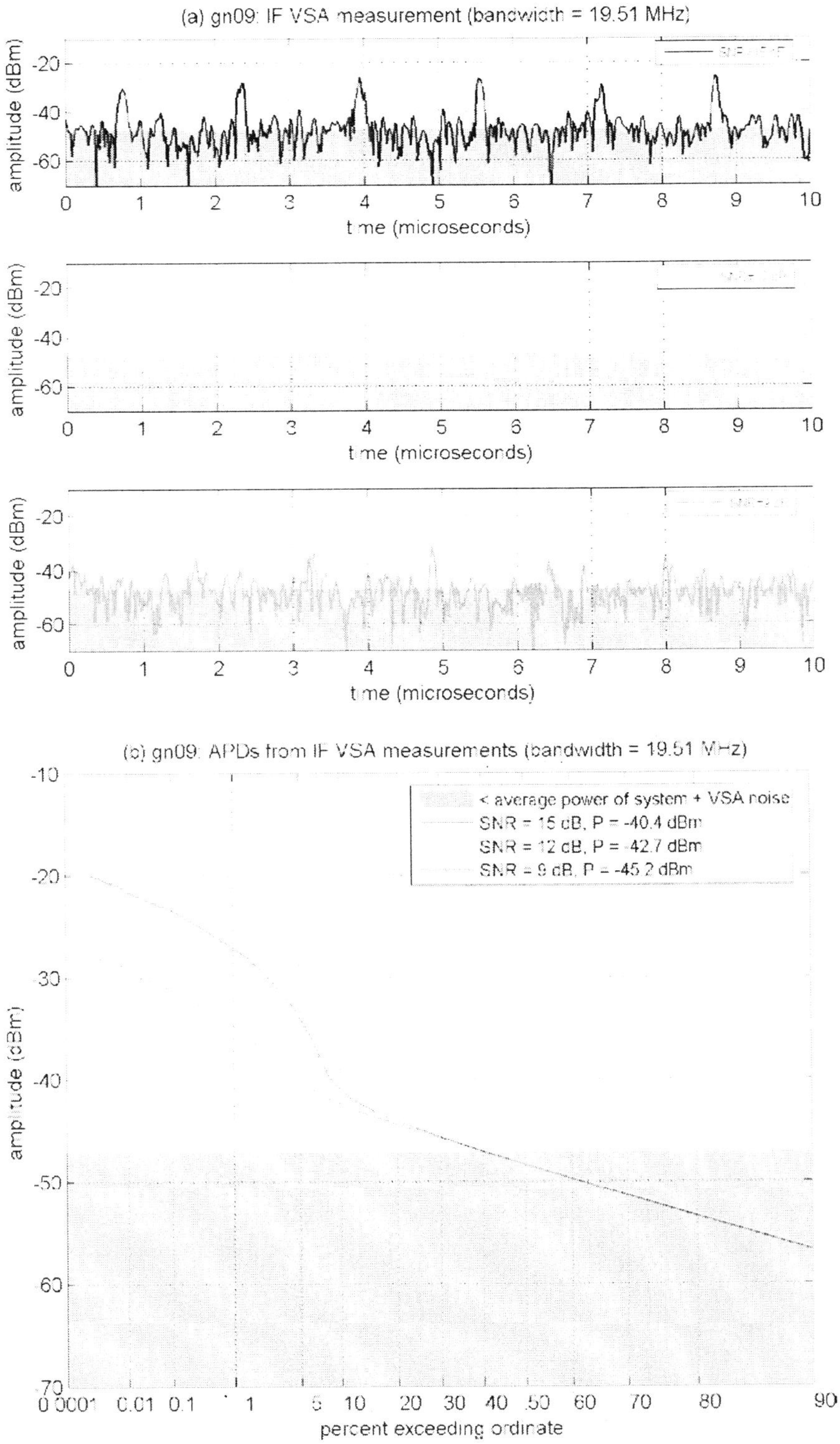

Figure D-20. IF measurements of GN-09 at INR_{TOV}.

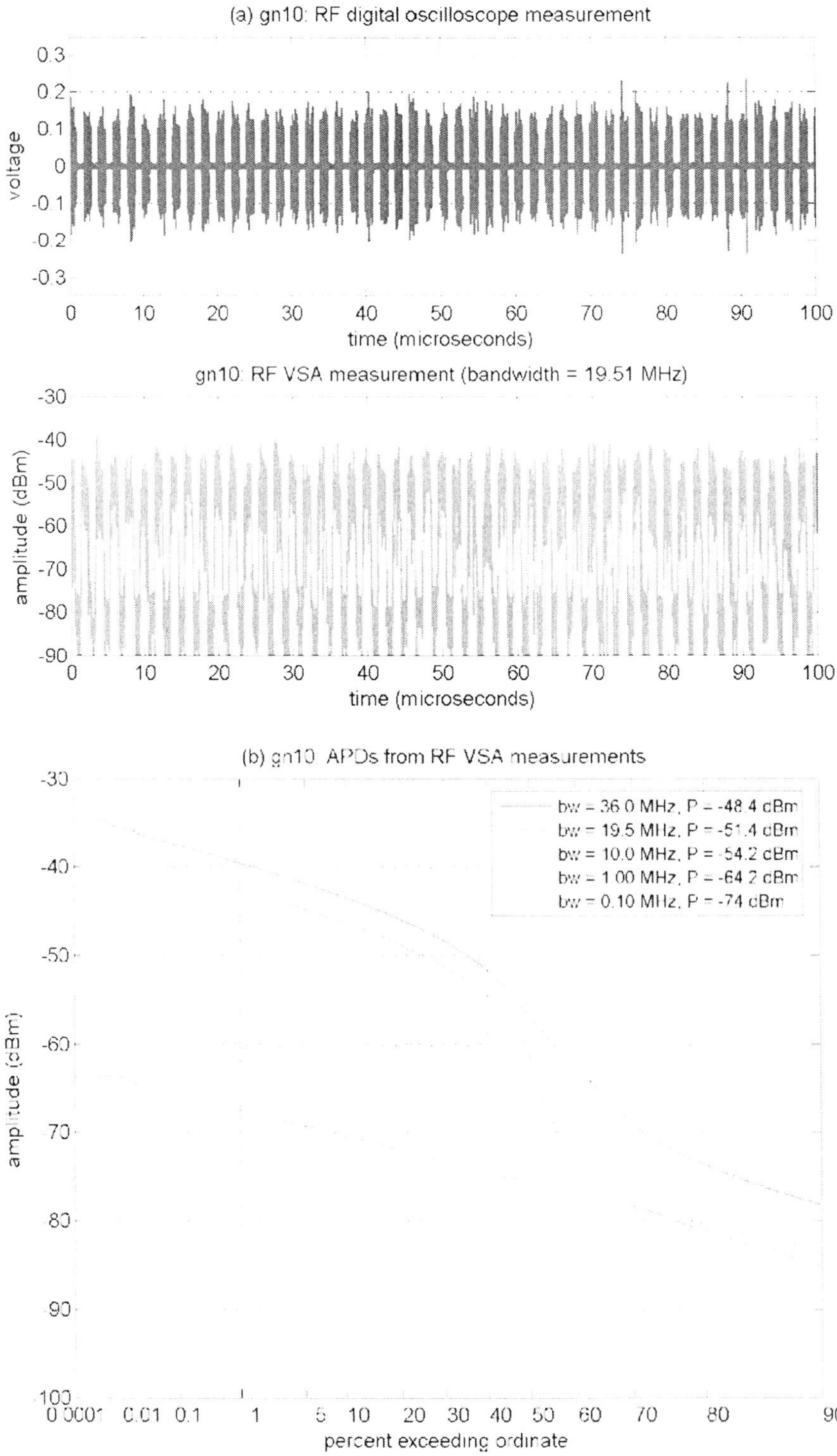

Figure D-21. RF measurements of GN-10 (τ_{on} = 1000 ns, ζ = 0.5000).

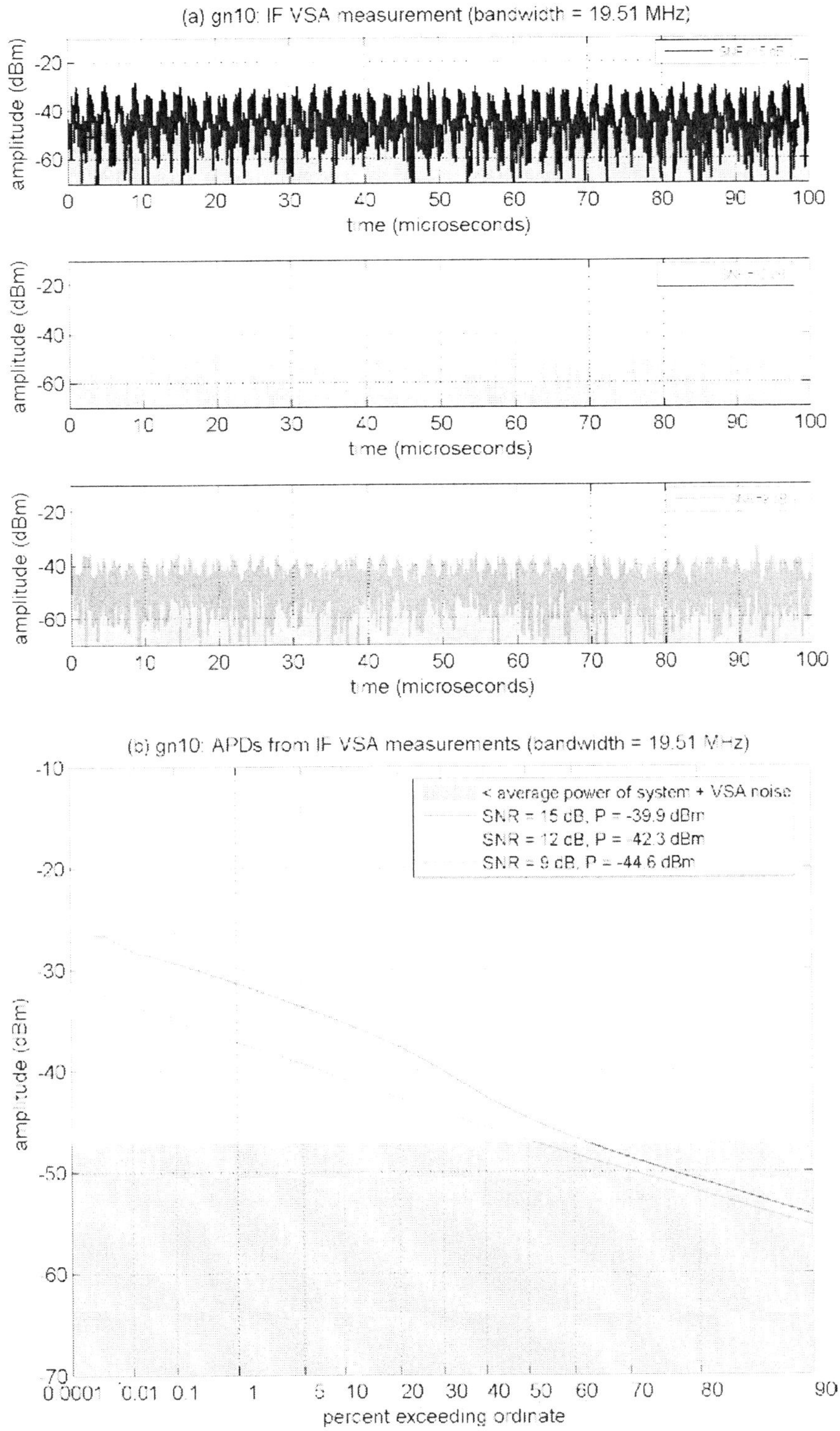

Figure D-22. IF measurements of GN-10 at INR_{TOV}.

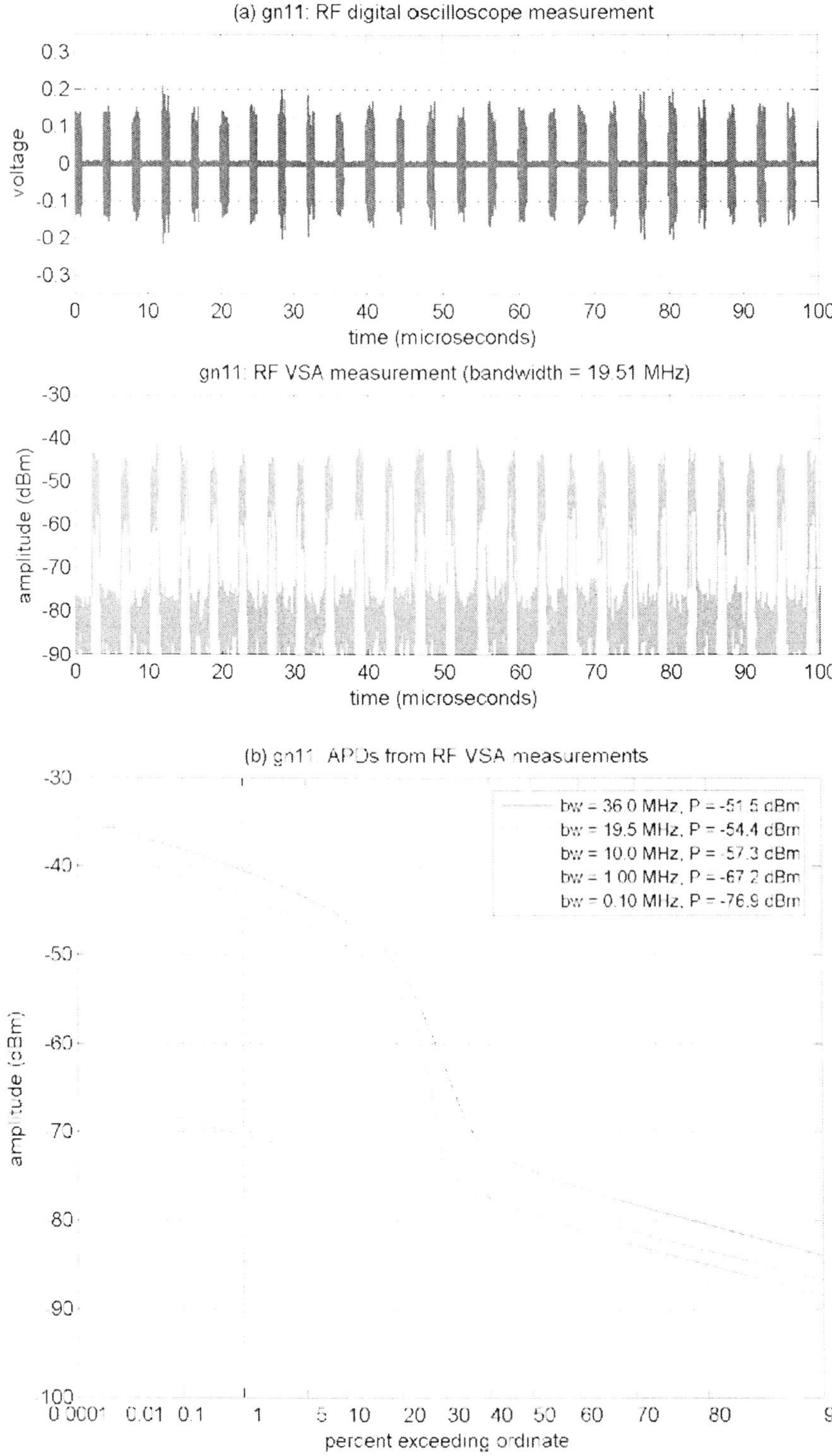

Figure D-23. RF measurements of GN-1 1 (τ_{on} = 1000 ns, ζ = 0.2500).

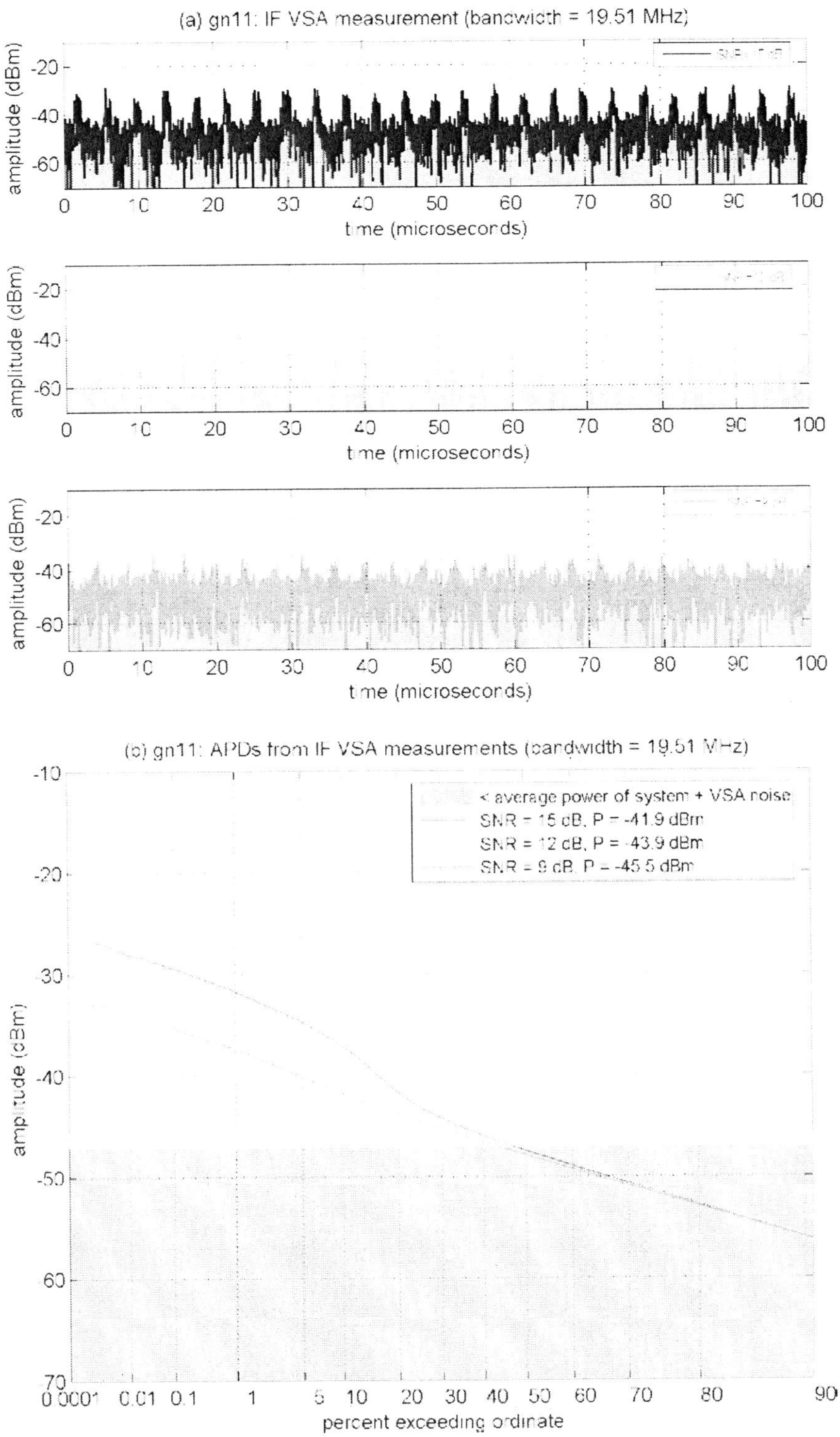

Figure D-24. IF measurements of GN- 11 at INR_{TOV}.

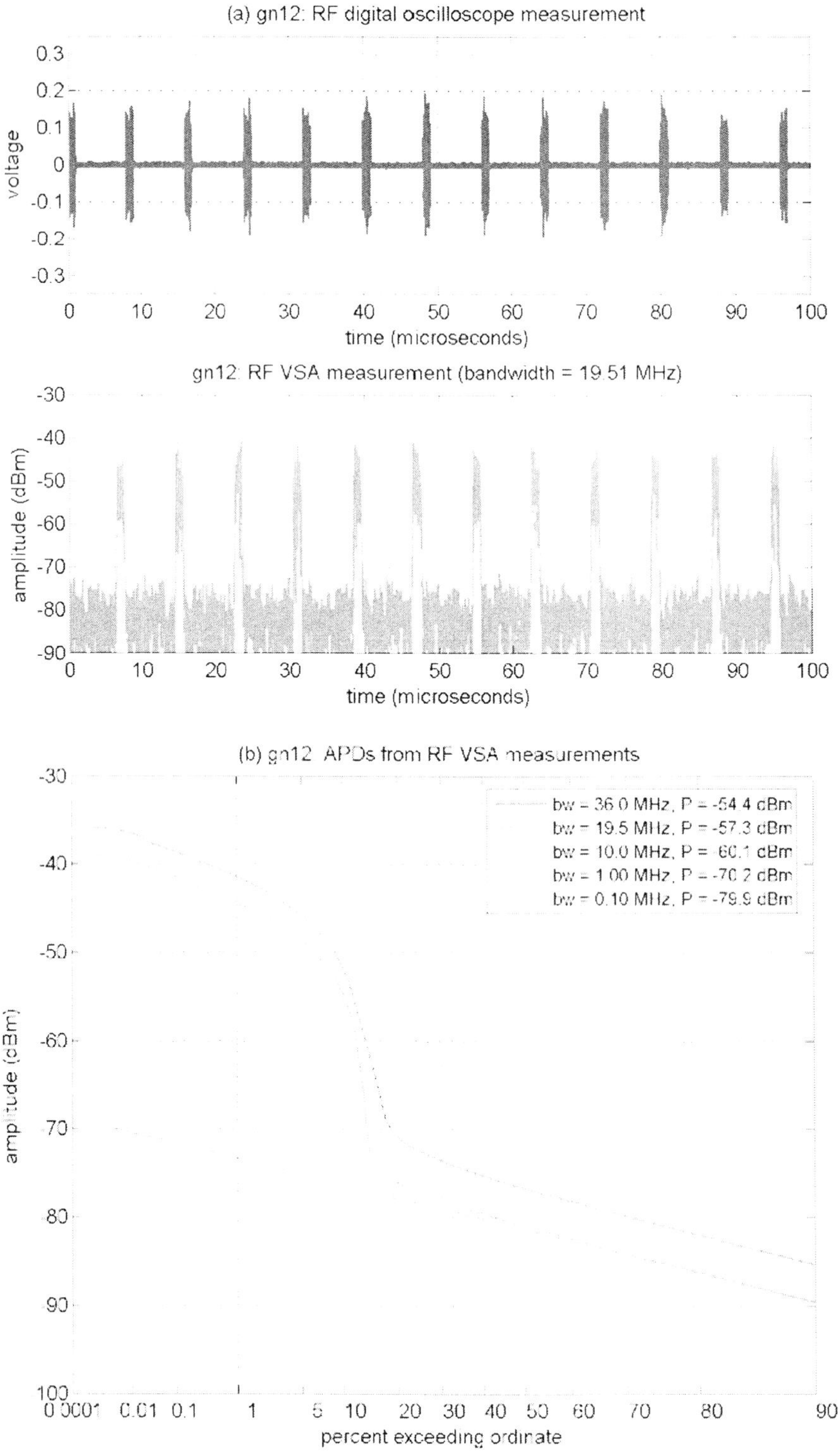

Figure D-25. RF measurements of GN-12 (τ_{on} = 1000 ns, ζ = 0.1250).

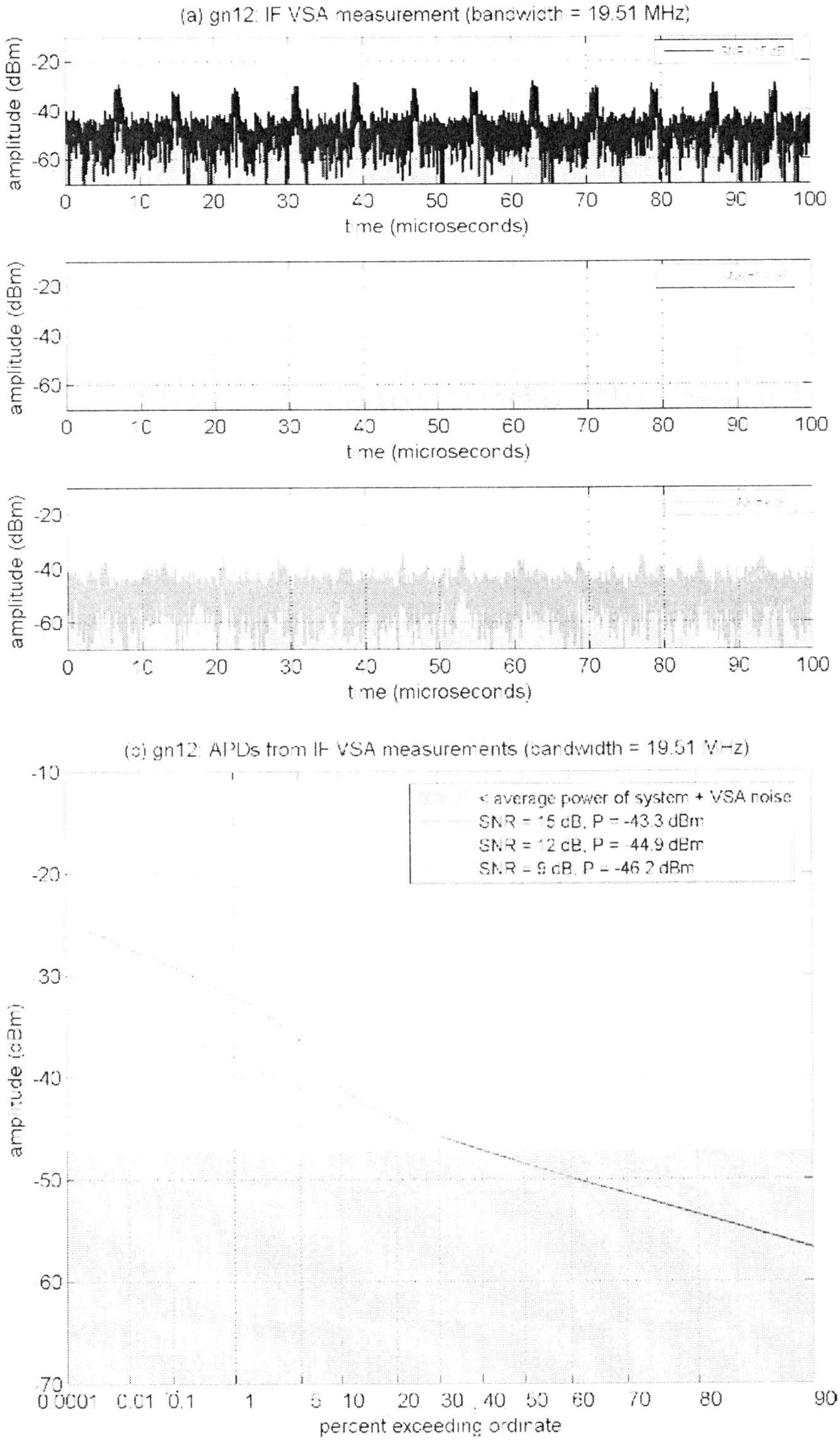

Figure D-26. IF measurements of GN- 12 at INR_{TOV}.

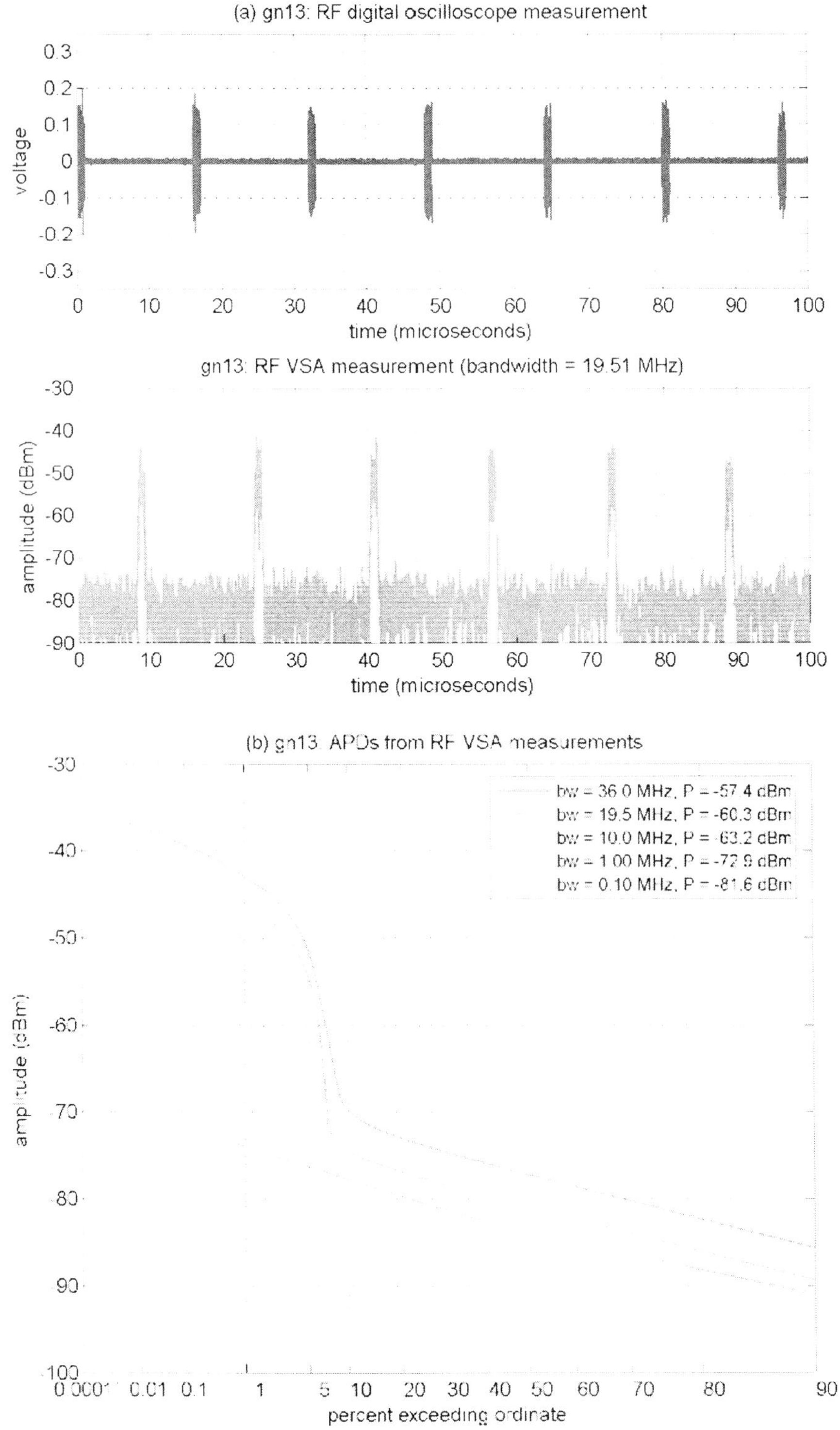

Figure D-27. RF measurements of GN-13 (τ_{on} = 1000 ns, ζ = 0.0625).

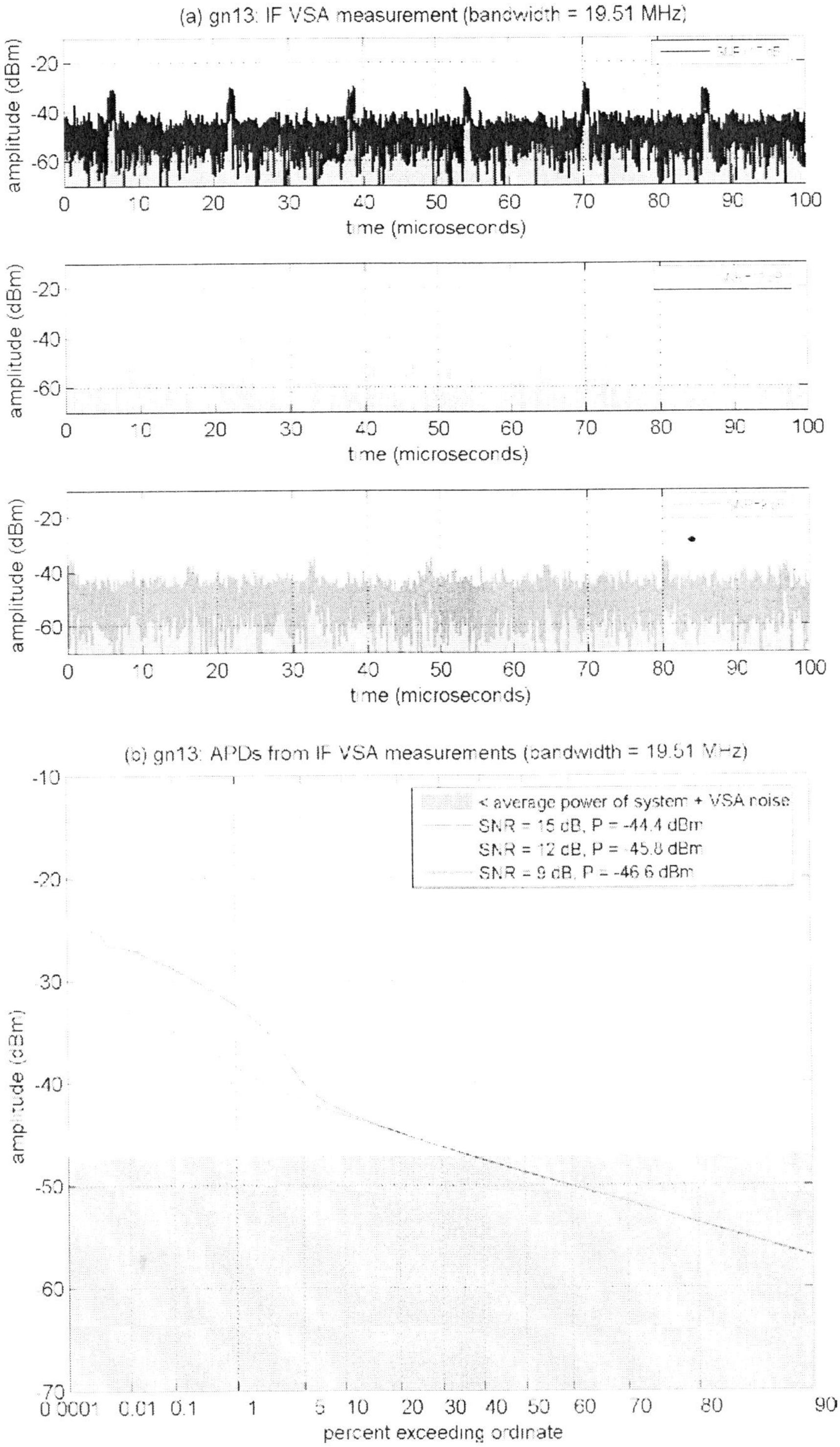

Figure D-28. IF measurements of GN-13 at INR_{TOV}.

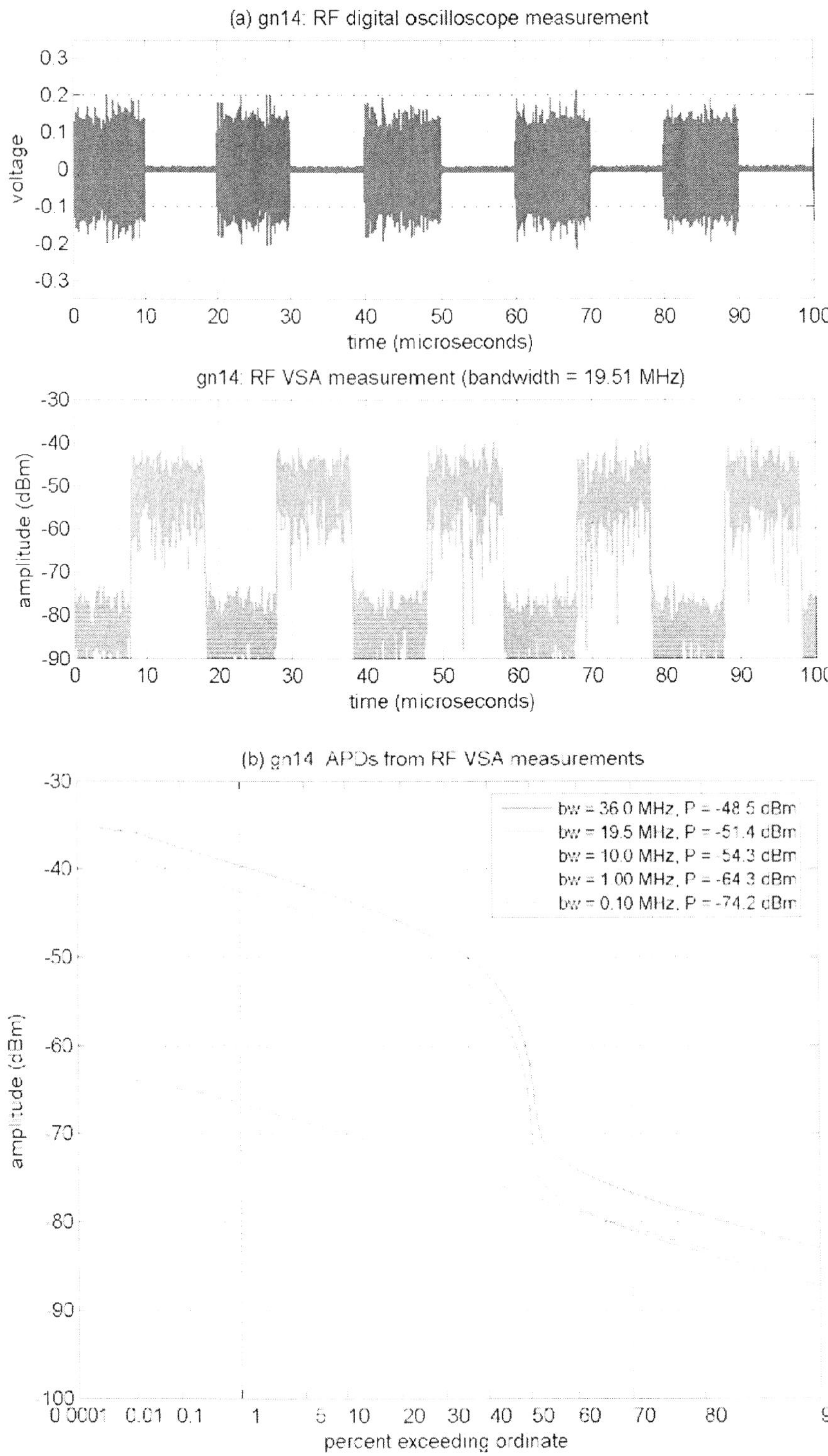

Figure D-29. RF measurements of GN-14 (τ_{on} = 10000 ns, ζ = 0.5000).

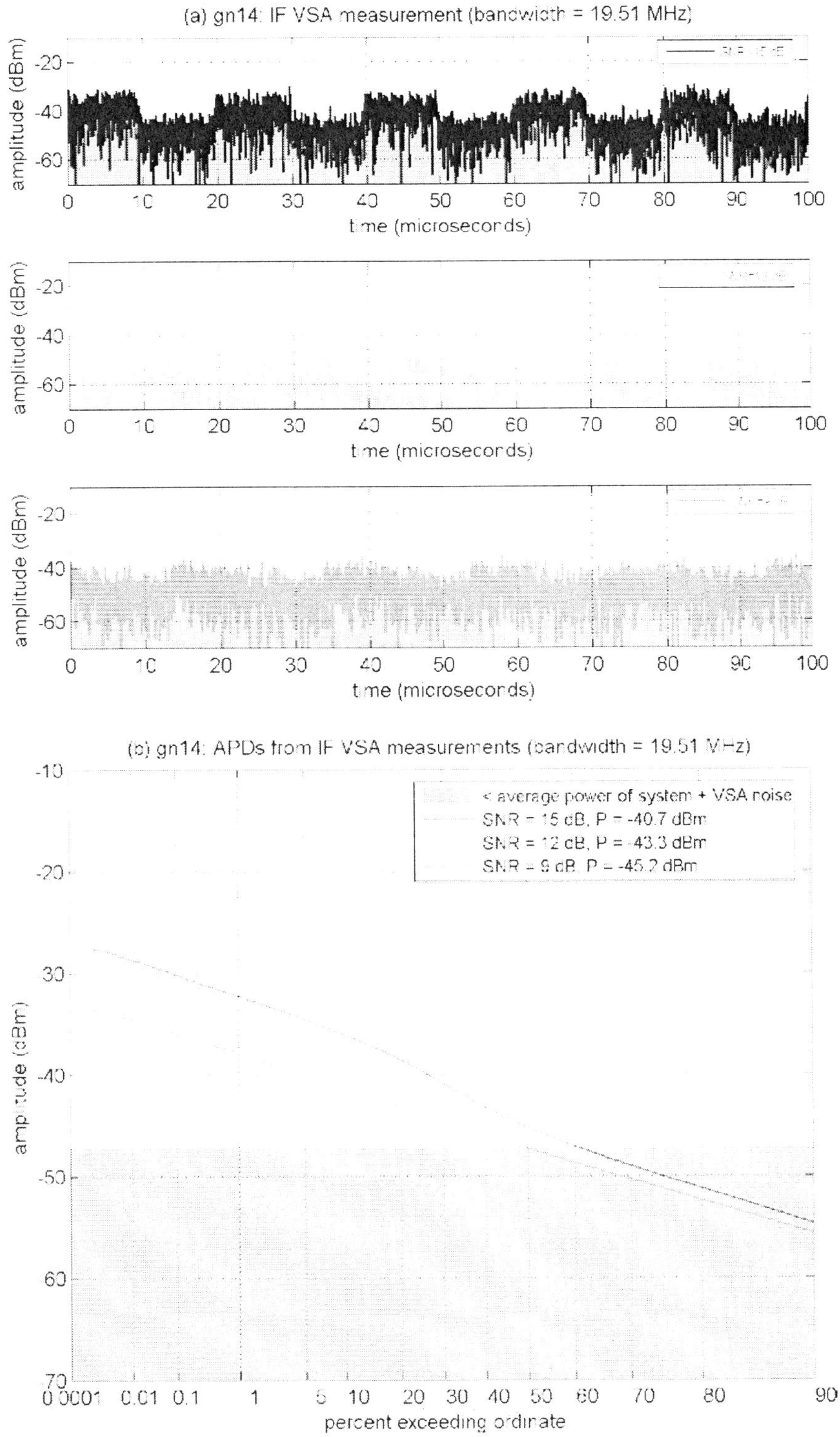

Figure D-30. IF measurements of GN-14 at INR_{TOV}.

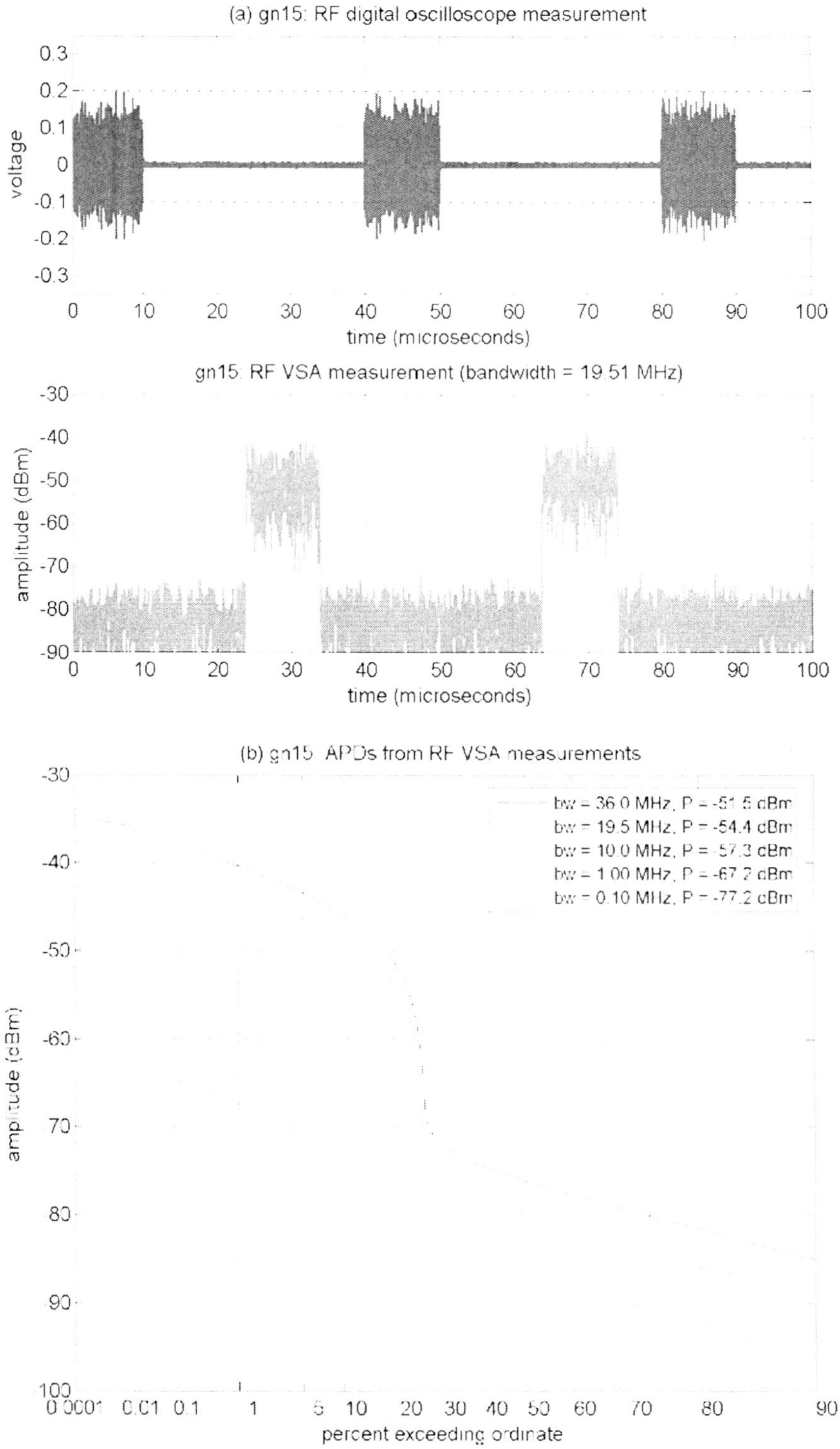

Figure D-31. RF measurements of GN-15 (τ_{on} = 10000 ns, ζ = 0.2500).

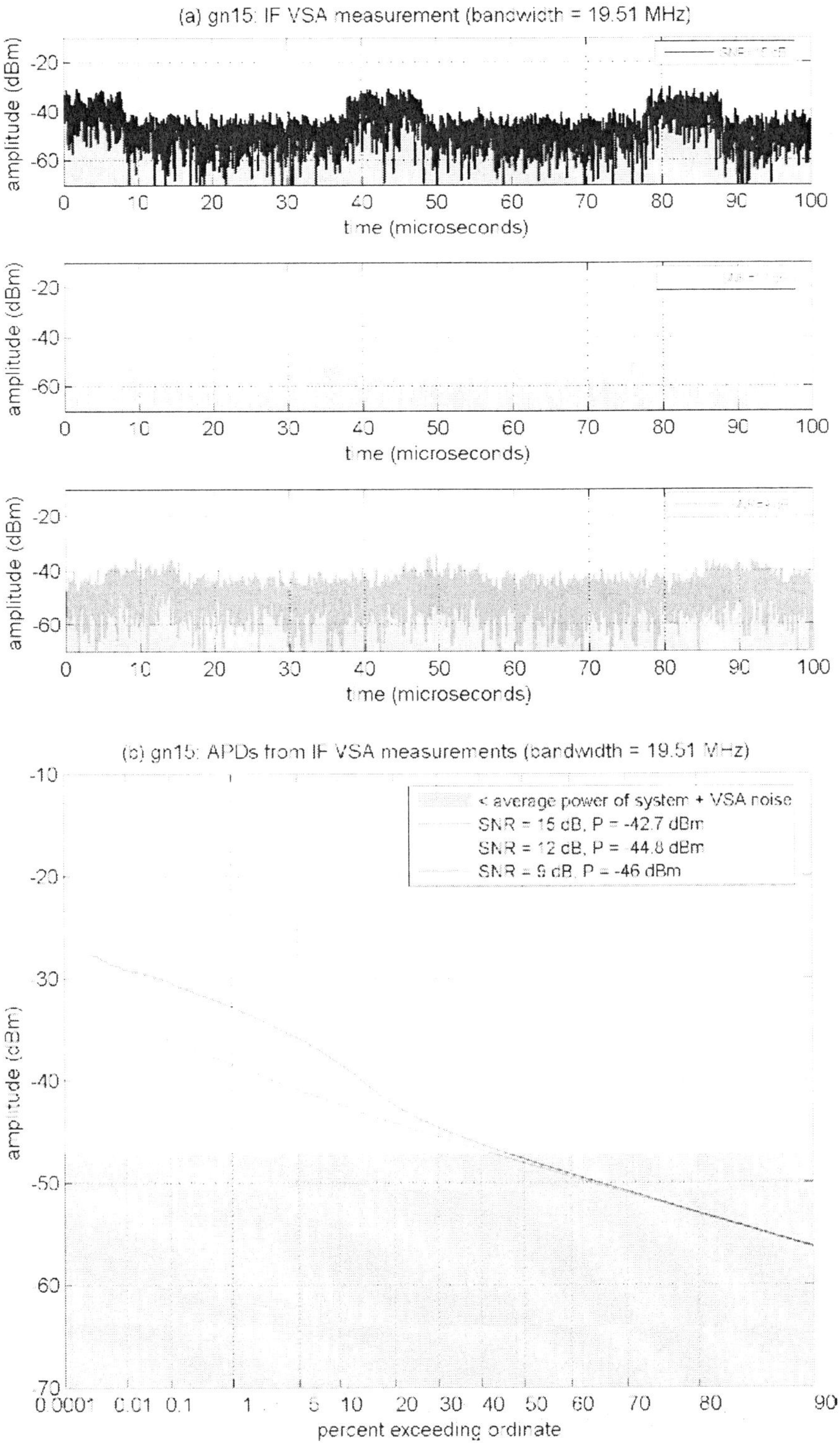

Figure D-32. IF measurements of GN-15 at INR_{TOV}.

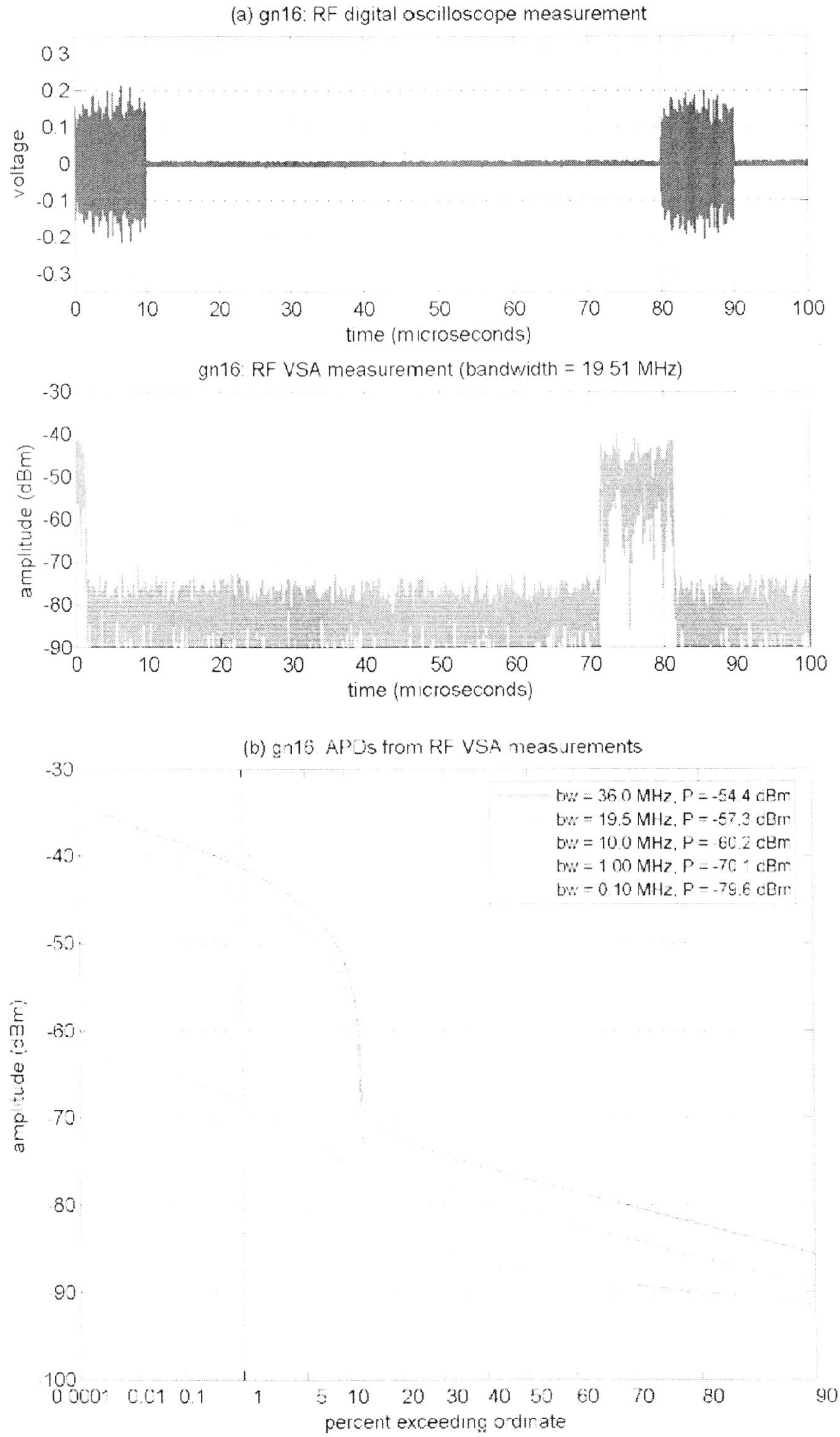

Figure D-33. RF measurements of GN-16 (τ_{on} = 10000 ns, ζ = 0.1250).

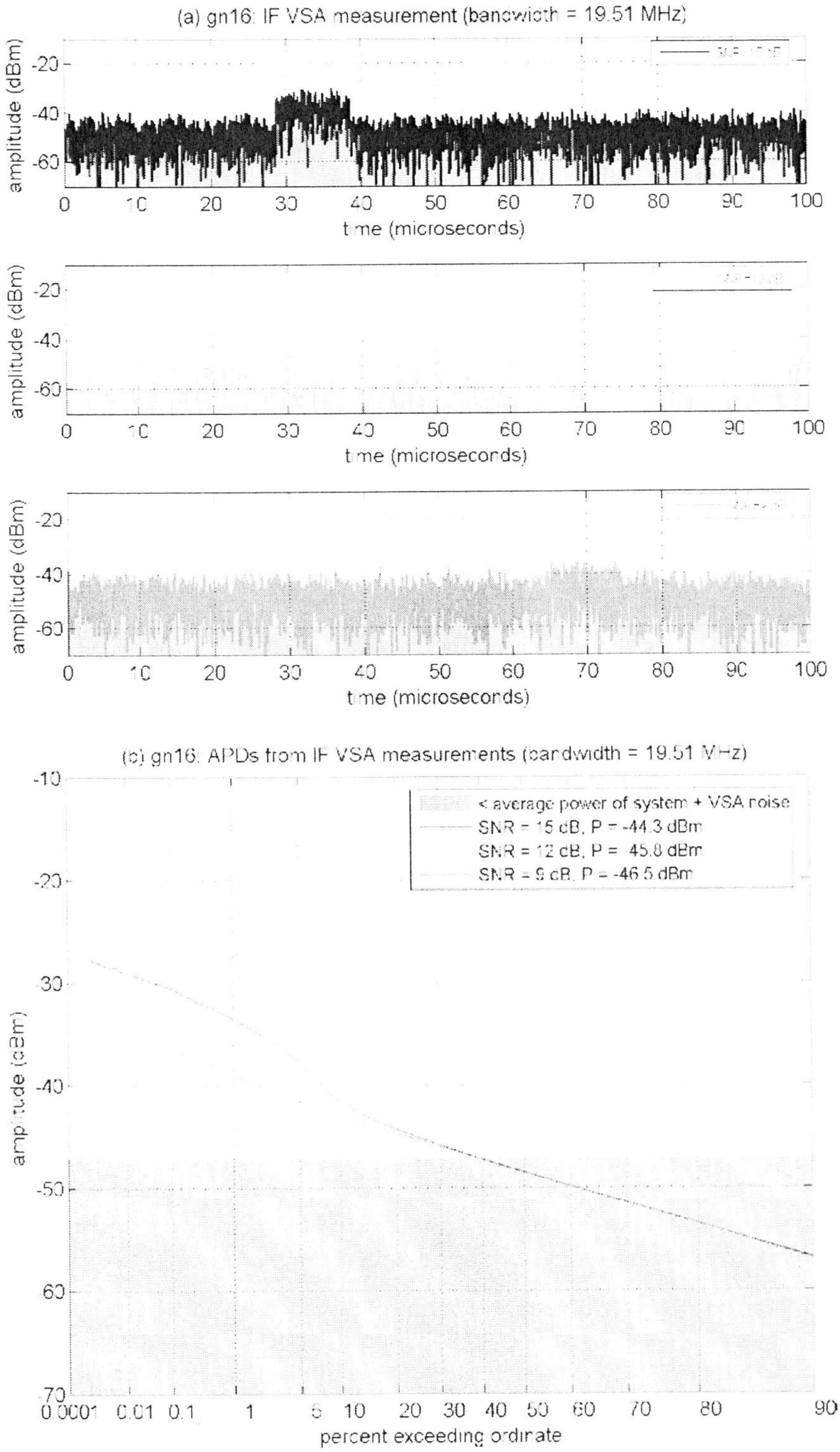

Figure D-34. IF measurements of GN- 16 at INR_{TOV}.

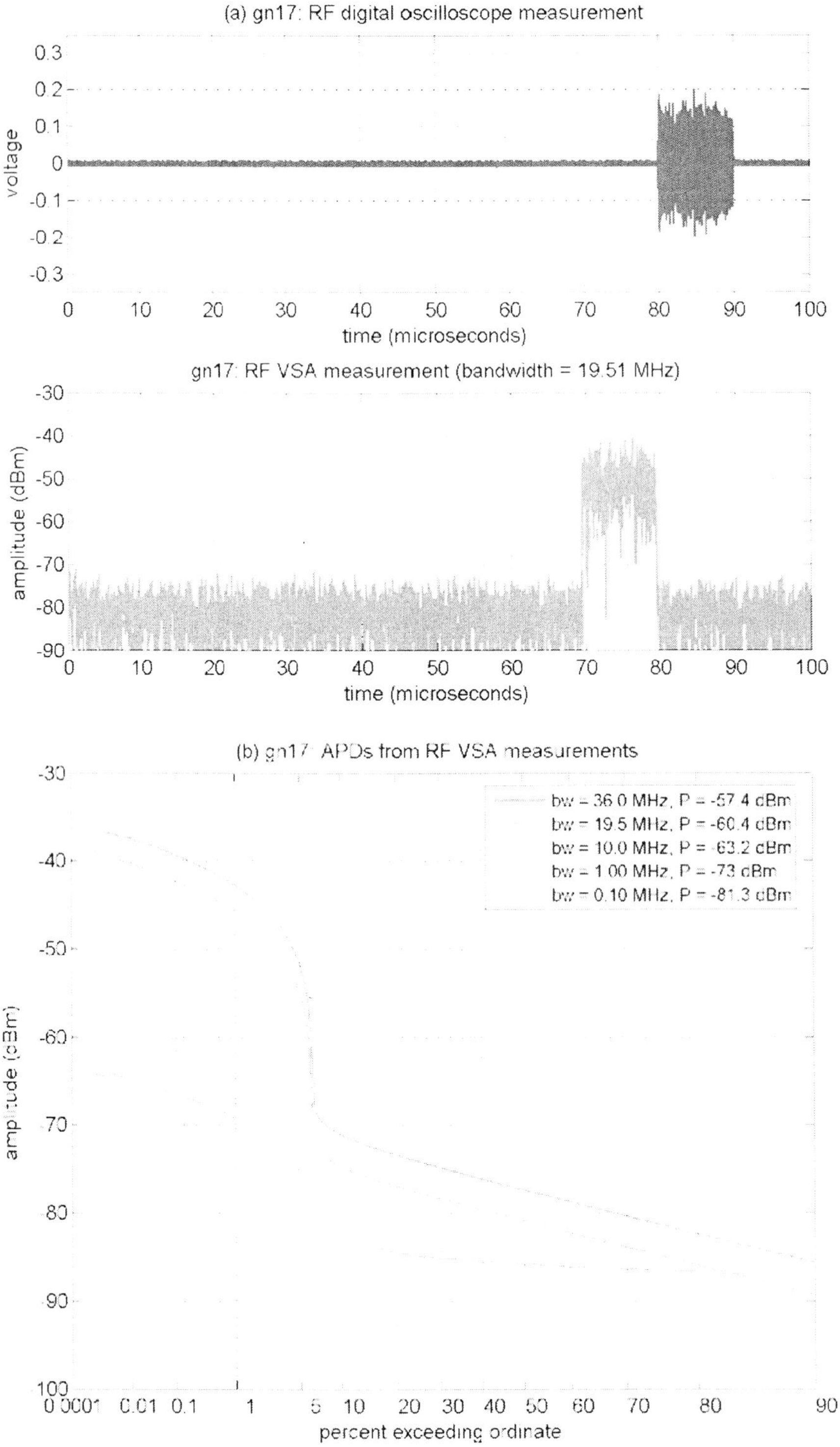

Figure D-35. RF measurements of GN-17 (τ_{on} = 10000 ns, ζ = 0.0625).

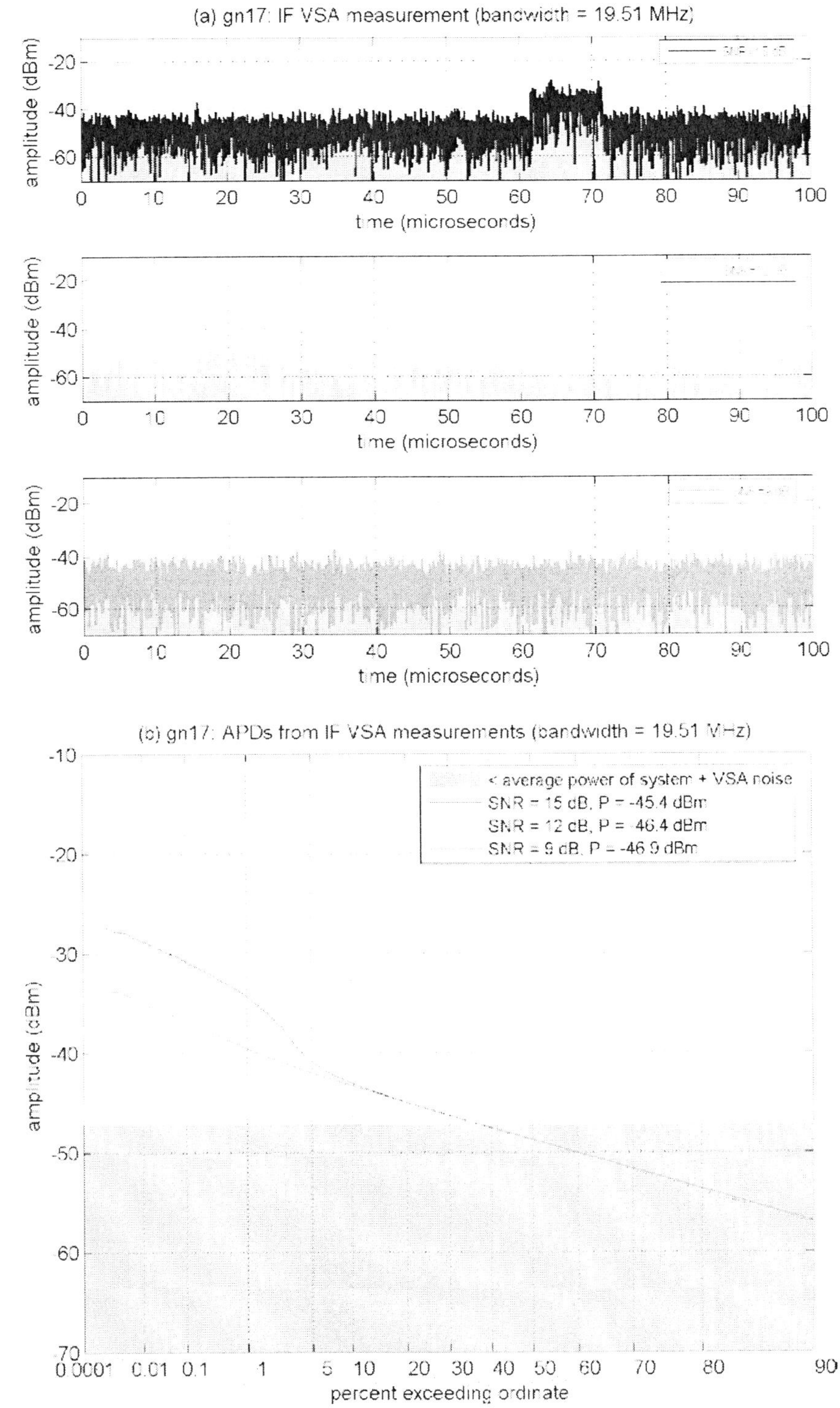

Figure D-36. IF measurements of GN-17 at INR_{TOV}.

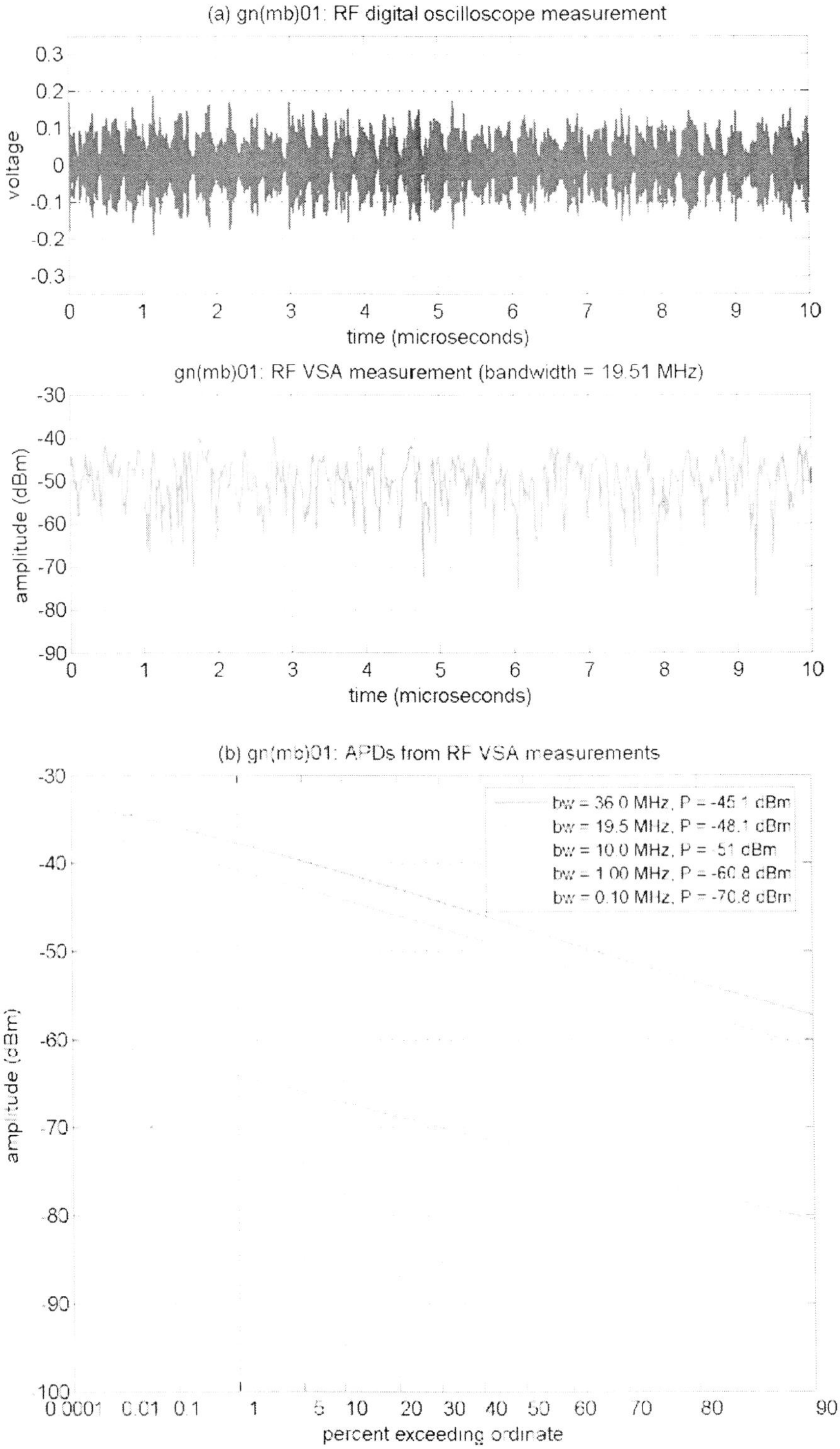

Figure D-37. RF measurements of GN(MB)-01 ($b = 1$, $d = 1$).

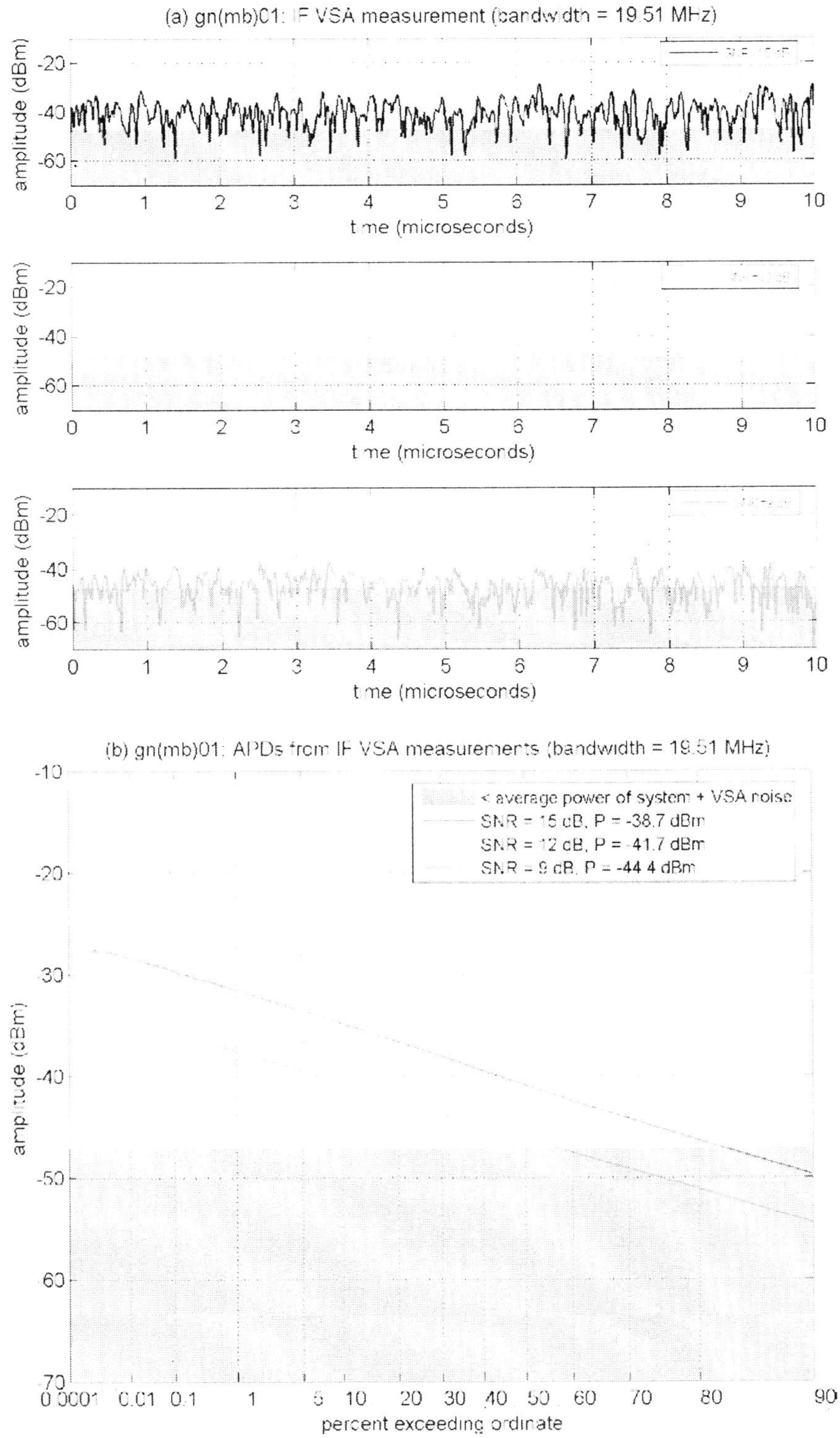

Figure D-38. IF measurements of GN(MB)-01 at INR_{TOV}.

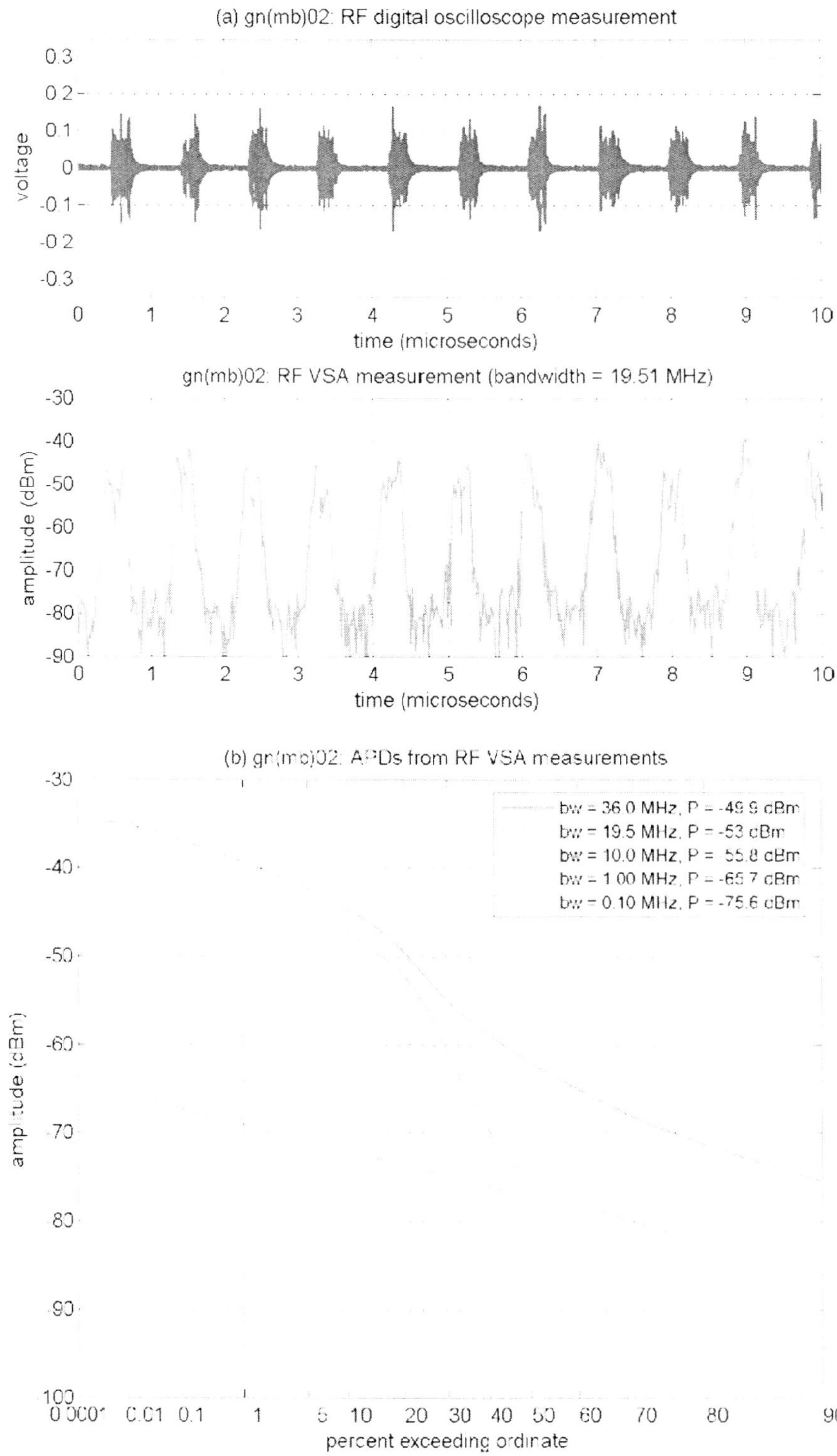

Figure D-39. RF measurements of GN(MB)-02 ($b = 3$, $d = 1$).

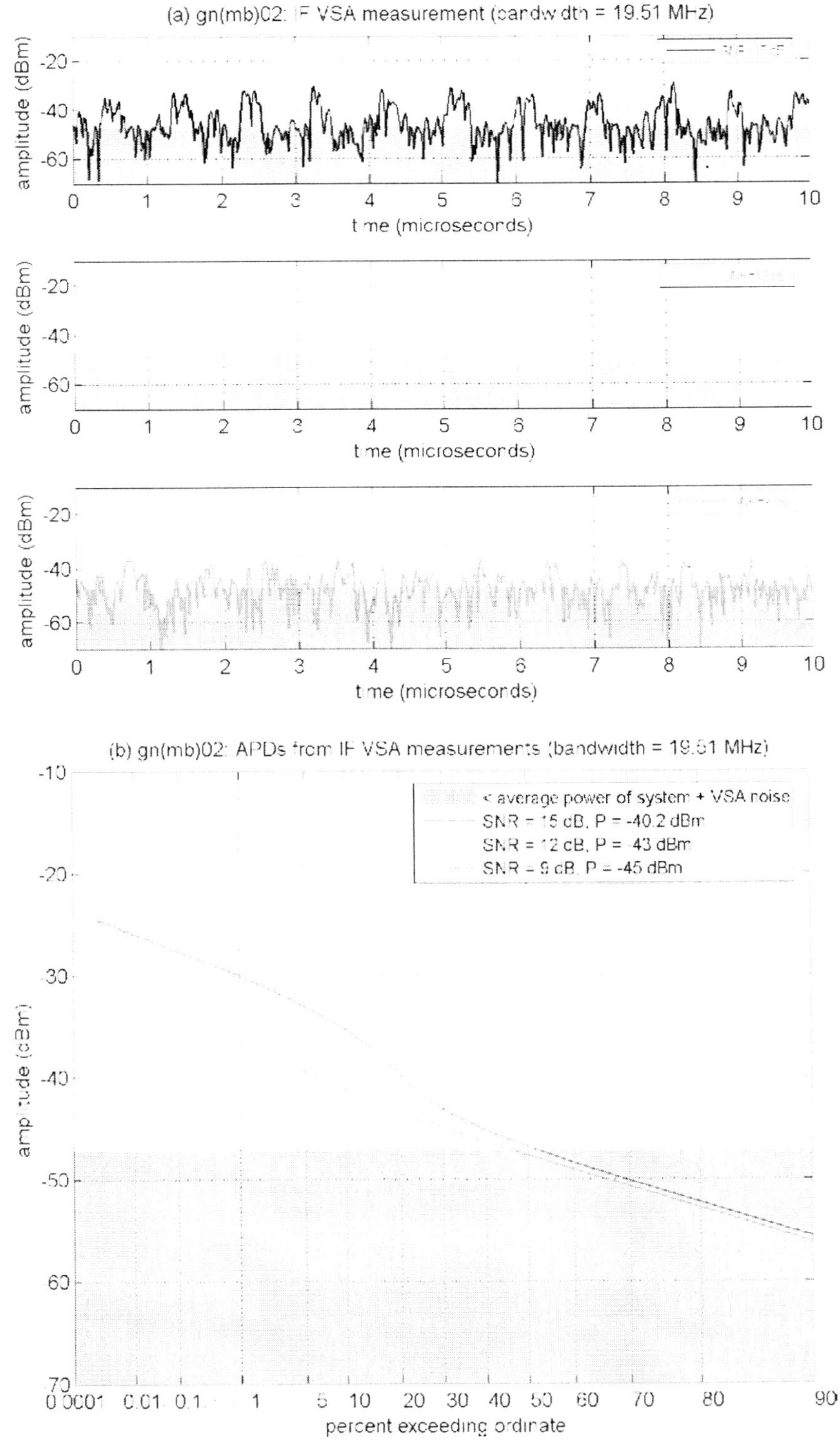

Figure D-40. IF measurements of GN(MB)-02 at INR_{TOV}.

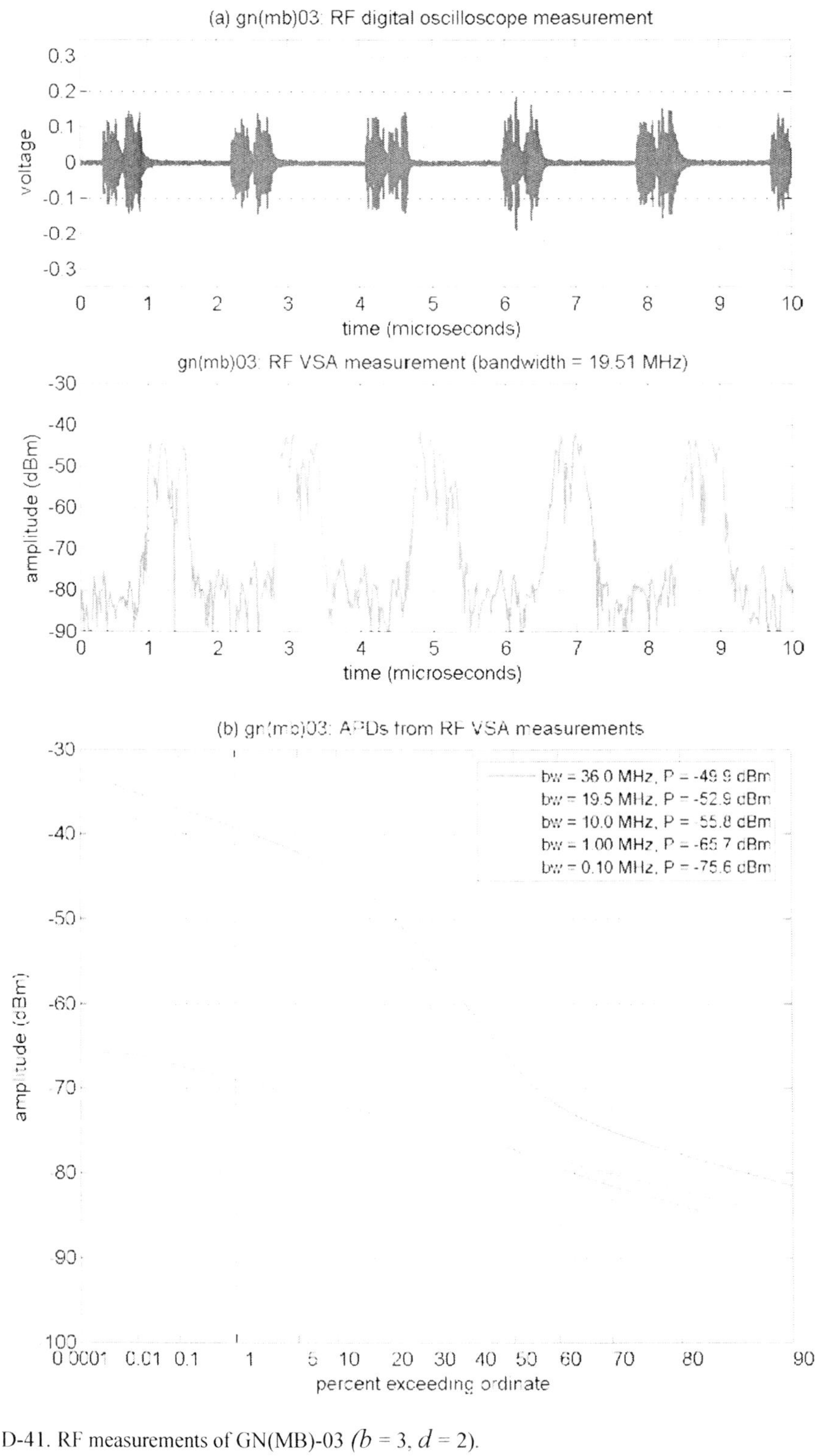

Figure D-41. RF measurements of GN(MB)-03 ($b = 3$, $d = 2$).

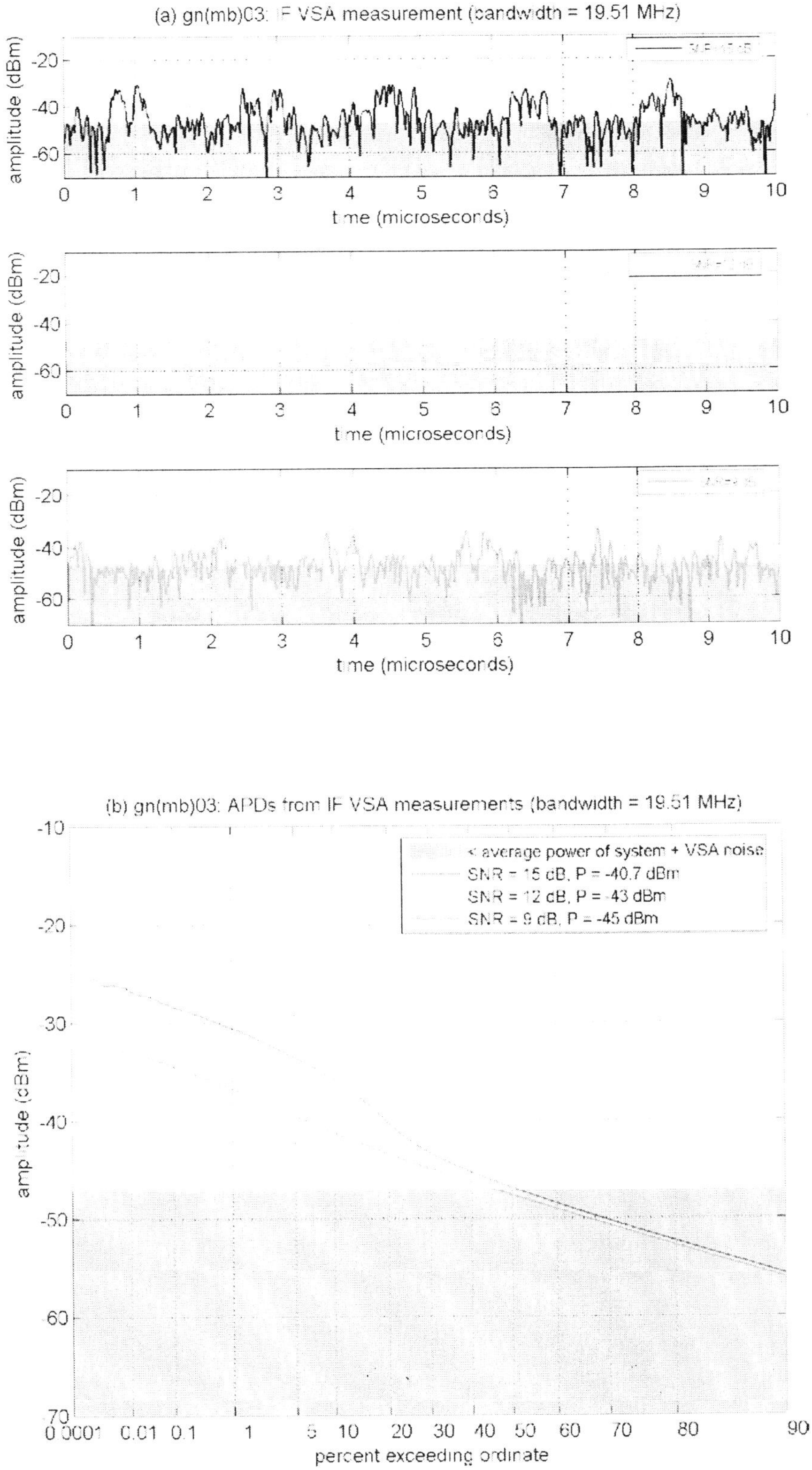

Figure D-42. IF measurements of GN(MB)-03 at INR_{TOV}

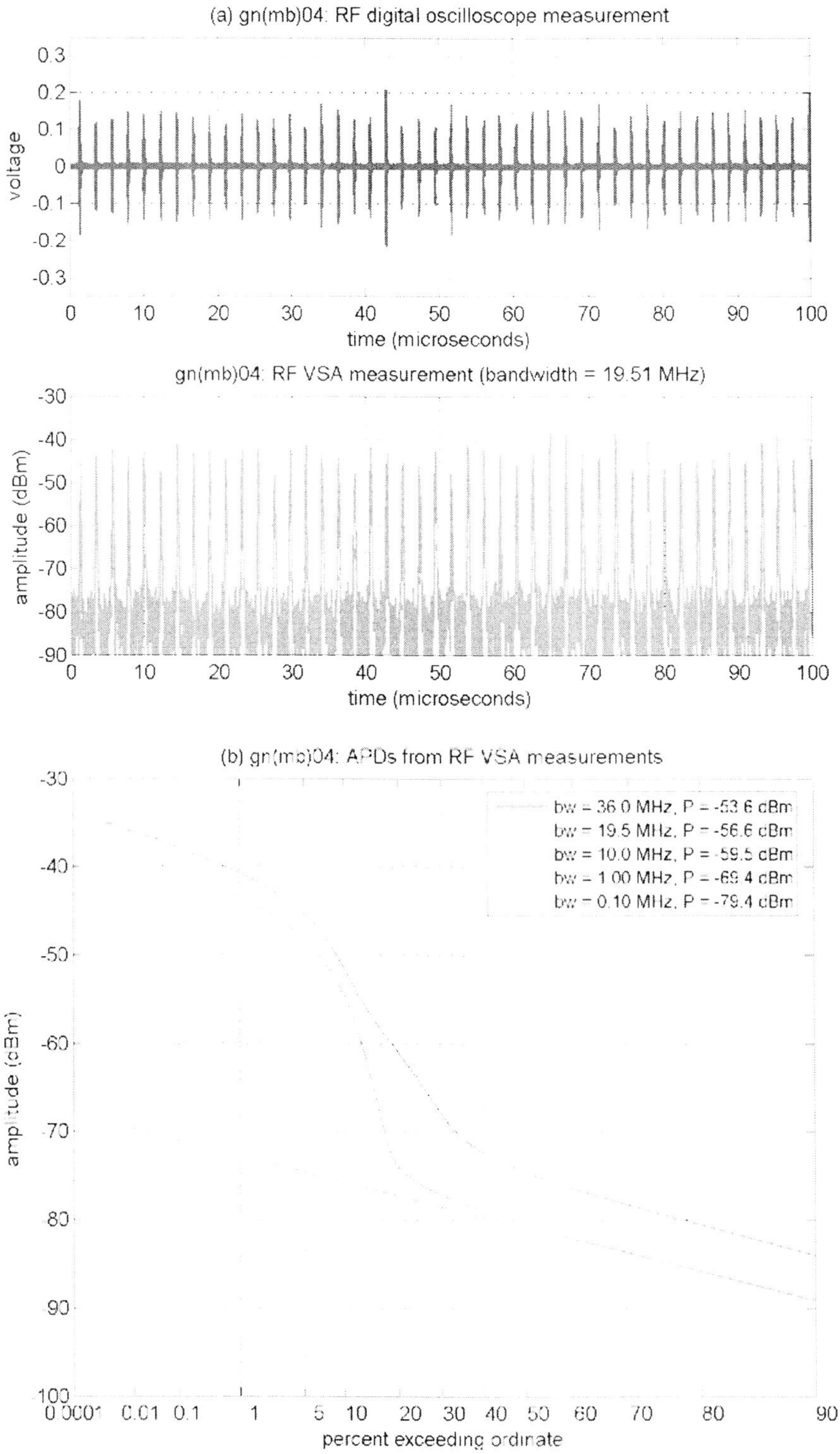

Figure D-43. RF measurements of GN(MB)-04 $(b = 7, d = 1)$.

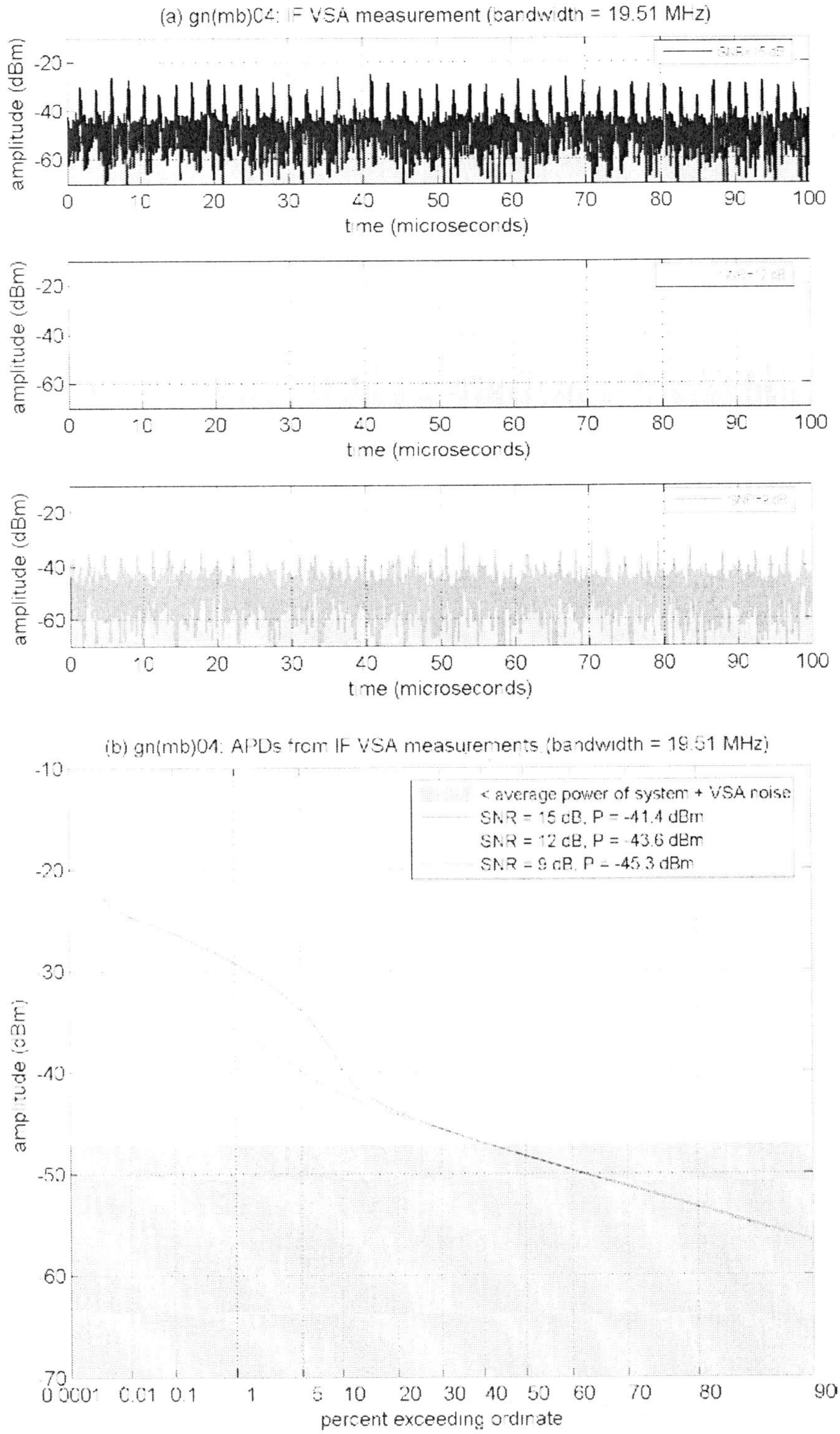

Figure D-44. IF measurements of GN(MB)-04 at INR_{TOV}

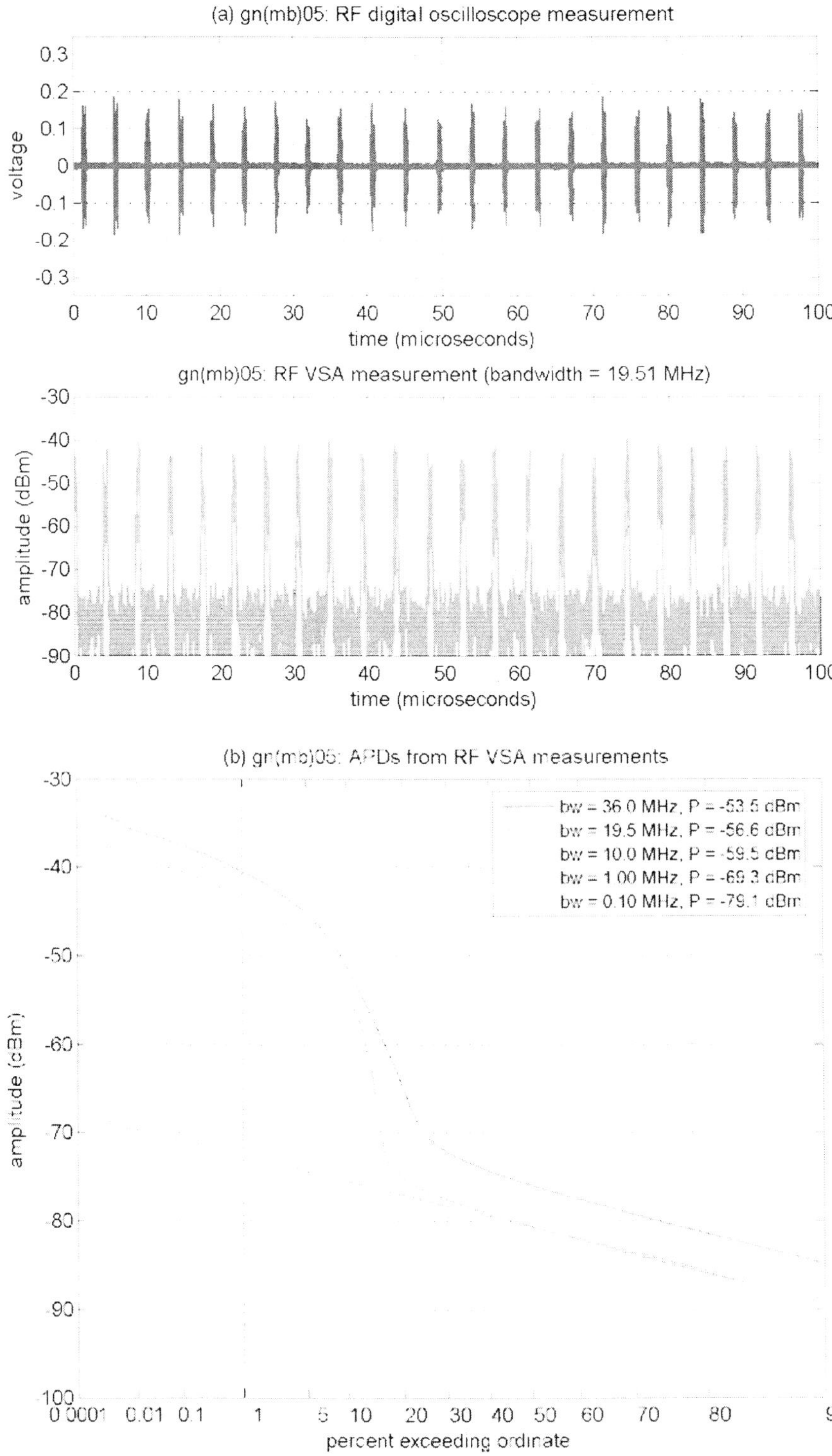

Figure D-45. RF measurements of GN(MB)-05 $(b = 7, d = 2)$.

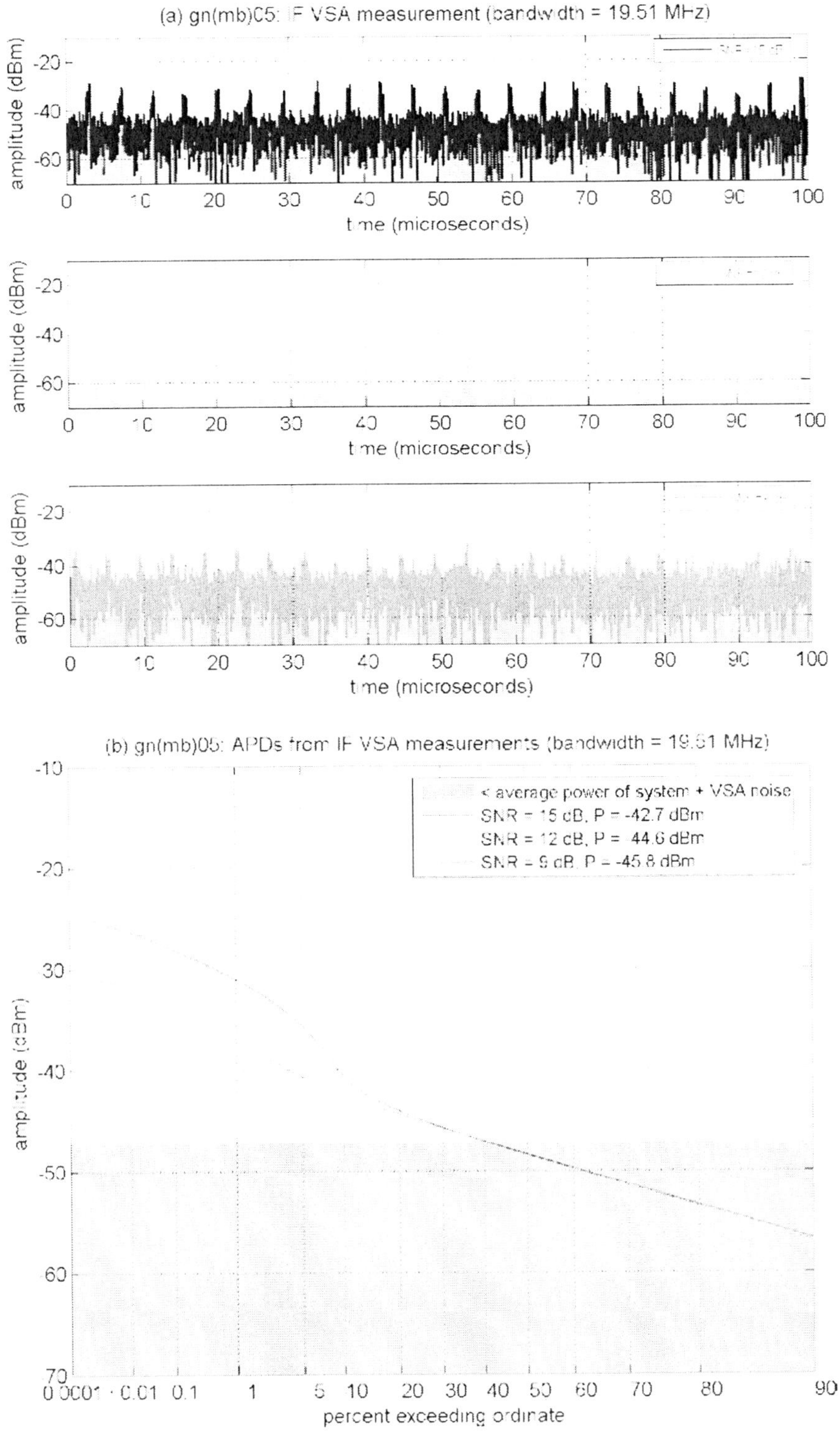

Figure D-46. IF measurements of GN(MB)-05 at INR_{TOV}

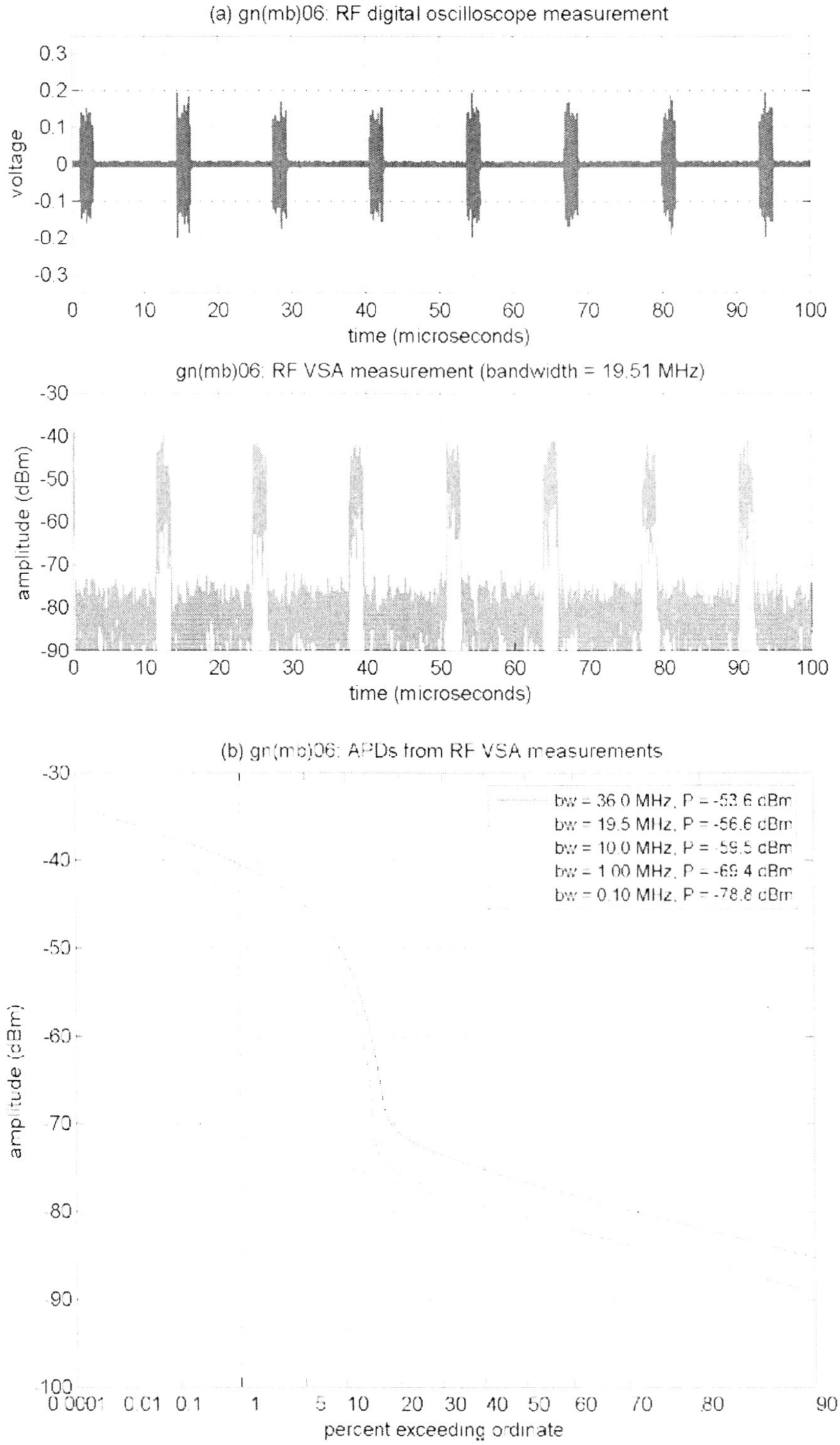

Figure D-47. RF measurements of GN(MB)-06 ($b = 7$, $d = 6$).

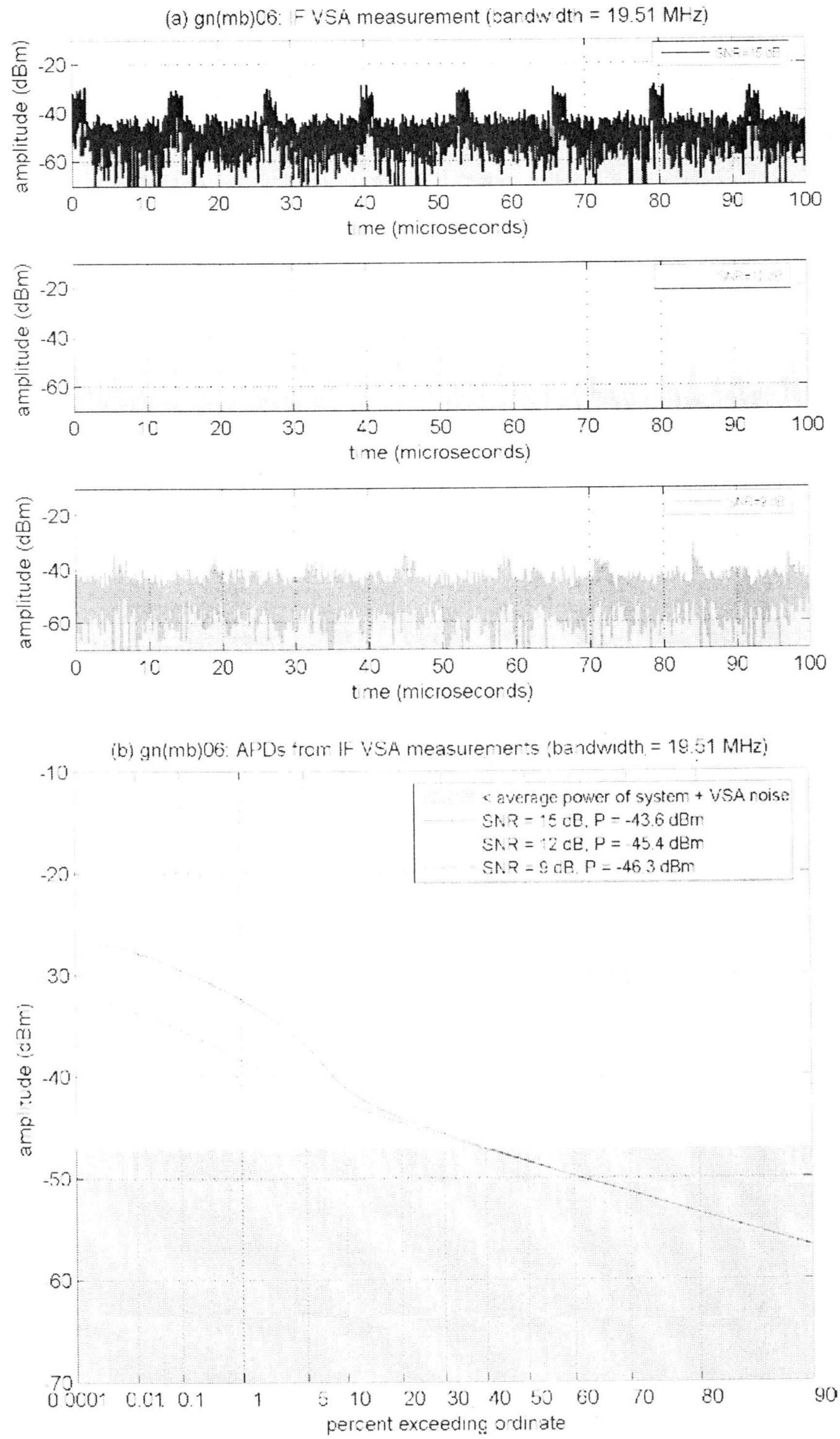

Figure D-48. IF measurements of GN(MB)-06 at INR_{TOV}.

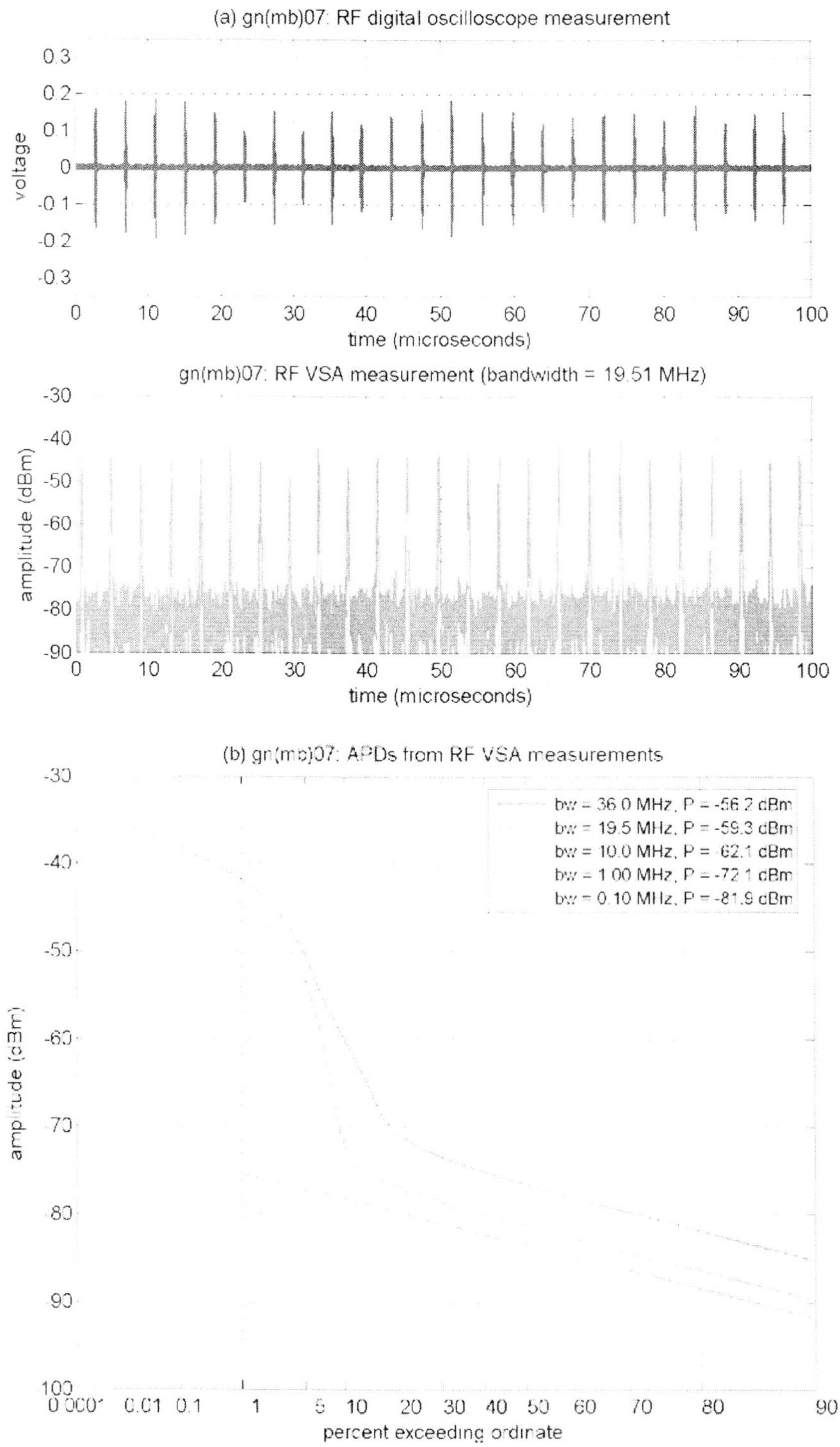

Figure D-49. RF measurements of GN(MB)-07 ($b = 13$, $d = 1$).

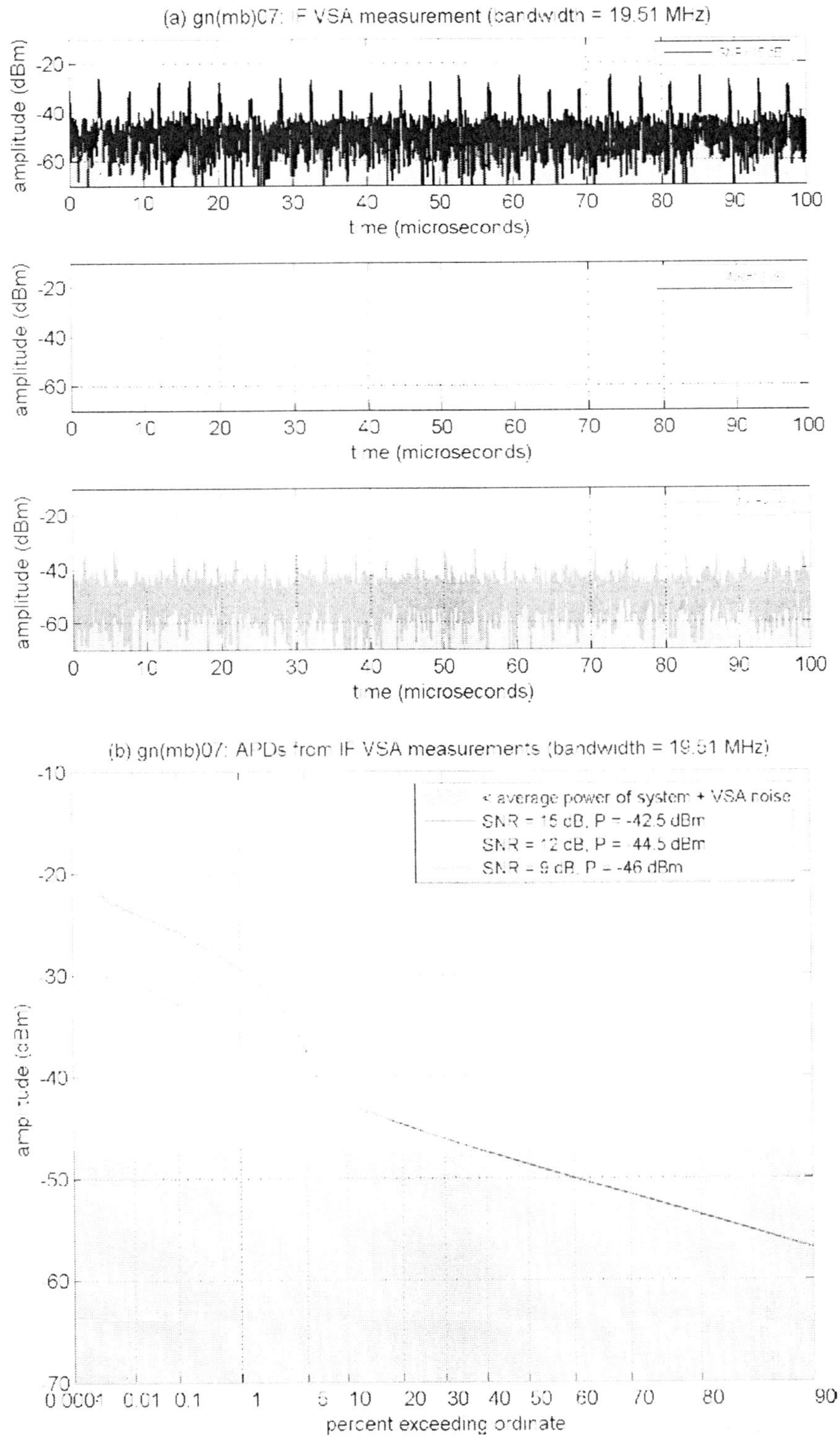

Figure D-50. IF measurements of GN(MB)-07 at INR_{TOV}.

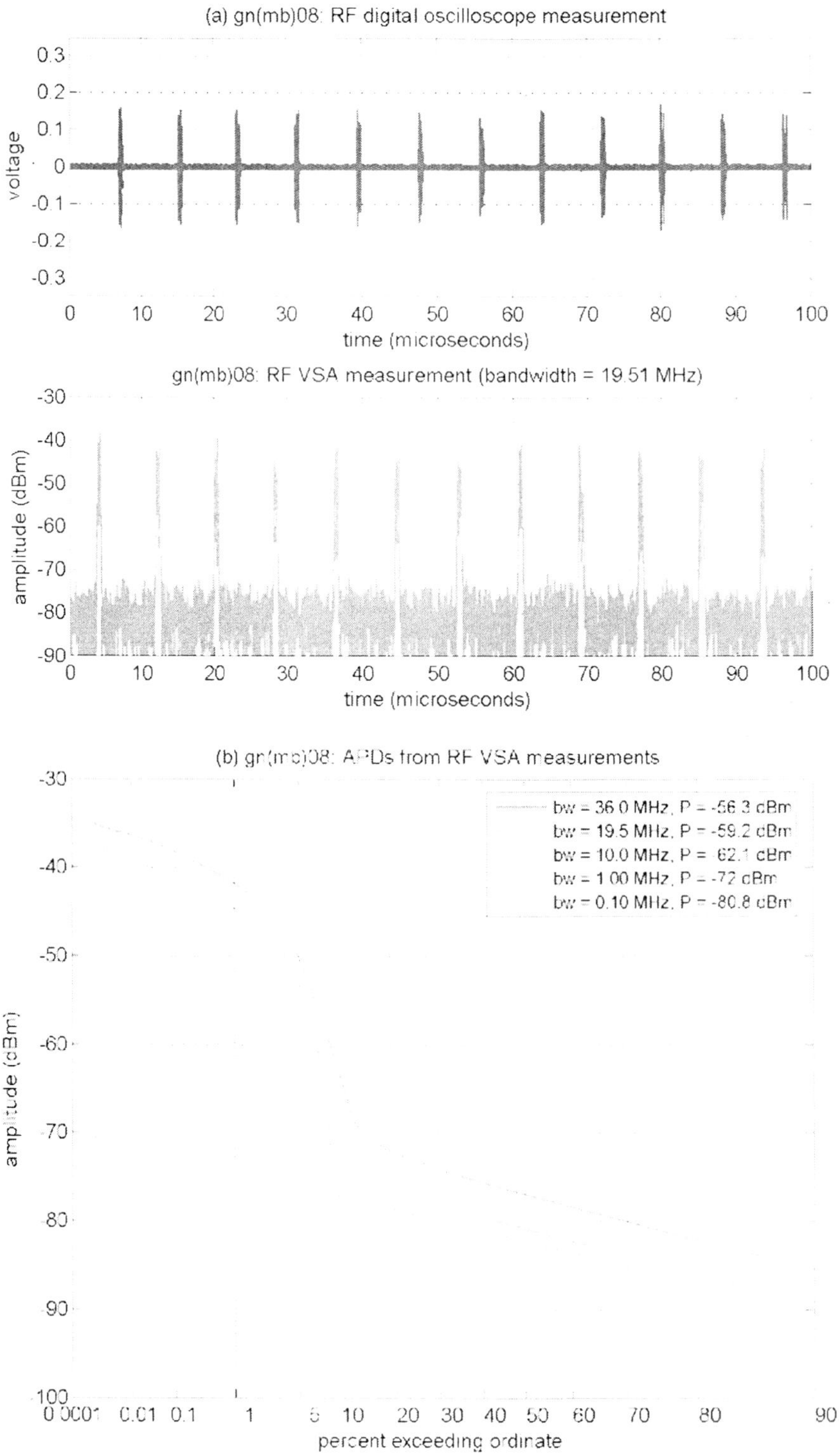

Figure D-51. RF measurements of GN(MB)-08 (b = 13, d = 2).

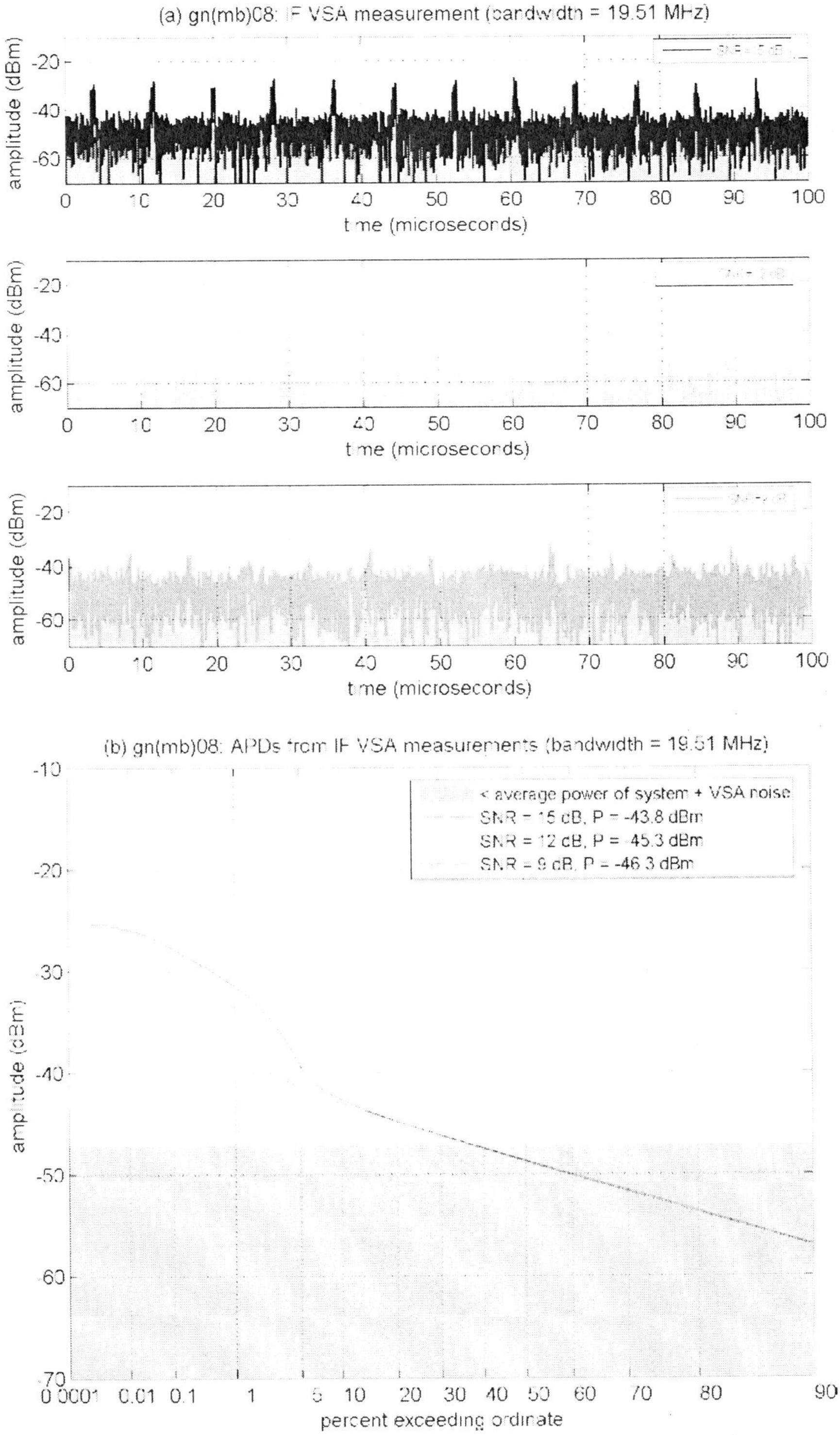

Figure D-52. IF measurements of GN(MB)-08 at INR_{TOV}

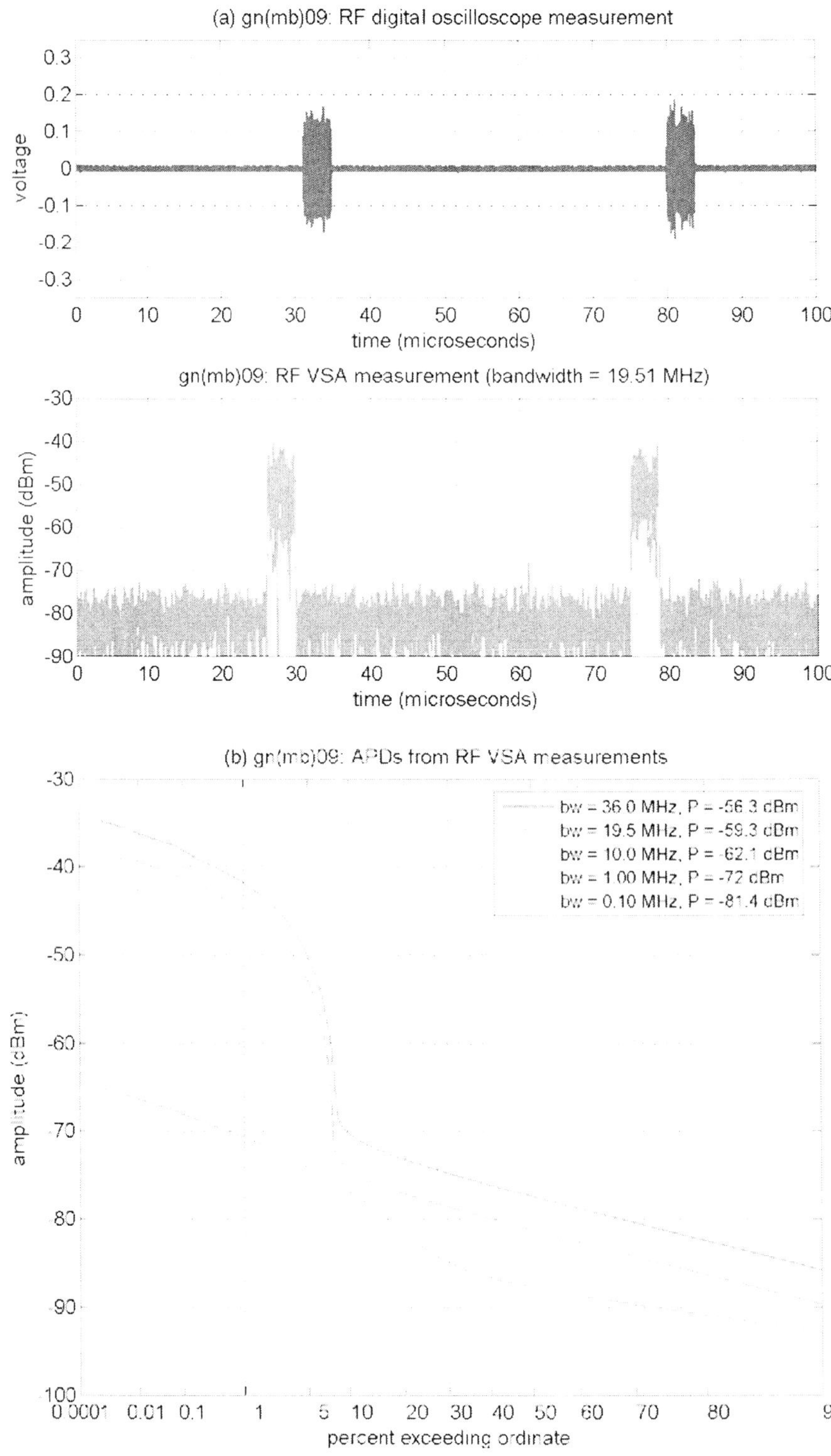

Figure D-53. RF measurements of GN(MB)-09 (b = 13, d = 12).

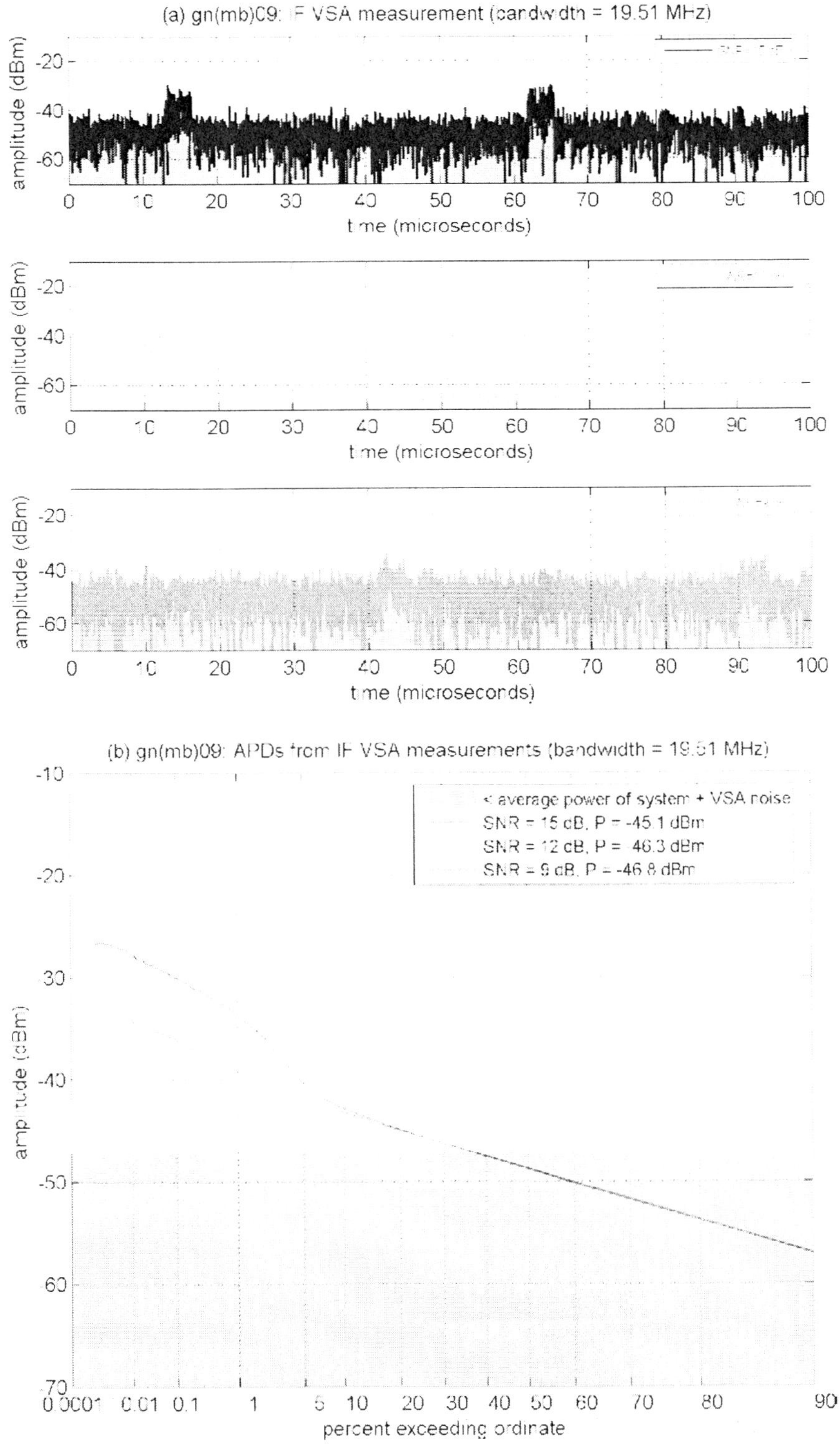

Figure D-54. IF measurements of GN(MB)-09 at INR_{TOV}

In: Interference Potential of Ultrawideband Signals
Editors: M. Cotton, R. Achatz, pp. 181-313

ISBN: 1-60021-359-6

PART 3: MEASUREMENT OF ULTRAWIDEBAND INTERFERENCE TO C-BAND SATELLITE DIGITAL TELEVISION RECEIVERS*

ABSTRACT

This chapter provides results from tests that measured digital television (DTV) susceptibility to ultrawideband (UWB) interference. A test system was developed to inject interference with known characteristics into a victim receiver and quantitatively measure susceptibility. In this experiment, a C-band satellite DTV victim receiver was injected with Dithered-Pulse (DP), Direct-Sequence (DS), and Multi-Band OFDM (MB) UWB interference. Results showed that the UWB signals could be categorized into three signal sets of common DTV susceptibility behavior. Interestingly, the categorized signals, band-limited by the DTV receiver filter, also had common characteristics. Set 1 consists of signals whose DTV susceptibility and band-limited signal characteristics resemble Gaussian noise.

Set 2 consists of signals more deleterious than Gaussian noise interference. Notably, these signals had a wide range of band-limited signal characteristics and susceptibilities. Set 3 consists of a signal that is relatively benign. Results also showed that measurable band-limited characteristics, e.g., burst duration (BD), burst interval (BI), fractional on-time (ζ_{DTV}), and peak-to-average ratio (P/A), of the interfering signal are useful for predicting susceptibility. Finally, it was determined that continuous and gated noise signals can be used to emulate the interference effects of DS and MB signals for the DTV victim receiver and operational scenarios tested in this study. This might not be true, however, for testing the susceptibility of other victim receivers operating in narrower bandwidths as indicated by amplitude probability distributions as a function of frequency for MB signals band-limited to relatively narrow bandwidths.

Key words: digital television; interference; satellite communications; ultrawideband.

1. INTRODUCTION

In April 2002, the Federal Communications Commission (FCC) released FCC 02-48 [1] legalizing intentional, low-power ultrawideband (UWB) emissions between 3.1 GHz and 10.6

GHz for communications devices operated indoors. UWB emissions were limited to -41 dBm average power in 1-MHz bandwidth and 0 dBm peak power in 50-MHz bandwidth, where average power is measured over a 1-millisecond integration time and peak power measurement duration is unspecified. The rules define a UWB device as one that emits signals with 10-dB bandwidth greater than 500 MHz or greater than 20% of the center frequency.

The FCC rules do not specify how the bandwidth requirement is achieved, consequently allowing industry considerable breadth in choosing a modulation. This breadth is exemplified by the development of Direct-Sequence (DS) and Multi-band Orthogonal Frequency-Division Multiplexing (MB) ultrawideband technologies. Proponents of DS and MB technologies have both sought standardization from IEEE (Institute of Electrical and Electronics Engineers) 802.15 working group 3a on high-rate (greater than 20 million bits per second) Wireless Personal Area Networks (WPAN). As the name implies, WPAN is intended for short-distance (nominally less than 10 meters) wireless networking of devices such as PCs, personal digital assistants, and mobile phones.

Both DS and MB transmitters are based on state-of-the-art integrated circuitry. DS modulation controls pulse polarity and hence supports phase shifting modulations. It achieves its ultra-wide bandwidth by transmitting sufficiently narrow pulses. An MB device simultaneously modulates 122 carriers spaced 4.125 MHz apart to achieve the ultra-wide bandwidth and frequency hops the modulated carriers between non-overlapping bands.

Since previous work performed at the Institute for Telecommunication Sciences (ITS) [2 – 5] did not specifically look at susceptibility of receivers to interference from DS or MB signals, and since there is little published information on this subject, ITS entered into a Cooperative Research and Development Agreement (CRADA) with the Freescale subsidiary of Motorola, Inc. to study how UWB interference might be predicted.

1.1. Experiment

Interference potential is a general concept where performance degradation of a victim receiver is predicted from interference signal characteristics. Interference potential is derived from numerous susceptibility tests on receivers with a variety of bandwidths and signal demodulation techniques. This 3-part chapter series describes one such susceptibility test on a C-band satellite DTV receiver. This victim receiver was chosen because it demodulates signals transmitted in the 3.7 to 4.2 GHz frequency range, which lies within the band allocated for UWB operation. It also uses a variety of bandwidths and signal demodulation techniques, which makes it an ideal victim receiver for the study of interference potential. Additionally, instruments capable of providing quantitative DTV signal quality data from various receiver subsystems are readily available.

Part 1 [6] of this chapter series describes procedures to characterize UWB signal emissions and measure DTV susceptibility. These procedures were executed on a test bed consisting of a satellite signal generator, a vector signal generator (VSG) emitting software-simulated UWB waveforms, a DTV signal monitor, and a vector signal analyzer (VSA). Specifically, the procedures describe how signal quality metrics from various receiver

* Excerpted from Report TR-06-437, dated February 2006.

subsystems within the DTV signal monitor can be used to quantify susceptibility and how signal characterization metrics can be obtained from the VSA.

The complexity of the UWB signals made experimental verification difficult. Part 2 [7] of this chapter series describes how continuous and gated noise signals were used in their place for verification purposes. These signals eliminated the complexities of the UWB signals and provided a simple parameterized theoretical basis for understanding DTV susceptibility results. These tests verified the pre-Viterbi bit error rate (BER) and post-Reed-Solomon segment error rate (SER) DTV signal quality metrics.

Beyond verification, insight into DTV susceptibility to UWB signals was gained from the gated- noise experiment. Results clearly showed that DTV susceptibility was dependent on the interfering signal gating parameters, such as on-time (τ_{on}) and off-time (τ_{off}), and the bandwidth of the DTV receiver in addition to the more commonly used average and peak power metrics. This dependence on τ_{on} and τ_{off} motivated an investigation into the effects of band-limiting on the temporal parameters and their subsequent correlation to DTV susceptibility. These band-limited metrics include burst duration (BD) and burst interval (BI), corresponding to τ_{on} and τ_{off}, respectively.

In this chapter, Part 3 of the chapter series, the experiment is extended to actual UWB signals. The primary objective of this chapter is to evaluate the susceptibility of the DTV receiver to the UWB signals. The secondary objective is to determine if DTV susceptibility can be predicted from UWB signal parameters and the bandwidth of the victim receiver as well as band-limited metrics such as BD and BI. The final objective is to determine if continuous and gated Gaussian noise can be used in place of DS and MB signals, respectively, in susceptibility tests.

1.2. Organization of Book

UWB signals addressed in this chapter can be divided into two categories: Dithered-Pulse (DP) and DS signals that are specified with pulse parameters, i.e., pulse width (w), pulse repetition period (T_{pulse}), and fractional dither (f_D), and MB signals that are specified with gating parameters, i.e., τ_{on} and τ_{off}. Accordingly, DTV susceptibility and signal characterization results for DP and DS signals are given in Section 2, while results for MB signals are given in Section 3.

DTV susceptibility results are plotted in two different ways. First, post-Reed-Solomon SER and pre-Viterbi BER of the DTV victim receiver are plotted as a function of interference-to-noise ratio (INR) to demonstrate how DTV performance depends on average power of the interfering signals. Second, INR and BER at the threshold of visibility (TOV), i.e., INR_{TOV} and BER_{TOV}, are plotted as a function of the reciprocal fractional on-time ($1/\zeta$) in dB. These plots demonstrate how DTV susceptibility and forward error correction (FEC) performance depend on pulse and gating parameters of the interfering signals. Signal characterization includes temporal and amplitude analyses of measured UWB signals, which demonstrate the effects of band-limiting by the victim receiver. The temporal analyses are based on crossing statistics. Measured BD and BI of the band-limited signals are compared to w and T_{pulse} of the DP and DS signals and to τ_{on} and τ_{off} of the MB signals. Amplitude analyses are based on the amplitude probability distribution (APD). Measured peak-to-average ratios (P/As) of the band-limited DP and DS signals are compared to P/A of band-limited Gaussian noise, measured P/As of the band-limited MB signals are compared to P/As of corresponding band-limited gated Gaussian noise signals. Also provided are spectral analyses based on the power spectral density (PSD).

Section 4 summarizes DTV susceptibility test results, identifies signals with common DTV susceptibility behavior, plots DTV susceptibility results as a function of band-limited metrics, i.e, BD, BI, ζ_{DTV}, and P/A, to illustrate correlations, and evaluates the significance and scope of the findings. Finally, the validity of using continuous and gated noise signals to replicate DS and MB signals in this and other susceptibility tests is discussed.

Appendices to this chapter contain information supporting the main body. Appendix A develops theoretical expressions for the PSD of the DP and DS signals. Appendix B provides measured crossing statistics of the UWB signals measured at the satellite radio frequency (RF) and band-limited to the bandwidth of the victim receiver (B_{DTV}), APDs of the UWB signals measured at the satellite RF in a variety of bandwidths, APDs of the UWB signals measured at the first intermediate frequency (IF) of the victim receiver in the presence of low-noise block downconverter (LNB) noise for interference powers corresponding to INR_{TOV}, and APD statistics as a function of frequency and bandwidth of the UWB signals measured at the satellite RF.

2. DP and DS Interference

This section provides DTV susceptibility test results and characterization measurements for DP and DS interference.

2.1. Pulse Parameters

DP and DS signals achieve their ultra-wide bandwidth by transmitting pulses of short duration. These signals are specified by pulse width (w), the nominal pulse period (T_{pulse} = 1/PRF, where PRF is the pulse repetition frequency), and fractional dither (f_D) to quantify pulse offset from T_{pulse}. Table 1 specifies parameters that define the DP signals considered in this chapter. Fractional on-time is approximated as $\zeta \approx w/T_{pulse}$. Figure 1 illustrates pulse parameters for DP-04.

Table 1. DP Pulse Parameters

Type	Index	w (ns)	T_{pulse} (ns)	ζ	f_D
DP	01	0.094	10,000	9.4×10^{-6}	0.5
	02		1000	9.4×10^{-5}	
	03		100	9.4×10^{-4}	
	04		10	9.4×10^{-3}	

Table 2 specifies parameters that define the DS signals considered in this chapter. Recall that DS devices multiply each data bit by a code word consisting of L chips. The chips have values of {-1, 0, 1} and are transmitted every $T_{chip} = 1/f_{chip}$, where f_{chip} = 1320 million chips per second. The coded data phase modulates pulses that are approximately 0.58 ns wide.

Sparse codes used for DS-02, DS-03, and DS-05 are all zeros except for a single one, which results in distinct pulses every $T_{pulse} = LT_{chip}$. DS codes used for DS-04 and DS-06 have relatively few zeros, which results in tightly-packed pulses every $T_{pulse} = T_{chip}$. Figure 2 illustrates pulse parameters for DS-03 and DS-04.

Table 2. DS Pulse Parameters[†]

Type	Index	DS Parameters		Pulse Parameters			
		L	Code Type	w (ns)	T_{pulse} (ns)	ζ	f_D
DS	01	1	N/A	0.58	0.758	0.77	0.0
	02	3	Sparse		2.27	0.26	
	03	6	Sparse		4.55	0.13	
	04	12	DS		0.758	0.77	
	05	12	Sparse		9.09	0.064	
	06	24	DS		0.758	0.77	

[†] The specific codes are [1] for DS-01, [0 1 0] for DS-02, [0 0 1 0 0 0] for DS-03, [0 -1 -1 -1 1 1 1 -1 1 1 -1 1] for DS-04, [0 0 0 0 0 0 0 0 1 0 0 0] for DS-05, and [-1 1 -1 -1 1 -1 -1 1 -1 0 -1 0 -1 -1 1 1 1 -1 1 1 1 -1 -1 -1] for DS-06.

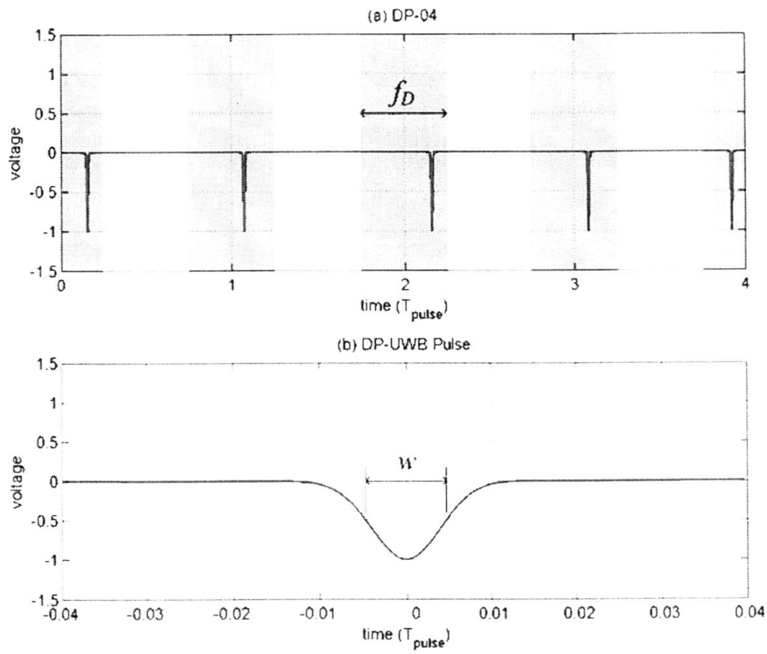

Figure 1. Simulated DP-04 (T_{pulse} = 10 ns).

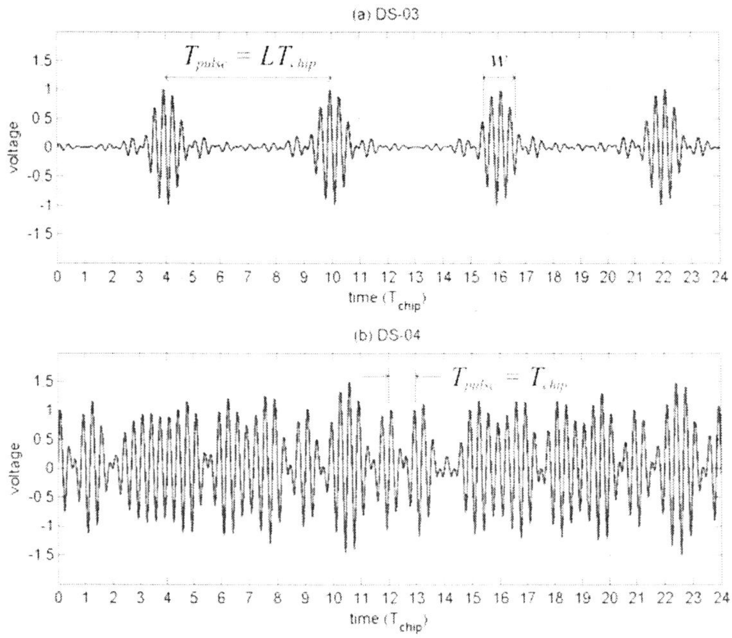

Figure 2. Simulated (a) DS-03 and (b) DS-04 transmitting random data (T_{chip} = 0.758 ns).

2.2. DTV Susceptibility to DP and DS Interference

Results of the DTV susceptibility tests for DP and DS interference are described in this section. Post-Reed-Solomon segment error rate and pre-Viterbi bit error rate are plotted versus the ratio of average interference power to average noise power. Additionally, DTV susceptibility and FEC performance metrics, i.e., INR_{TOV} and BER_{TOV}, are plotted versus $1/\zeta$ in dB. Figure 3 illustrates reference points for signal powers and DTV signal quality metrics acquired during the DTV susceptibility tests.

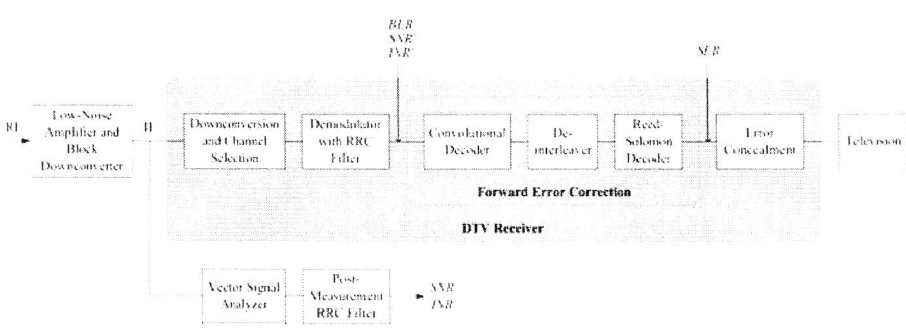

Figure 3. SNR, SNR′, INR, INR′, BER, and SER reference points.

2.2.1. DTV Signal Quality as a Function of Interference Average Power

Figures 4 – 9 provide composite plots of measured SER and BER as a function of INR for measured SNR = {9, 12, 15} dB corresponding to SNR′ = {8.2, 10.5, 12.5} dB at the output of the demodulator. Each page is dedicated to a single SNR, each plot is dedicated to a specific signal type, and each curve represents a single T_{pulse}. Figures 4, 6, and 8 provide composite plots of SER, while figures 5, 7, and 9 provide composite plots of BER. For each plot, Gaussian noise (GN-01) results are provided as a reference.

The following are general comments regarding the shift, separation, and slope of these SER and BER curves. Both BER and SER shifted toward greater INR with increasing SNR. This occurred because more interference was needed to degrade stronger satellite signals. The shift of DP-01 BER and SER was significantly greater than the other interference signals. SER due to DS signals was similar to SER due to GN-01. SER due to DP signals showed separation from GN-01 only when T_{pulse} was greater than 100 ns. Relative to Gaussian-noise degradation, the DTV receiver was more susceptible to DP-02 and less susceptible to DP-01. With the exception of DP-01 and DP-02, all DP and DS SER curves were steeper than corresponding BER curves, which indicates that SER was more sensitive than BER to changes in average interference power as expected with forward error correction. When T_{pulse} > 100 ns, the SER slope flattened with increasing T_{pulse}. In fact, the slope of DP-01 SER was so low that oscillations indicative of the systematic error mentioned in Appendix B of Part 2 of this chapter series were revealed.

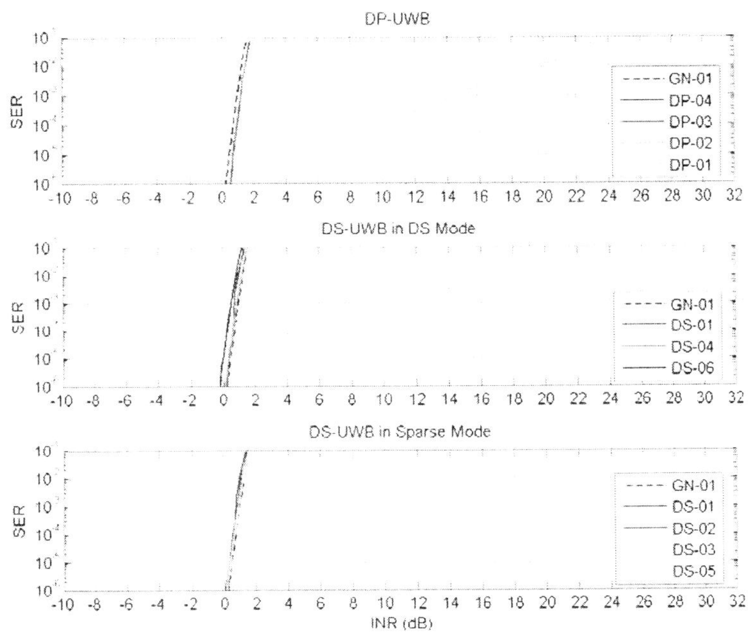

Figure 4. SER versus INR for a DTV channel operating at SNR = 9 dB and exposed to DP and DS interference.

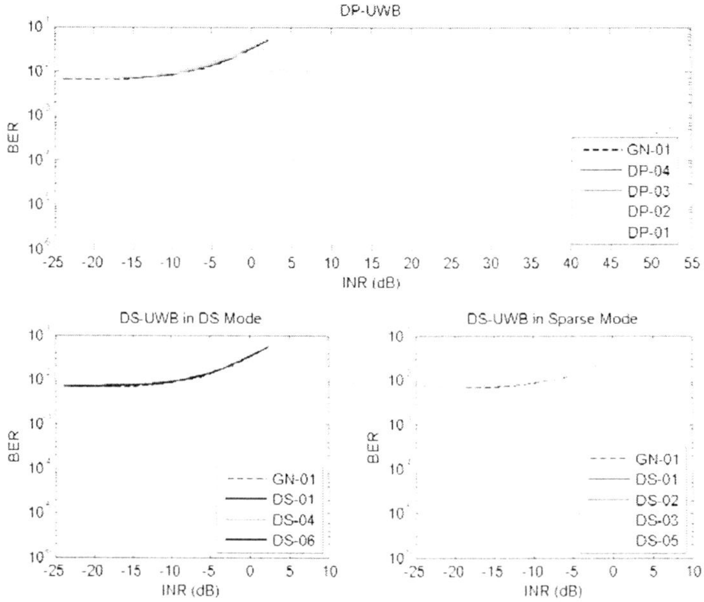

Figure 5. BER versus INR for a DTV channel operating at SNR = 9 dB and exposed to DP and DS interference.

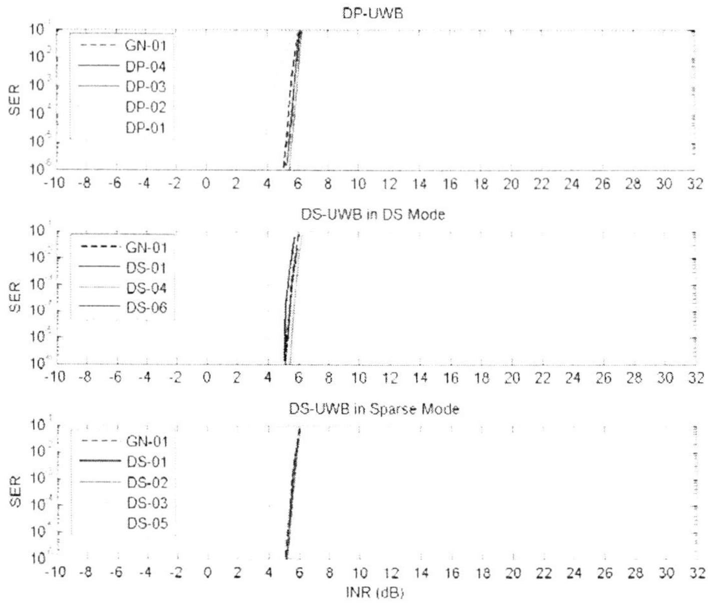

Figure 6. SER versus INR for a DTV channel operating at SNR = 12 dB and exposed to DP and DS interference.

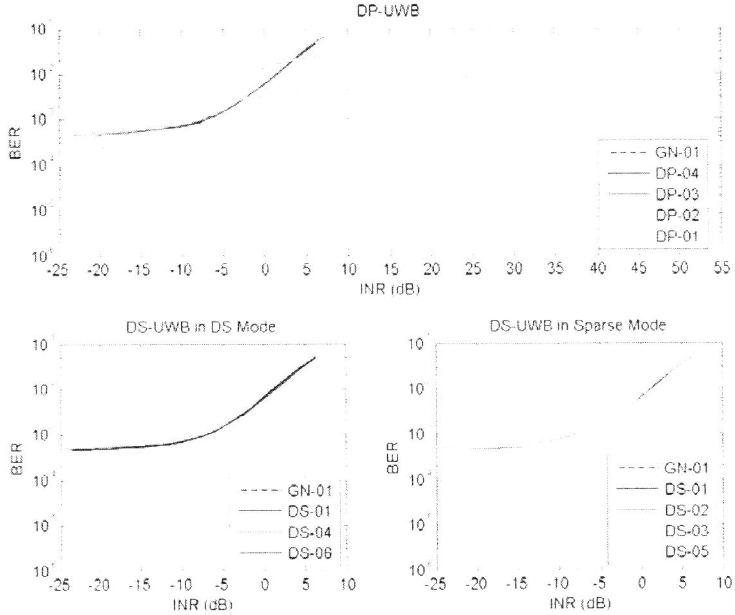

Figure 7. BER versus INR for a DTV channel operating at SNR = 12 dB and exposed to DP and DS interference.

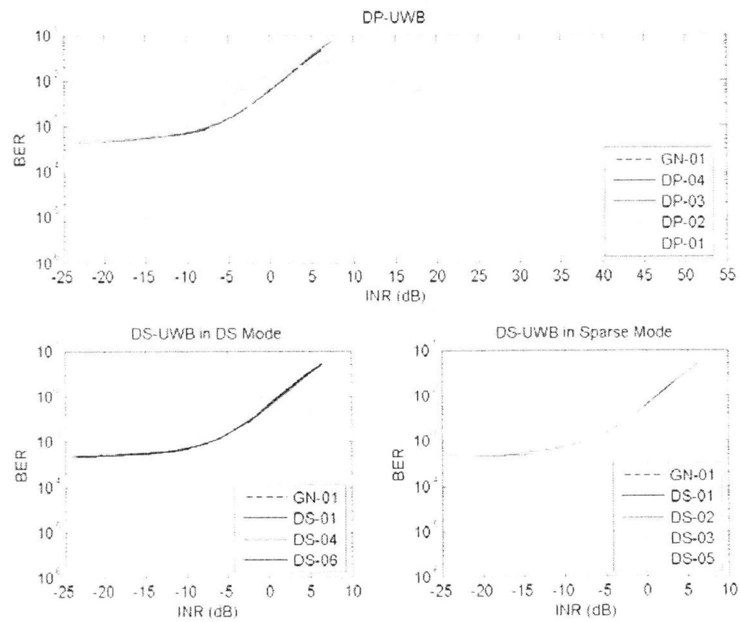

Figure 8. SER versus INR for a DTV channel operating at SNR = 15 dB and exposed to DP and DS interference.

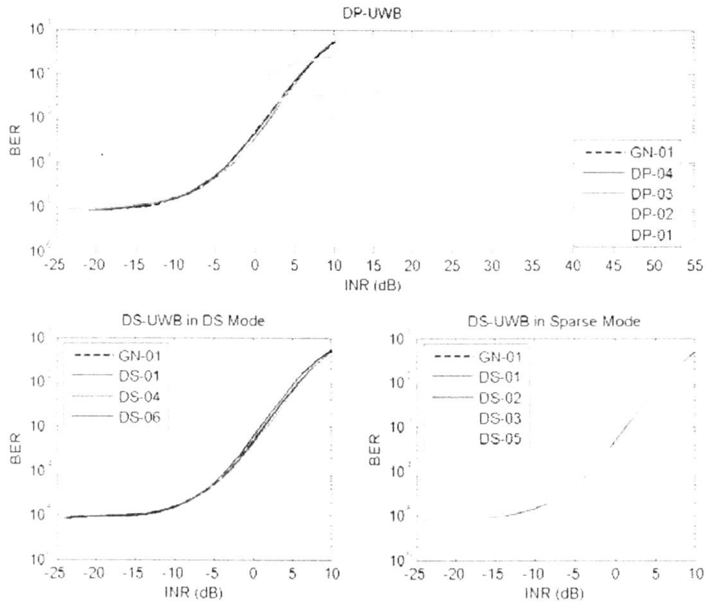

Figure 9. BER versus INR for a DTV channel operating at SNR = 15 dB and exposed to DP and DS interference.

2.2.2. DTV Susceptibility as a Function of Fractional on-Time

Figure 10 displays measured SER and BER curves for a DTV victim receiver operating at SNR = 15 dB and exposed to increasing levels of Gaussian noise interference. This figure also identifies important DTV susceptibility and FEC performance metrics. Importantly, $SER_{TOV} = 10^{-4}$ is a level identified by video quality studies [8] as the threshold of visibility (TOV) where video quality degradation is first evident.

The INR that corresponds to SER_{TOV}, INR_{TOV}, is our primary DTV susceptibility metric. That is, receiver susceptibility increases as INR_{TOV} decreases. The BER that corresponds to SER_{TOV}, BER_{TOV}, is our FEC performance metric. This metric quantifies the ability of the FEC to overcome bit errors caused by interference to achieve SER_{TOV}. In other words, larger BER_{TOV} indicates more corrected bits and better FEC performance at the threshold of visibility.

Table 3 gives INR_{TOV} and BER_{TOV} results for DP and DS interference; GN-01 results from Part 2 of this chapter series are also included for comparison purposes. Differences between INR_{TOV} of DS, DP-03, DP-04, and GN-01 results were less than 0.5 dB.

Recall from Part 2 of this chapter series that high DTV susceptibility, i.e., low INR_{TOV}, correlated with poor FEC performance, i.e., low BER_{TOV}. This demonstrated that DTV susceptibility was related to how effective the FEC was at overcoming bit errors caused by the interference at TOV. DP-01 demonstrates a distinct contrast with this trend, i.e., high DTV susceptibility did not correspond to poor FEC performance based on the chosen metric BER_{TOV}.

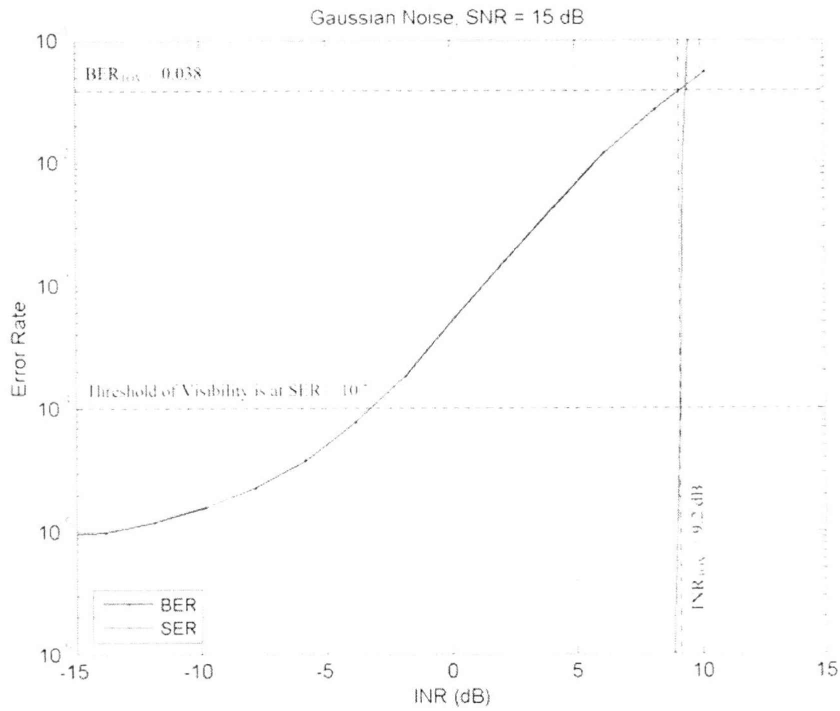

Figure 10. Illustration of threshold-of-visibility metrics.

Table 3. Measured DTV Susceptibility and FEC Performance for DP and DS Interference

Type	Index	SNR ≈ 9 dB		SNR ≈ 12 dB		SNR ≈ 15 dB	
		INR_{TOV} (dB)	BER_{TOV}	BER_{TOV}	INR_{TOV} (dB)	INR_{TOV} (dB)	BER_{TOV}
DP	01	10.2	0.012	20.4	0.007	25.2	0.006
	02	-0.7	0.025	4.2	0.021	8.0	0.022
	03	1.0	0.041	5.8	0.045	9.6	0.049
	04	0.9	0.038	5.6	0.039	9.0	0.040
DS	01	0.6	0.038	5.4	0.038	9.1	0.038
	02	0.5	0.038	5.5	0.038	9.1	0.039
	03	0.9	0.038	5.7	0.039	9.0	0.040
	04	0.6	0.039	5.7	0.041	9.2	0.042
	05	0.6	0.039	5.7	0.041	9.2	0.042
	06	0.2	0.037	5.2	0.038	8.8	0.039
GN	01	0.7	0.038	5.4	0.038	9.2	0.038

Figures 11 – 16 provide plots of INR_{TOV} and BER_{TOV} as a function of $10\log(1/\zeta)$, where DP pulse parameters w = 0.094 ns and $T_{pulse} = \{10^4, 10^3, 10^2, 10\}$ ns correspond to $\zeta = \{9.4\times10^{-6}, 9.4\times10^{-5}, 9.4\times10^{-4}, 9.4\times10^{-3}\}$ and $10\log(1/\zeta) = \{50, 40, 30, 20\}$ dB, and DS pulse parameters w = 0.58 ns and $T_{pulse} = \{0.758, 2.27, 4.55, 9.09\}$ ns correspond to $\zeta = \{0.77, 0.26,$

0.13, 0.064} and 10log(1/ζ) = {1.6, 5.9, 8.9, 12} dB. The independent variable 10log(1/ζ) was chosen because average power is proportional to ζ. Data points for DS-04 and DS-06 were excluded from the plots because they have the same ζ as DS-01. The horizontal dashed reference line in each INR_{TOV} plot corresponds to DTV susceptibility to Gaussian noise interference. Results demonstrate that DTV susceptibility can be related to particular pulse parameters. Generally speaking, DTV susceptibility to DS interference and DP interference with 10log(1/ζ) ≤ 30 dB was comparable to that of Gaussian noise interference. At larger 10log(1/ζ), however, the DTV receiver was relatively more susceptible to DP-02 and less susceptible to DP-01.

2.3. Characterization of DP and DS Signals

Characterization of the DP and DS signals was accomplished via temporal, amplitude, and spectral analyses. The UWB signals were measured with the vector signal analyzer under optimal measurement conditions at the RF output of the vector signal generator. Appendix B discusses the measurement setup and provides an extensive set of results. This section provides band-limited signal characterization results for the B_{DTV} = 19.51-MHz DTV receiver root-raised-cosine (RRC) filter.

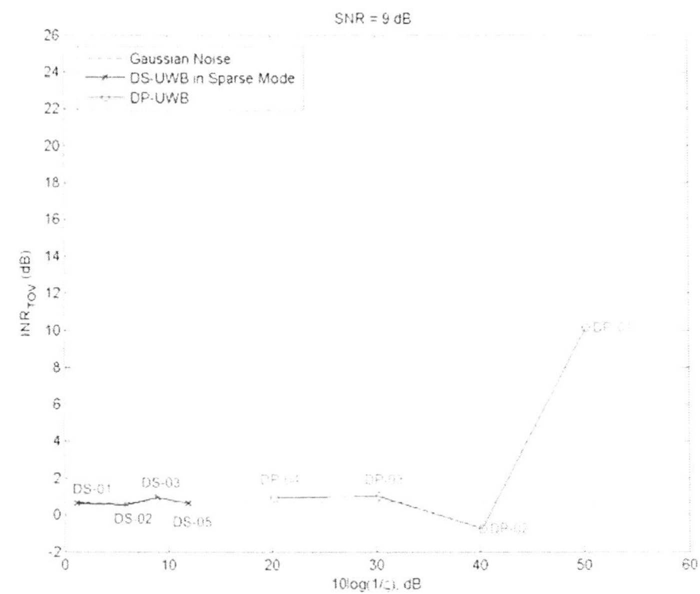

Figure 11. INR_{TOV} versus 10log(1/ζ) for a DTV channel operating at SNR = 9 dB and exposed to DP and DS interference.

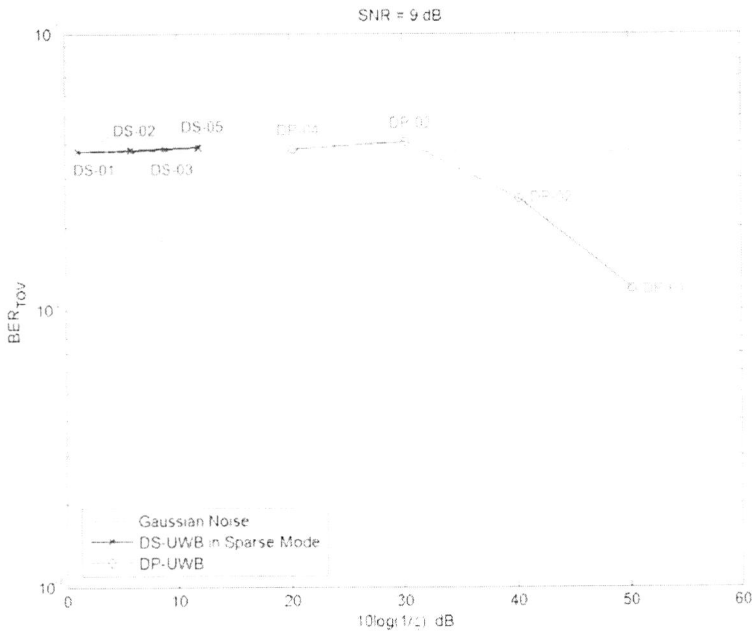

Figure 12. BER_{TOV} versus $10\log(1/\zeta)$ for a DTV channel operating at SNR = 9 dB and exposed to DP and DS interference.

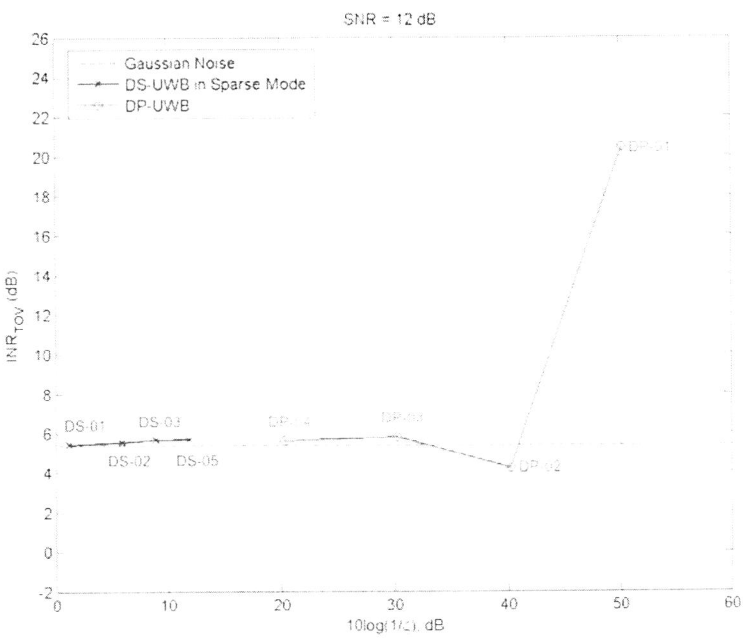

Figure 13. INR_{TOV} versus $10\log(1/\zeta)$ for a DTV channel operating at SNR = 12 dB and exposed to DP and DS interference.

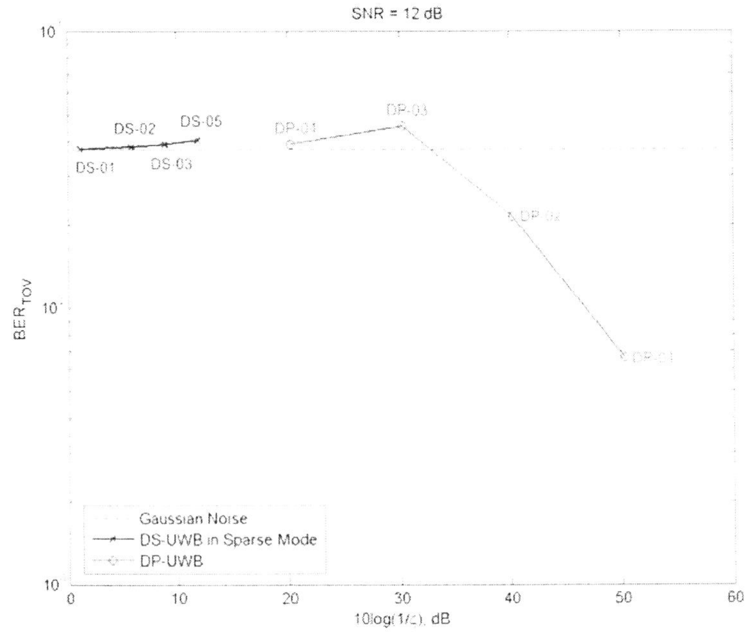

Figure 14. BER_{TOV} versus 10log(1/ζ) for a DTV channel operating at SNR = 12 dB and exposed to DP and DS interference.

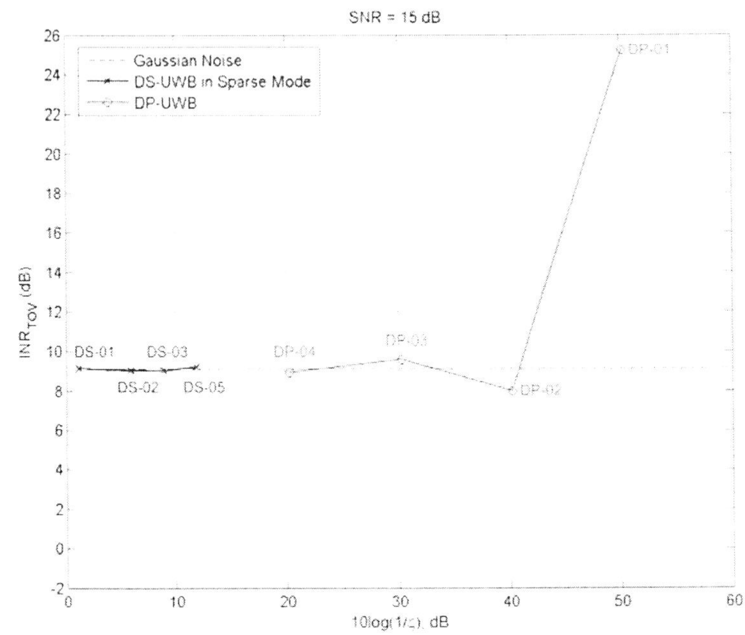

Figure 15. INR_{TOV} versus 10log(1/ζ) for a DTV channel operating at SNR = 15 dB and exposed to DP and DS interference.

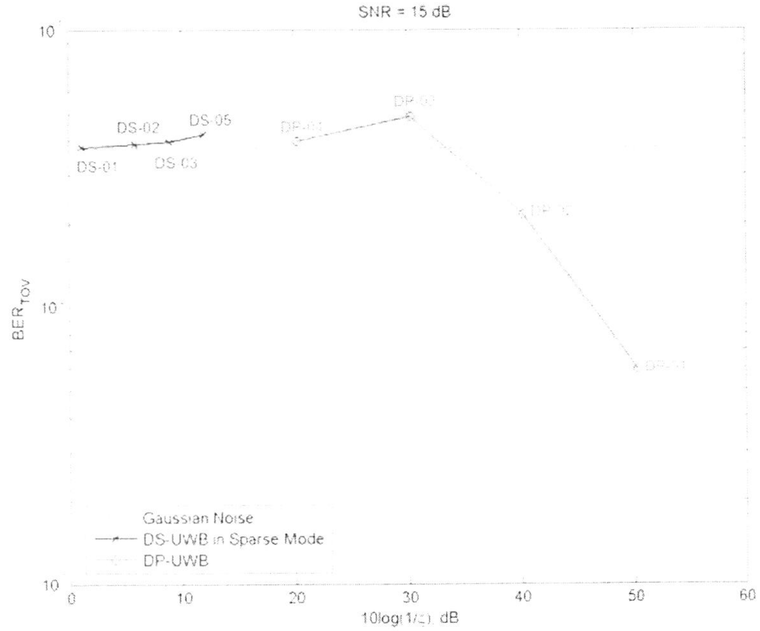

Figure 16. BER_{TOV} versus $10\log(1/\zeta)$ for a DTV channel operating at SNR = 15 dB and exposed to DP and DS interference.

2.3.1. Temporal Analysis

Temporal characteristics of the DP and DS signals were evaluated with crossing statistics, i.e., probabilities of the time duration that commences when a signal crosses an amplitude threshold and stays either above or below that threshold. Measured cumulative distribution functions for pulse duration (PD) and pulse interval (PI) statistics are given in Section B.2. Burst duration (BD) and burst interval (BI) are defined as the 10th percentile of PD and PI, respectively, using a crossing threshold equal to -70 dBm. This crossing threshold was useful for quantifying the effects of band-limiting on temporal characteristics of UWB signals, because it is high enough to avoid significant contribution from the measurement system noise floor and low enough to discern pulses and gated signals given the described measurement conditions.

Table 4 summarizes the effects of band-limiting on the temporal characteristics of the DP and DS signals. DP-01 and DP-02 pulses were resolved because T_{pulse} was significantly greater than $1/B_{DTV}$. As expected for these signals, band-limiting lengthened pulse widths, BD > w, and increased fractional on-times, $\zeta_{DTV} = BD/(BD + BI) > \zeta$. Notice that BI for DP-01 was greater than T_{pulse} due to dithering. BD and BI are less meaningful estimates when pulses are not resolved in the limiting bandwidth B_{DTV}, because there are no on- and off-times to distinguish. This was the case for DS signals, DP-03, and DP-04, whose crossing statistics resembled those of GN-01, i.e., BI was invariably 0.02 μs, and the variability in BD had little effect on $\zeta_{DTV} \approx 1$.

Table 4. Measured Temporal Characteristics of DP and DS Signals Band-Limited to B_{DTV}

Type	Index	Pulse Parameters			Band-Limited Metrics		
		w (μs)	T_{pulse} (μs)	ζ	BD (μs)	BI (μs)	ζ_{DTV}
DP	01	0.000094	10.0	9.4×10^{-6}	0.22	11.24	0.02
	02		1.0	9.4×10^{-5}	0.20	0.91	0.18
	03		0.1	9.4×10^{-4}	2.26	0.02	0.99
	04		0.01	9.4×10^{-3}	8.46	0.02	1.00
DS	01	0.00058	0.000758	0.77	7.40	0.02	1.00
	02		0.00227	0.26	8.36	0.02	1.00
	03		0.00455	0.13	8.10	0.02	1.00
	04		0.000758	0.77	9.03	0.02	1.00
	05		0.00909	0.064	11.22	0.02	1.00
	06		0.000758	0.77	3.71	0.02	0.99
GN	01	N/A	N/A	1.00	6.92	0.02	1.00

2.3.2. Amplitude Analysis

Amplitudes of the DP and DS signals were characterized with amplitude probability distributions, which give the probability that the signal amplitude exceeds a certain value. In this chapter, APDs are plotted on Rayleigh graphs where Rayleigh-distributed amplitudes of complex-Gaussian noise appear as a negatively-sloped, straight line. The sorting method, documented in Appendix D of Part 1 of this chapter series, was used to estimate and plot APDs of 131,072 samples that are free of correlation introduced by the measurement system bandwidth and band-limiting in post-measurement processing.

Figures 17 and 18 provide composite APD plots of measured DP and DS signals, respectively. The upper bound of the shaded region is the average power of the VSA noise, approximately -79.7 dBm in B_{DTV}. Notice that measured average powers, indicated in the legends, were proportional to ζ. Band-limiting had profound effects on the APDs of the DP and DS signals. For signals with $T_{pulse} > 1/B_{DTV}$, a step-like APD occurred, where higher amplitudes corresponded to elongated pulses and lower amplitudes corresponded to system noise observable during the shortened off-times between pulses. DP-01 and DP-02 band-limited to B_{DTV} have step-like APDs. For signals with $T_{pulse} \leq 1/B_{DTV}$ a negatively-sloped, straight APD can occur, corresponding to Rayleigh-distributed amplitudes due to elongated pulses overlapping each other. DS-01 and DS-02 band-limited to B_{DTV} have Rayleigh-distributed APDs.

Table 5 summarizes measured peak-to-average ratio and APD characteristics for the band-limited DS and DP signals. APD characteristics are labeled RAYL for Rayleigh-distributed amplitudes or RP for resolved pulses. More specifically, RP designates a step-like APD whose amplitude falls below the average power of the VSA noise at least 10% of the time. Notice that measured P/As of the DS signals were 0 – 3 dB less than P/A of GN-01.

Table 5. Measured Amplitude Characteristics of DP and DS Signals Band-Limited to B_{DTV}

Type	Index	P/A (dB)	APD
DP	01	23.5	RP
	02	13.8	RP
	03	5.4	
	04	8.1	
DS	01	9.5	RAYL
	02	9.0	RAYL
	03	8.1	
	04	6.6	
	05	6.6	
	06	6.8	
GN	01	9.6	RAYL

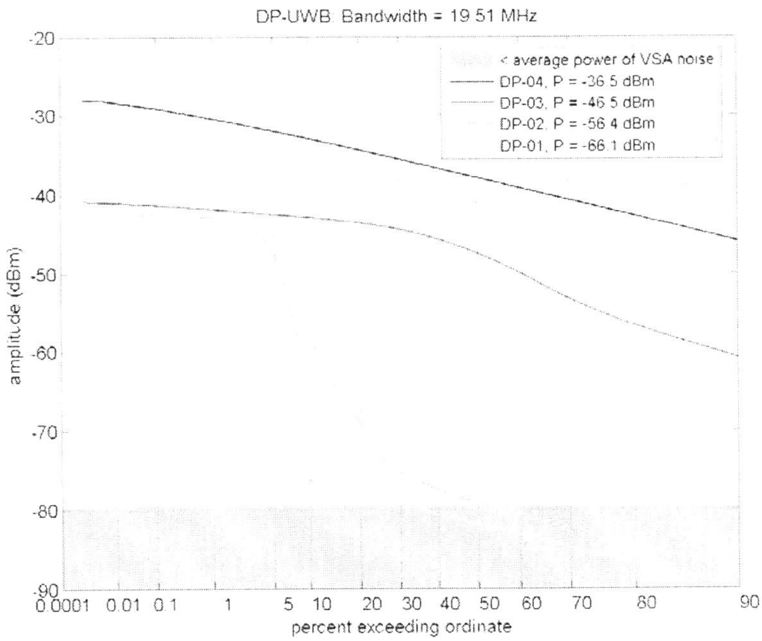

Figure 17. APDs of DP signals band-limited to B_{DTV}.

2.3.3. Spectral Analysis

Spectral characteristics of the DP and DS signals were characterized with the power spectral density, which represents the average power present per unit bandwidth as a function of frequency. Appendix A provides theoretical analyses of the PSD for DP and DS signals. In short, PSDs of the DP signals have discrete and continuous components that depend on the pulse spectrum and the dither function, while PSDs of the DS signals have only a continuous component that depends on the pulse spectrum and the code word spectrum.

PSDs were calculated from VSA measurements by averaging squared spectrum magnitudes of 500 rectangular-windowed, non-overlapping blocks of signal data. The periodic nature of the signals required that the block size be an integer multiple of the pulse period. To have the same Δf, the block size must also be a common integer multiple of all pulse periods. For the DP signals, these requirements were satisfied with a block size equal to 100 μs, which corresponds to Δf = 10.0 kHz. For the DS signals, these requirements were satisfied with a block size equal to 90.9 μs, which corresponds to Δf = 11.0 kHz.

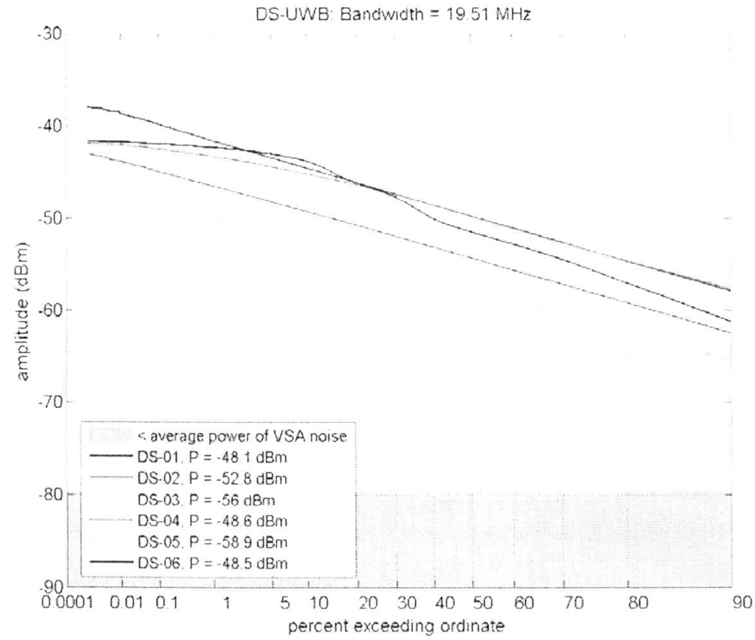

Figure 18. APDs of DS signals band-limited to B_{DTV}.

Figures 19 – 20 provide PSDs of the DP and DS signals, respectively, in the frequency band of the DTV victim receiver, i.e., 0 Hz on the plots corresponds to 3820 MHz. The victim receiver RRC filter was not applied to the data in post-measurement processing. The measured PSDs of DP signals, DS-01, DS-02, DS-03, DS-04, and DS-05 were flat over the DTV frequency band. In contrast and in agreement with figure A-6, the PSD of DS-06 was not flat due to the code word spectrum. Notice that the measured power densities of the DP and DS signals were proportional to ζ. Also, the spectral line at 0 Hz, most prevalent in DP-01, was due to VSG local-oscillator feed-through. The influence of the VSG local-oscillator feed-through on test results was discussed in Appendix C of Part 1 of this chapter series.

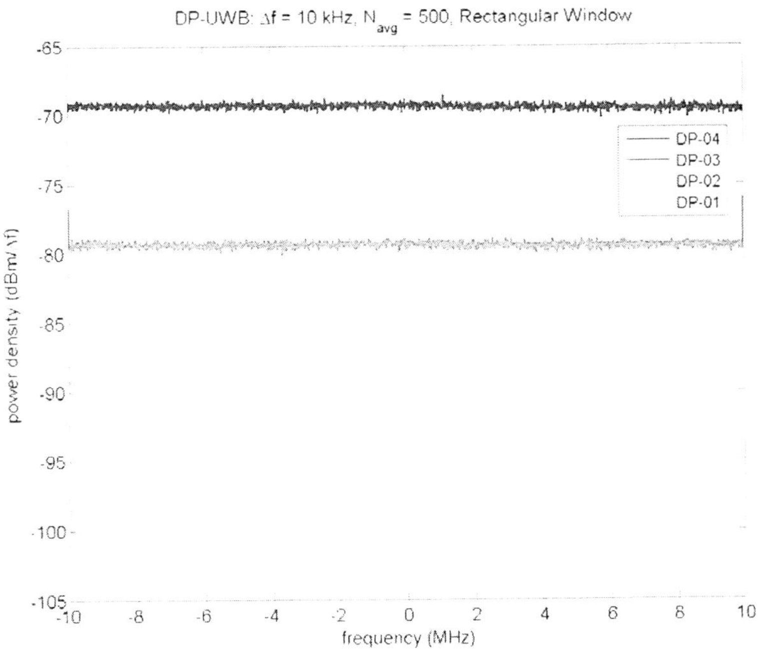

Figure 19. PSDs of DP signals.

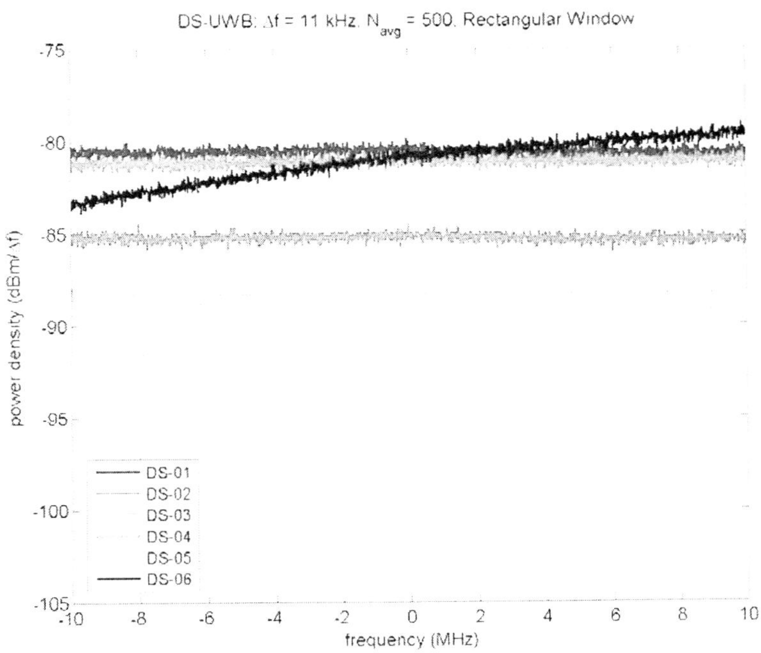

Figure 20. PSDs of DS signals.

3. MB INTERFERENCE

This section provides DTV susceptibility test results and characterization measurements for MB interference.

3.1. Gating Parameters

MB-OFDM achieves its ultra-wide bandwidth by simultaneously modulating 122 tones spaced 4.125 MHz apart. MB signals frequency-hop between bands according to a specified time-frequency code, which defines the number of bands (b) and the number of dwells (d) the signal stays in each band. From the perspective of a victim receiver operating within a single band, MB signals have two gating processes associated with (1) hopping and (2) insertion of zero-prefix and guard-intervals in each MB symbol. Each MB symbol is comprised of a zero-prefix (ZP), data block, and guard interval (GI); the MB symbol period is the sum of its constituents, i.e., $T_{OFDM} = T_{ZP} + T_{DATA} + T_{GI}$, where $T_{ZP} = 60.6$ ns, $T_{DATA} = 242.4$ ns, and $T_{GI} = 9.5$ ns.

Table 6 provides the MB parameters and corresponding gating parameters, i.e., on-time (τ_{on}), off-time (τ_{off}), and fractional on-time (ζ), considered in this chapter; these parameters also apply to GN(MB) gated-noise signals. Fractional on-time is calculated exactly as $\zeta = T_{DATA} / bT_{OFDM}$. On- and off-times are calculated as $\tau_{on} = dT_{OFDM} - T_{ZP} - T_{GI}$ and $\tau_{off} = (b - 1)dT_{OFDM} + T_{ZP} + T_{GI}$, respectively. These expressions ignore $T_{ZP} + T_{GI}$ between consecutive MB symbols and therefore are approximate in the multiple-dwell, i.e., $d > 1$, cases. This is necessary for later comparisons to BD and BI metrics and is justified for this experiment because $T_{ZP} + T_{GI} = 70.1$ ns is comparable to the reciprocal bandwidth of the victim receiver, i.e., $1/B_{DTV} = 51.3$ ns. Figure 21 illustrates gating parameters for MB-03.

Table 6. MB Gating Parameters

Type	Index	MB Parameters		Gating Parameters		
		b	d	τ_{on} (μs)	τ_{off} (μs)	ζ
MB, GN(MB)	01	1	N/A	0.24	0.07	0.776
	02	3	1	0.24	0.70	0.259
	03	3	2	0.55	1.32	0.259
	04	7	1	0.24	1.95	0.111
	05	7	2	0.55	3.82	0.111
	06	7	6	1.80	11.32	0.111
	07	13	1	0.24	3.82	0.060
	08	13	2	0.55	7.57	0.060
	09	13	12	3.68	45.07	0.060

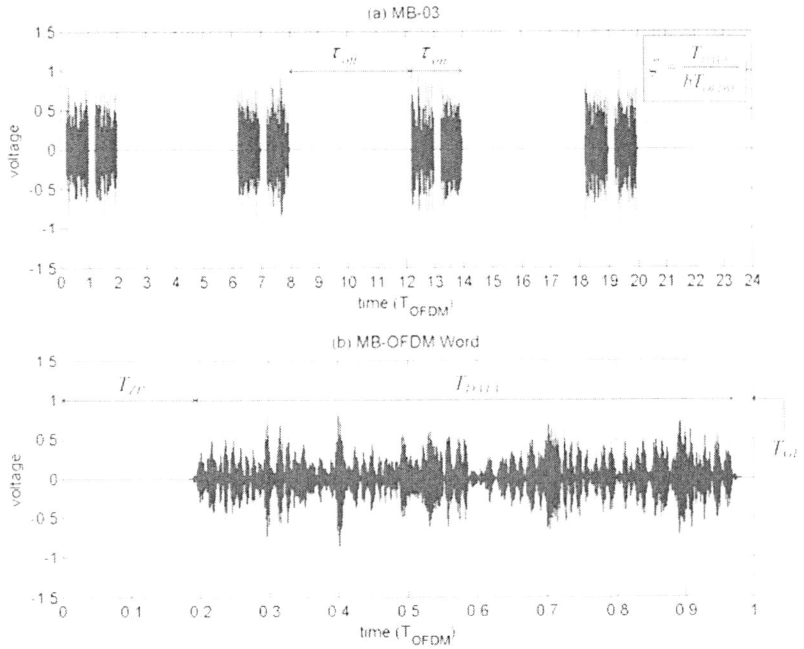

Figure 21. Simulated MB-03 transmitting random data (T_{OFDM} = 312.5 ns).

3.2. DTV Susceptibility to MB Interference

Results of the DTV susceptibility tests for MB interference are described in this section. Post-Reed-Solomon segment error rate and pre-Viterbi bit error rate are plotted versus the ratio of average interference power to average noise power. Additionally, DTV susceptibility and FEC performance metrics, i.e., INR_{TOV} and BER_{TOV}, are plotted versus $1/\zeta$ in dB.

3.2.1. DTV Signal Quality as a Function of Interference Average Power

In this subsection, SER and BER are plotted versus INR for DTV susceptibility to MB interference. Figures 22 – 27 provide composite plots of measured SER and BER as a function of INR. Each page is dedicated to a single SNR, each plot is dedicated to a specific range of d, and each curve represents a specific value of b. Figures 22, 24, and 26 provide composite plots of SER, while figures 23, 25, and 27 provide composite plots of BER. For each plot, GN-01 and MB-01 results are provided as a reference.

The following are general comments regarding the shift, separation, and slope of SER and BER curves. Both SER and BER shifted toward greater INR with increasing SNR. SER curves with fixed SNR and τ_{on} show increased separation (from GN-01) with decreasing ζ. Likewise, all SER curves with fixed SNR and ζ exhibit increased separation with increasing τ_{on}. SER curves were generally steeper than corresponding BER curves. For sufficiently small ζ, slopes of the SER curves were not as steep as higher-ζ cases. These observations are similar to those discussed in Appendix A of Part 2 of this chapter series for GN(MB) gated-noise interference.

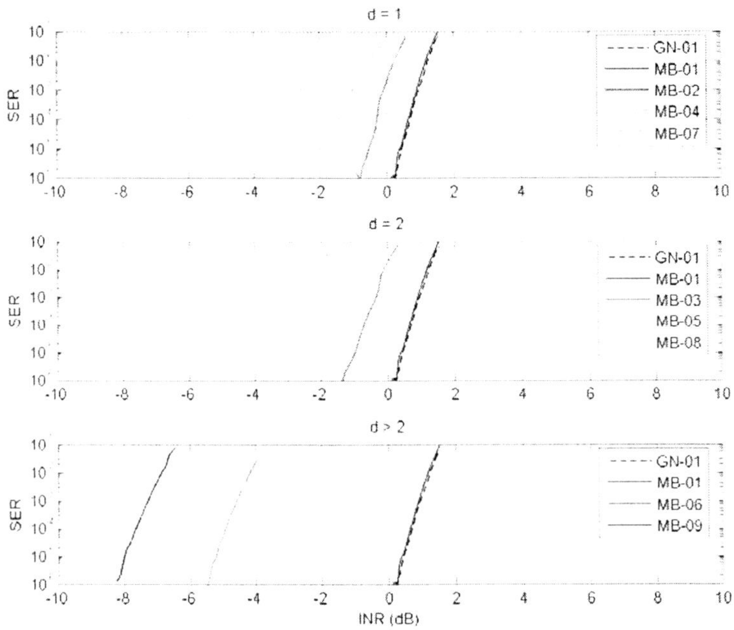

Figure 22. SER versus INR for a DTV channel operating at SNR = 9 dB and exposed to MB interference.

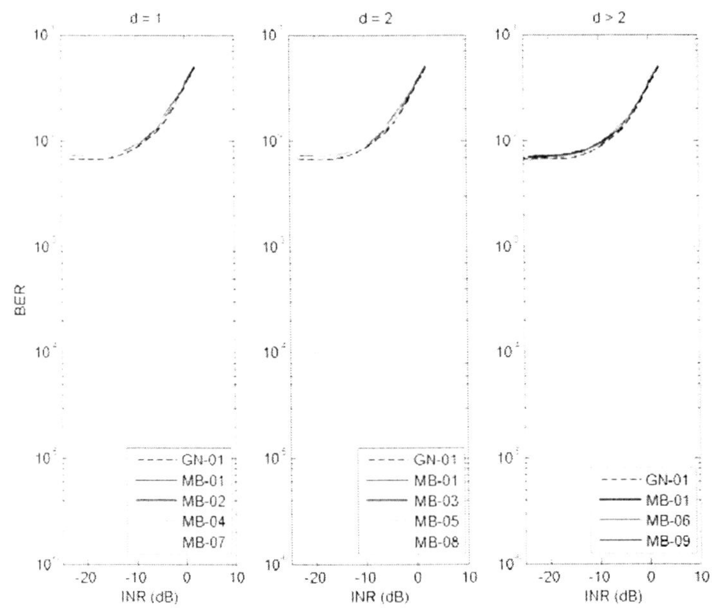

Figure 23. BER versus INR for a DTV channel operating at SNR = 9 dB and exposed to MB interference.

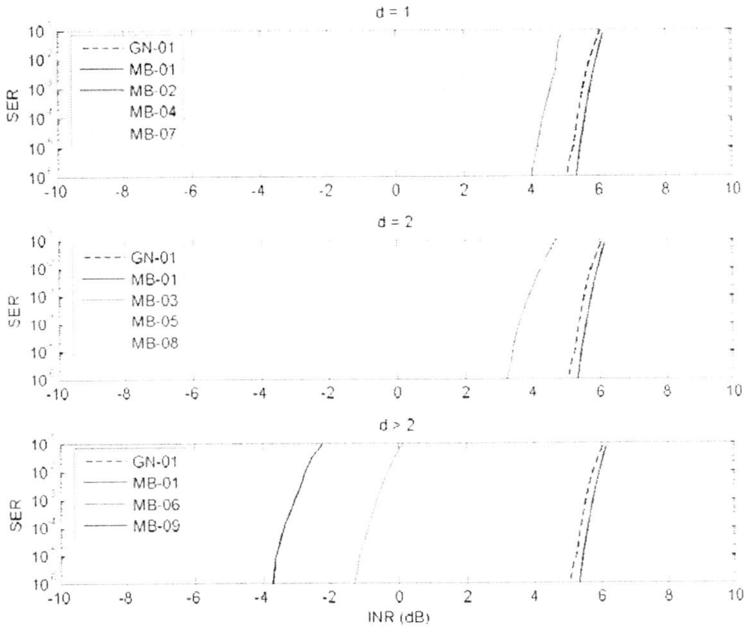

Figure 24. SER versus INR for a DTV channel operating at SNR = 12 dB and exposed to MB interference.

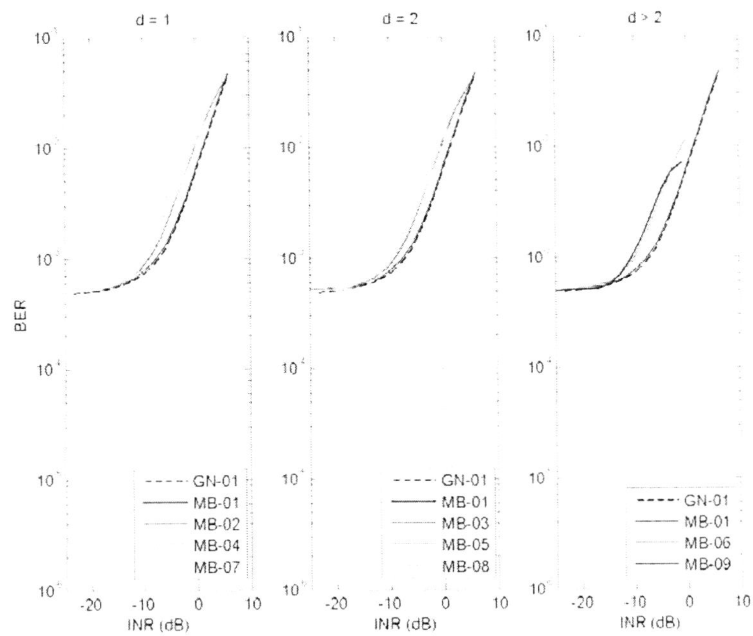

Figure 25. BER versus INR for a DTV channel operating at SNR = 12 dB and exposed to MB interference.

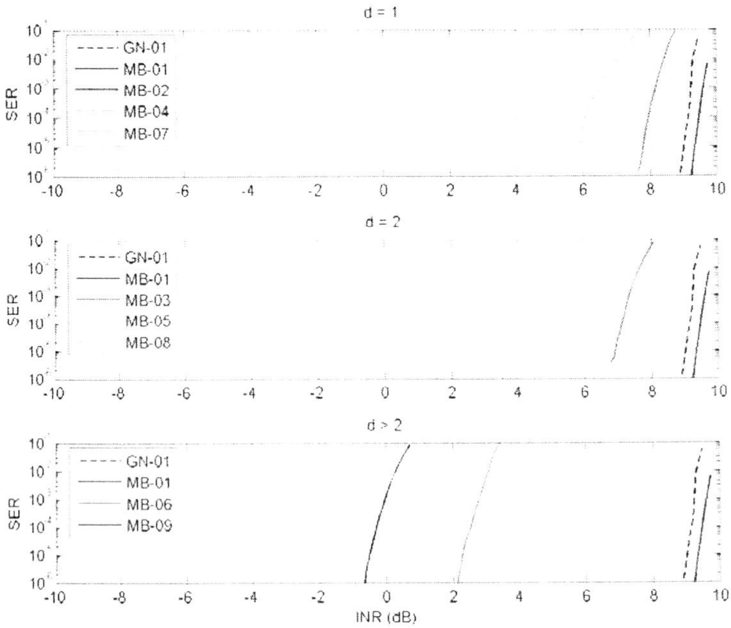

Figure 26. SER versus INR for a DTV channel operating at SNR = 15 dB and exposed to MB interference.

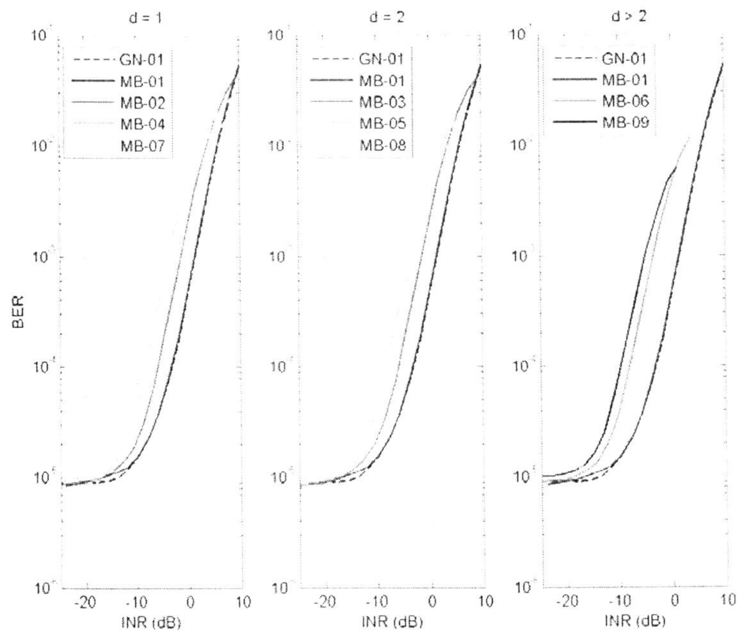

Figure 27. BER versus INR for a DTV channel operating at SNR = 15 dB and exposed to MB interference.

3.2.2. DTV Susceptibility as a Function of Fractional on-Time

Table 7 summarizes INR_{TOV} and BER_{TOV} results for MB interference; GN(MB) results from Part 2 of this chapter series are included for comparison purposes. Differences between INR_{TOV} of corresponding MB and GN(MB) results were less than 1 dB.

Figures 28 – 33 plot INR_{TOV} and BER_{TOV} versus $1/\zeta$ in dB, where b = {1, 3, 7, 13} correspond to ζ = {0.776, 0.259, 0.111, 0.060} and $10\log(1/\zeta)$ = {1.10, 5.87, 9.55, 12.2} dB. The horizontal dashed reference line in each INR_{TOV} plot corresponds to DTV susceptibility to Gaussian noise interference. In terms of average interference power, the DTV receiver was generally more susceptible to MB signals than Gaussian noise. This is observed in figures 28, 30, and 32 where INR_{TOV} was lower for MB-02 – MB-09 than for GN-01. Additionally, the MB results demonstrated that DTV susceptibility was dependent on temporal parameters of the interfering signal. For MB-01, τ_{off} was comparable to $1/B_{DTV}$ = 51.3 ns, and DTV susceptibility was similar to that of Gaussian noise and independent of τ_{on} and ζ. For MB-02 – MB-09, however, τ_{off} was significantly greater than $1/B_{DTV}$, and DTV susceptibility was clearly dependent on τ_{on} and ζ. More specifically, for fixed SNR and τ_{on}, DTV susceptibility increased with decreasing ζ. Likewise, for fixed SNR and ζ, DTV susceptibility increased with increasing τ_{on}.

Table 7. Measured DTV Susceptibility and FEC Performance for MB Interference

Type	Index	SNR = 9 dB		SNR = 12 dB		SNR = 15 dB	
		INR_{TOV} (dB)	BER_{TOV}	INR_{TOV} (dB)	BER_{TOV}	INR_{TOV} (dB)	BER_{TOV}
MB	01	0.6	0.039	5.6	0.041	9.5	0.043
	02	-0.3	0.032	4.4	0.032	8.0	0.034
	03	-0.7	0.030	3.6	0.028	7.1	0.029
	04	-1.4	0.024	3.0	0.021	6.1	0.021
	05	-3.0	0.019	1.2	0.015	4.2	0.014
	06	-4.9	0.015	-0.9	0.010	2.6	0.010
	07	-3.1	0.018	0.9	0.013	4.1	0.021
	08	-4.9	0.015	-1.1	0.009	2.1	0.009
	09	-7.7	0.011	-3.4	0.006	-0.3	0.005
GN(MB)	01	0.5	0.037	5.2	0.036	8.5	0.036
	02	-0.7	0.030	3.7	0.027	7.0	0.028
	03	-0.9	0.028	3.5	0.025	6.5	0.025
	04	-1.9	0.022	2.3	0.018	5.8	0.019
	05	-3.2	0.018	0.8	0.013	4.2	0.013
	06	-5.0	0.015	-1.2	0.009	2.3	0.009
	07	-3.3	0.016	0.7	0.012	4.3	0.011
	08	-5.1	0.014	-1.1	0.009	2.3	0.008
	09	-7.4	0.011	-3.3	0.005	-0.3	0.005
GN	01	0.7	0.038	5.4	0.038	9.2	0.038

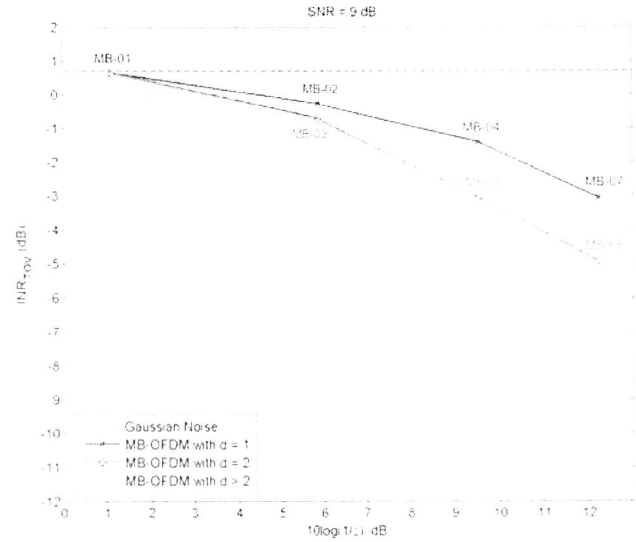

Figure 28. INR_{TOV} versus 10log(1/ζ) for a DTV channel operating at SNR = 9 dB and exposed to MB interference.

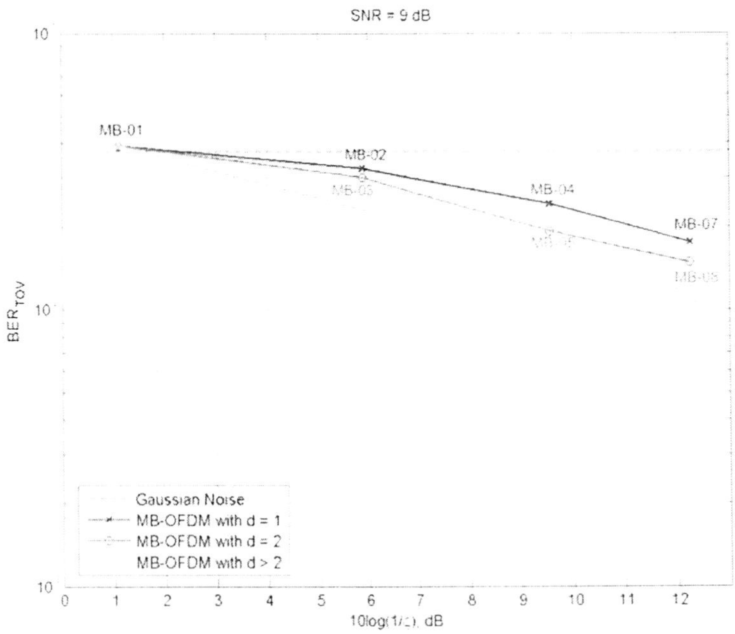

Figure 29. BER_{TOV} versus 10log(1/ζ) for a DTV channel operating at SNR = 9 dB and exposed to MB interference.

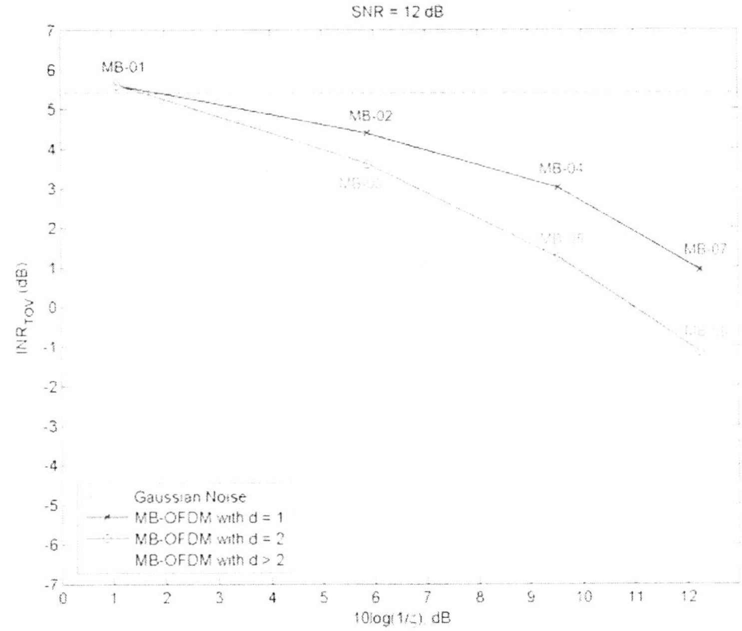

Figure 30. INR_{TOV} versus 10log(1/ς) for a DTV channel operating at SNR = 12 dB and exposed to MB interference.

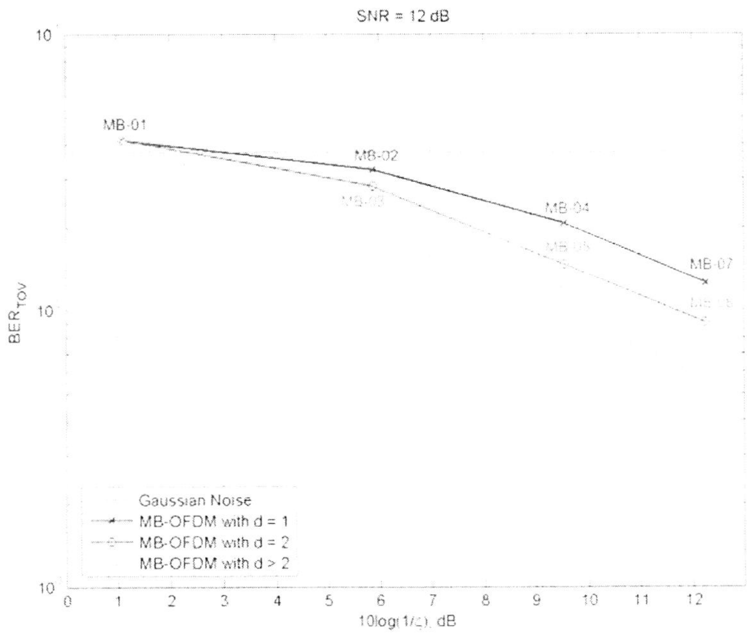

Figure 31. BER_{TOV} versus 10log(1/ς) for a DTV channel operating at SNR = 12 dB and exposed to MB interference.

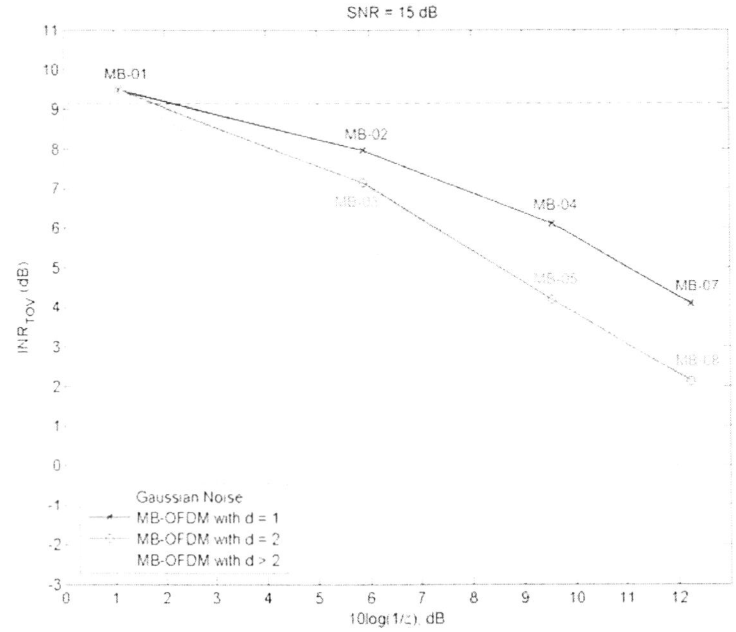

Figure 32. INR_{TOV} versus 10log(1/ζ) for a DTV channel operating at SNR = 15 dB and exposed to MB interference.

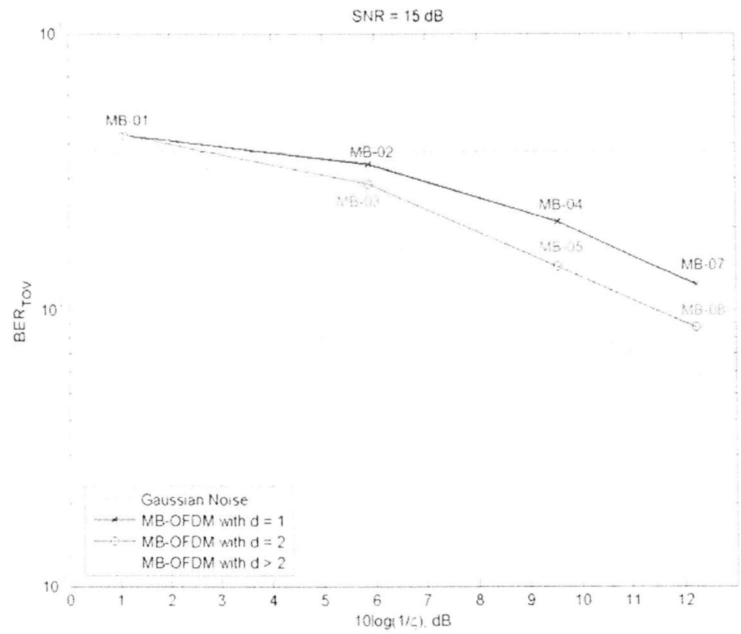

Figure 33. BER_{TOV} versus 10log(1/ζ) for a DTV channel operating at SNR = 15 dB and exposed to MB interference.

3.3. Characterization of MB Signals

Characterization of MB signals was accomplished via temporal, amplitude, and spectral analyses. This section provides band-limited signal characterization results for the B_{DTV} = 19.51-MHz DTV receiver RRC filter.

3.3.1. Temporal Analysis

Table 8 summarizes the effects of band-limiting on the temporal characteristics of MB signals. As expected, band-limiting lengthened on-times, BD > τ_{on}, shortened off-times, BI < τ_{off}, and increased fractional on-times, $\zeta_{DTV} > \zeta$. BD and BI approached the respective gating parameters only when τ_{on} and τ_{off} were significantly larger than $1/B_{DTV}$. Notice that measured MB temporal characteristics are identical to those of the GN(MB) signals.

Table 8. Measured Temporal Characteristics of MB Signals Band-Limited to B_{DTV}

Index	Gating Parameters			MB			GN(MB)		
	τ_{on} (μs)	τ_{off} (μs)	ζ	BD (μs)	BI (μs)	ζ_{DTV}	BD (μs)	BI (μs)	ζ_{DTV}
01	0.24	0.07	0.776	3.7	0.0	1.00	3.7	0.0	1.00
02	0.24	0.70	0.259	0.4	0.6	0.40	0.4	0.6	0.40
03	0.55	1.32	0.259	0.7	1.2	0.37	0.7	1.2	0.37
04	0.24	1.95	0.111	0.4	1.8	0.18	0.4	1.8	0.18
05	0.55	3.82	0.111	0.7	3.7	0.16	0.7	3.7	0.16
06	1.80	11.32	0.111	1.9	11.2	0.15	1.9	11.2	0.15
07	0.24	3.82	0.060	0.4	3.7	0.10	0.4	3.7	0.10
08	0.55	7.57	0.060	0.7	7.4	0.09	0.7	7.4	0.09
09	3.68	45.07	0.060	2.5	44.9	0.05	2.5	44.9	0.05

3.3.2. Amplitude Analysis

Figures 34 and 35 provide composite APD plots of measured single- and multi-dwell MB signals, respectively. Notice that measured average powers, indicated in the legends, were proportional to ζ. Band-limiting had profound effects on the APDs of the MB signals. For MB-01, τ_{off} was less than or comparable to $1/B_{DTV}$ and a negatively-sloped, relatively straight APD occurred. For MB-02 – MB-09, τ_{off} exceeded $1/B_{DTV}$ and a step-like APD occurred where higher amplitudes corresponded to MB interference during elongated on-times and lower amplitudes corresponded to system noise observable during shortened off-times. This was most pronounced for MB-08 and MB-09.

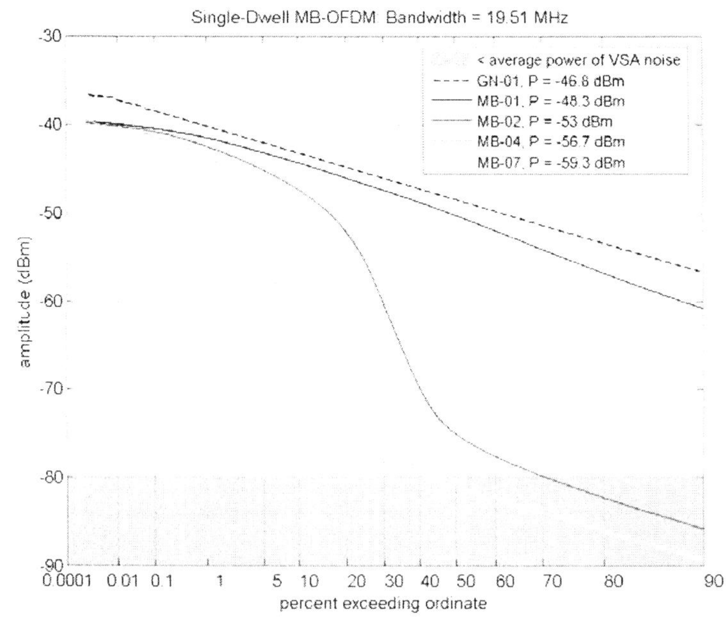

Figure 34. APDs of single-dwell MB signals band-limited to B_{DTV}.

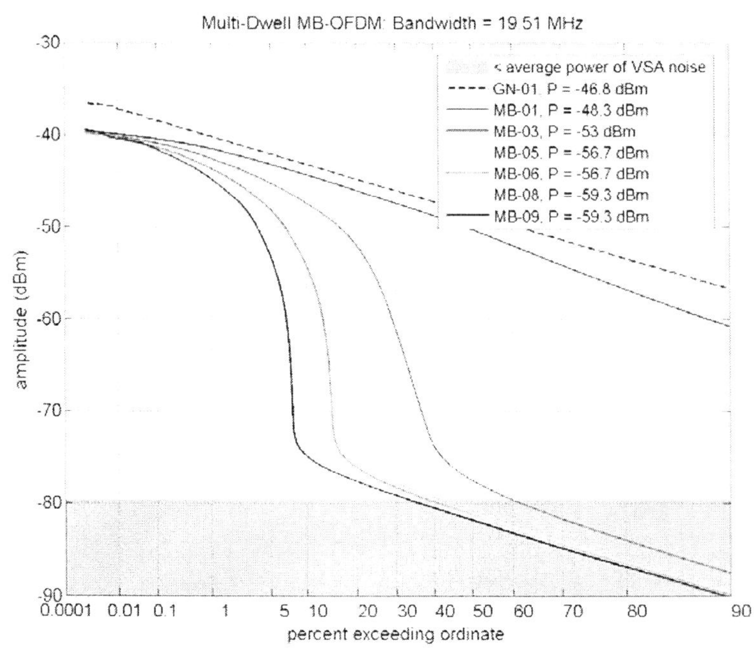

Figure 35. APDs of multi-dwell MB signals band-limited to B_{DTV}.

Table 9 summarizes measured P/A and APD characteristics for the MB signals band-limited to B_{DTV}. APD characteristics are labeled RG for resolved gating, i.e., step-like APD whose amplitude falls below the average power of the VSA noise at least 10% of the time.

Notice that measured P/As of the MB signals were 1 – 2 dB less than the P/A of corresponding GN(MB) signals.

Table 9. Measured Amplitude Characteristics of MB Signals Band-Limited to B_{DTV}

Index	MB		GN(MB)	
	P/A (dB)	APD	P/A (dB)	APD
01	8.3		10.5	
02	12.8	RG	14.7	RG
03	12.8	RG	14.6	RG
04	16.2	RG	18.0	RG
05	16.4	RG	17.9	RG
06	16.3	RG	17.8	RG
07	18.7	RG	20.2	RG
08	18.9	RG	20.1	RG
09	18.8	RG	19.9	RG

3.3.3. Spectral Analysis

Figures 36 and 37 provide measured PSDs of the single- and multi-dwell MB signals, respectively, over the VSA bandwidth. In figure 37, the PSD of MB-05 is directly beneath MB-06. To compute the PSD, 500 blocks of length = 170.6 μs (Δf = 5.9 kHz) were averaged. Measured MB PSDs were flat across the band of interest, and measured power density was proportional to ζ. Differences between measured MB and GN(MB) PSDs were negligible. Also, the spectral line at 0 Hz was due to VSG local-oscillator feed-through. The influence of the VSG local-oscillator feed-through on test results was discussed in Appendix C of Part 1 of this chapter series.

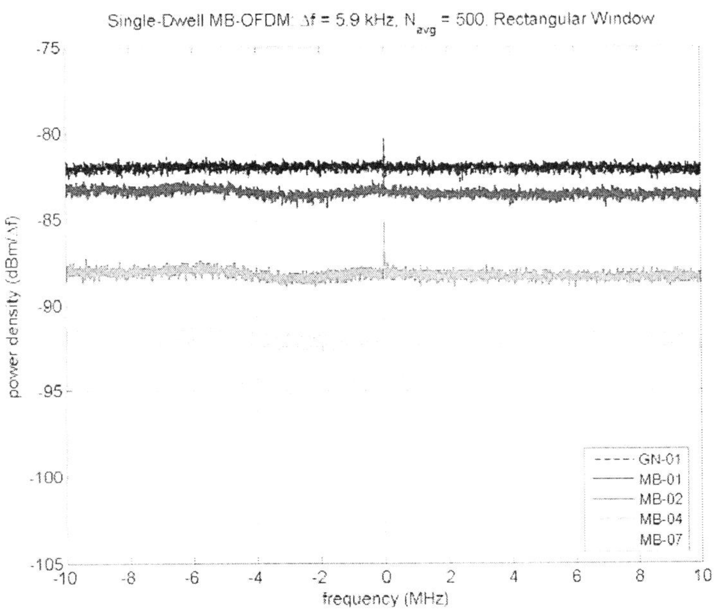

Figure 36. PSDs of single-dwell MB signals.

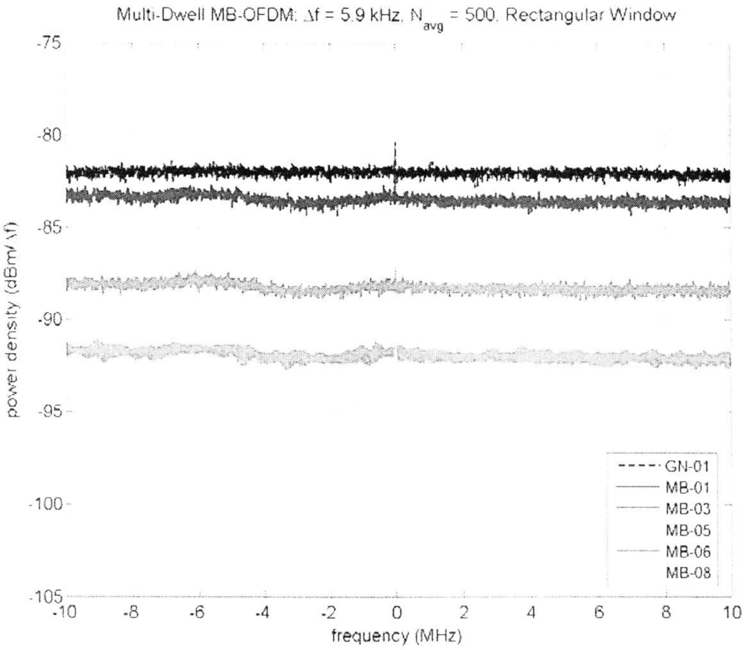

Figure 37. PSDs of multi-dwell MB signals.

4. Conclusion

The primary objective of this experiment was to measure and evaluate the susceptibility of a

C-band satellite DTV receiver to UWB interference. The secondary objective was to determine if DTV susceptibility could be predicted from measured signal characteristics. A final objective was to determine if continuous and gated Gaussian noise could be used to emulate DS and MB signals, respectively, in susceptibility tests. Each of these objectives are addressed in these concluding remarks.

4.1. Signal Sets of Common DTV Susceptibility Behavior

The primary objective of this experiment was to measure and evaluate the susceptibility of a DTV victim receiver to UWB interference. INR_{TOV} is our primary DTV susceptibility metric, i.e., DTV susceptibility increases as INR_{TOV} decreases and conversely decreases as INR_{TOV} increases. Table 10 provides ΔINR_{TOV} for the UWB signals ranked from least to most susceptible, where ΔINR_{TOV} is normalized to the INR_{TOV} of Gaussian noise. Notice that variation with SNR does not affect ranking, and for all signals except DP-01 SNR variation is minimal. Superimposed upon this ranking is categorization into three signal sets distinguished by their distance from the INR_{TOV} of Gaussian noise, i.e., Set 1 is defined as $-0.5 \text{ dB} \leq \Delta INR_{TOV} \leq 0.5 \text{ dB}$, Set 2 is defined as $\Delta INR_{TOV} < -0.5$ dB, and Set 3 is defined as $\Delta INR_{TOV} >$ 0.5 dB.

Signals in Set 1 are for all practical purposes equivalent to Gaussian noise in terms of DTV susceptibility. Signals in Set 2 have INR_{TOV} measurably below that of Gaussian noise and consequently are more deleterious to DTV reception. DP-01 is the only signal in Set 3; its INR_{TOV} is considerably greater than that of Gaussian noise and therefore is the most benign to DTV reception. Demarcation of Sets 1 and 3 is relatively straightforward as they are separated by a wide margin. Demarcation of Sets 1 and 2, however, is slightly more subjective. For example, INR_{TOV} of MB-02 is merely 1.0 – 1.2 dB below that of Gaussian noise.

Table 10. Ranked DTV Susceptibility relative to Gaussian Noise Degradation

Rank	Signal	ΔINR_{TOV} (dB)			Set
		SNR = 9 dB	SNR = 12 dB	SNR = 15 dB	
1	DP-01	9.5	15.0	16.1	3
2	DP-03	0.3	0.4	0.5	
3	MB-01	-0.1	0.2	0.3	
4	DS-03	0.2	0.3	-0.1	
5	DS-04	-0.1	0.3	0.1	
6	DS-05	-0.1	0.3	0.1	1
7	DP-04	0.2	0.2	-0.2	
8	GN-01	0.0	0.0	0.0	
9	DS-01	-0.1	0.1	0.0	
10	DS-02	-0.2	0.2	-0.1	
11	DS-06	-0.5	-0.2	-0.3	
12	MB-02	-1.0	-1.0	-1.2	
13	DP-02	-1.5	-1.2	-1.1	
14	MB-03	-1.4	-1.8	-2.0	
15	MB-04	-2.1	-2.4	-3.0	
16	MB-05	-3.7	-4.2	-5.0	2
17	MB-07	-3.8	-4.5	-5.1	
18	MB-06	-5.6	-6.3	-6.6	
19	MB-08	-5.6	-6.5	-7.0	
20	MB-09	-8.4	-8.8	-9.4	

During this experiment, there was interest in understanding the effects of UWB interference on the DTV forward error correction, consisting of a Viterbi decoder, convolutional byte de-interleaver, and Reed-Solomon decoder. Having access only to pre-Viterbi BER and post-Reed-Solomon SER, the FEC was regarded as a black box whose performance was evaluated by the average number of bit errors it corrected. In particular, FEC performance was deemed good when it achieved $SER_{TOV} = 10^{-4}$ from a relatively high BER_{TOV}. Interestingly in Sets 1 and 2, poor FEC performance, i.e., low BER_{TOV}, was correlated with high DTV susceptibility, i.e., low INR_{TOV}. Inspection of MB-01 and MB-09 in table 7 (SNR = 12 dB), for example, shows that BER_{TOV} decreased from 0.041 to 0.006 as INR_{TOV} decreased from 5.6 dB to -3.4 dB. As was pointed out in Section 2.2.2, DP-01 in Set 3 did not correlate in the same way. Disregarding this exception, the correlation between BER_{TOV} and INR_{TOV} indicates that FEC plays a strong role in susceptibility. The effects of UWB interference on FEC performance is a topic for further study.

4.2. Summary of Measured Signal Characteristics

UWB signal characterization measurements were performed in the DTV victim signal bandwidth. Measured UWB signals were subjected to temporal, amplitude, and spectral analyses. Temporal analyses used BD and BI statistics and corresponding ζ_{DTV} estimates. Amplitude analyses used the APD and P/A. Finally, spectral analyses focused on the PSD.

Temporal analyses of DP and DS signals showed that band-limiting elongated pulse width, w. This was evident in table 4, which demonstrates that BD > w and $\zeta_{DTV} > \zeta$. For signals with T_{pulse} less than or comparable to $1/B_{DTV}$, elongated pulses overlapped and the subsequent APD and P/A approached Rayleigh amplitude statistics. For signals with T_{pulse} significantly greater than $1/B_{DTV}$, elongated pulses remained distinct and a step-like APD occurred with correspondingly high P/A.

Similar results were observed with the MB signals; in these cases, however, band-limiting elongated on-time, τ_{on}, and shortened off-time, τ_{off}. These changes were evident in table 8, which demonstrated BD > τ_{on}, BI < τ_{off}, and $\zeta_{DTV} > \zeta$. For signals with τ_{off} less than or comparable to $1/B_{DTV}$, elongated on-times overlapped and the subsequent APD and P/A approached Rayleigh amplitude statistics. For signals with τ_{off} significantly greater than $1/B_{DTV}$, elongated on-times remained distinct and a step-like APD occurred with correspondingly high P/A.

Spectral analyses of DP, DS, and MB signals revealed flat power spectral densities like white Gaussian noise. A notable exception to this was DS-06 whose PSD sloped several dB over the DTV signal bandwidth. Theoretical analysis demonstrated that this was due to the DS code word spectral characteristics. Further comments on spectral analyses for bandwidths other than B_{DTV} can be found in Section 4.4.

4.3. Correlations between DTV Susceptibility and Signal Characteristics

The second objective of this experiment was to determine if DTV susceptibility could be predicted from measured signal characteristics. Indeed, in Part 2 of this chapter series it was found that DTV susceptibility to gated noise interference could not be predicted by interference power alone. It was demonstrated that DTV susceptibility was also dependent on the temporal parameters of the gated noise signals, i.e., τ_{on}, τ_{off}, and ζ. In this section, the analysis is taken one step further by plotting DTV susceptibility versus the band-limited metrics BD, BI, ζ_{DTV}, and P/A.

Figures 38 and 39 summarize the temporal metrics of the UWB signals and illustrate how DTV susceptibility, i.e., ΔINR_{TOV}, depends on these metrics. Figure 38 provides BI versus BD of the UWB signals, and figure 39 provides ΔINR_{TOV} as a function of ζ_{DTV}. In both plots, contours are drawn around the DTV susceptibility sets identified in Section 4.1. Signals within Set 1 have short BI < $1/B_{DTV}$ and a wide range of longer BD. For these signals, the pulse period or gating off-time is less than $1/B_{DTV}$, and band-limiting caused overlapping. This either preserved or created signal characteristics that resembled continuous noise as evidenced by ζ_{DTV} near unity. Hence, INR_{TOV} for signals within Set 1 resembles that of Gaussian noise. Signals within Set 2 have a wide range of BI > $1/B_{DTV}$ and BD. For these signals, the pulse period or gating off-time is greater than or comparable to $1/B_{DTV}$, and band-limiting did not create a continuous interference signal as evidenced by ζ_{DTV} ranging from 0.05 to 0.4. The wide range of BD and BI for signals within Set 2 caused a correspondingly wide range of ΔINR_{TOV} below that of Gaussian noise, where the DTV receiver was most susceptible to signals with long BD and BI, e.g., MB-09, and least susceptible to signals with short BD and BI, e.g., MB-02 and DP-02. DP-01 in Set 3 has long BI >> $1/B_{DTV}$. For this signal, BD is much less than BI making ζ_{DTV} small. Hence, the victim receiver was exposed to interference the least amount of time, and ΔINR_{TOV} is significantly greater than that of Gaussian noise.

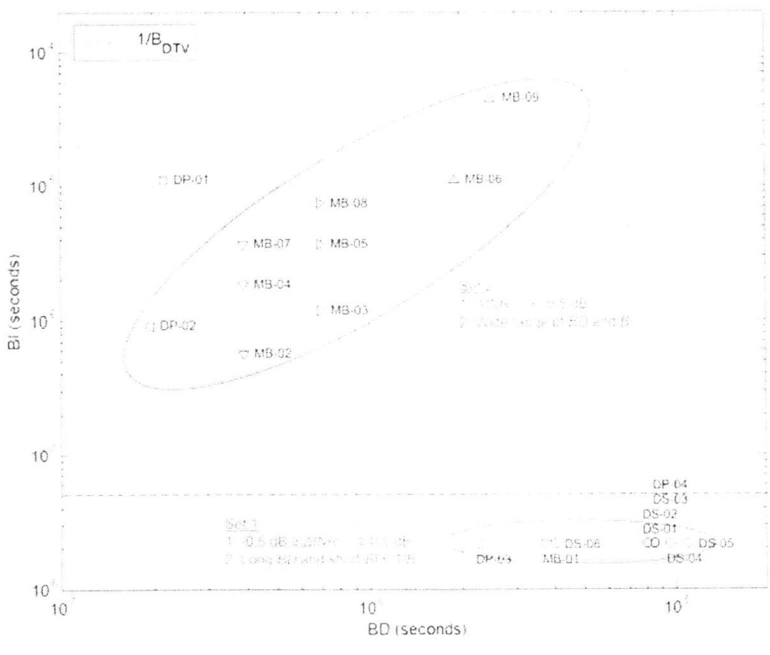

Figure 38. BI versus BD of UWB interference. Contours are drawn around signals within the same DTV susceptibility set.

Contours drawn around signals within the DTV susceptibility sets, defined by ΔINR_{TOV} alone, persist as regions when plotted against the temporal metrics BD, BI, and ζ_{DTV}. These findings suggest that DTV susceptibility categorization can be predicted from a priori knowledge of these band-limited metrics. Results suggest that UWB signals with ζ_{DTV} close to unity will generate interference much like Gaussian noise. Further, results suggest that UWB signals with ζ_{DTV} approaching zero will generate benign interference. Finally, results suggest that UWB signals with $0.05 \leq \zeta_{DTV} \leq 0.4$ are likely to generate a range of interference effects more deleterious than Gaussian noise interference, which requires more knowledge to determine the specific impact on the victim receiver. For example, figure 39 shows the signals clustering horizontally with those having the same number of bands. Within these clusters, signals are differentiated vertically by their dwell. Understanding this dependence on dwell required the in-depth study of the effects of gated noise on-time given in Part 2 of this chapter series.

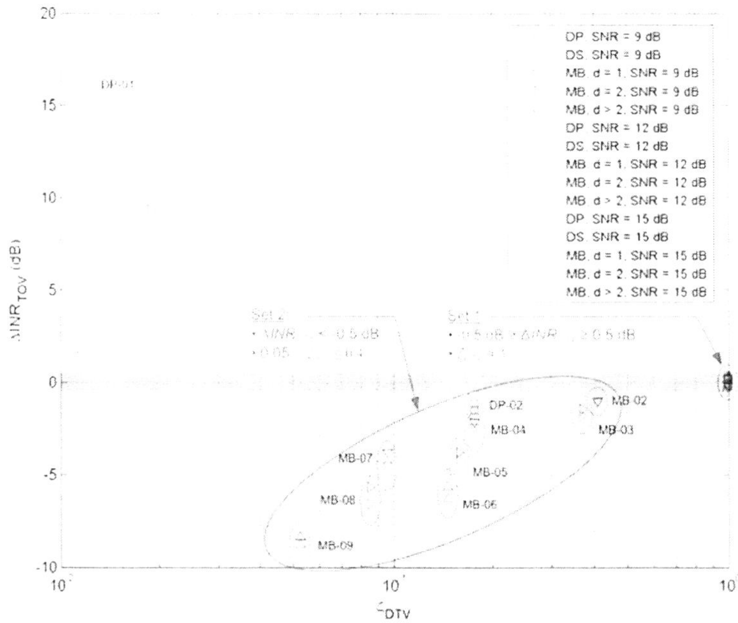

Figure 39. ΔINR_{TOV} versus ζ_{DTV} of UWB interference. Contours are drawn around signals within the same DTV susceptibility set.

Figure 40 provides ΔINR_{TOV} versus P/A with contours drawn around the DTV susceptibility sets. Signals within Set 1 are in a horizontal line in the range 5.4 dB $\leq$ P/A $\leq$ 9.6 dB, which indicates that the peaks of these signals have negligible effect on ΔINR_{TOV}. Signals within Set 2 follow a downward trend in the range 12.8 dB $\leq$ P/A $\leq$ 18.9 dB, which indicates that DTV susceptibility increases with increasing P/A. Exact determination of the P/A transitions between Sets 1 and 2 and Sets 2 and 3 requires further investigation.

Contours drawn around the signals within the DTV susceptibility signal sets, defined by ΔINR_{TOV} alone, persist as regions when plotted against P/A. This suggests that DTV susceptibility categorization can be predicted from a priori knowledge of the P/A metric. Results suggest that UWB signals with 5.4 dB $\leq$ P/A $\leq$ 9.6 dB will generate interference much like Gaussian noise. Further, results suggest that UWB signals with exceedingly high P/A will generate benign interference. Finally, results suggest that UWB signals with 12.8 dB $\leq$ P/A $\leq$ 18.9 dB are likely to generate a range of interference effects more deleterious than Gaussian noise interference, which requires more knowledge to determine the specific impact on the victim receiver.

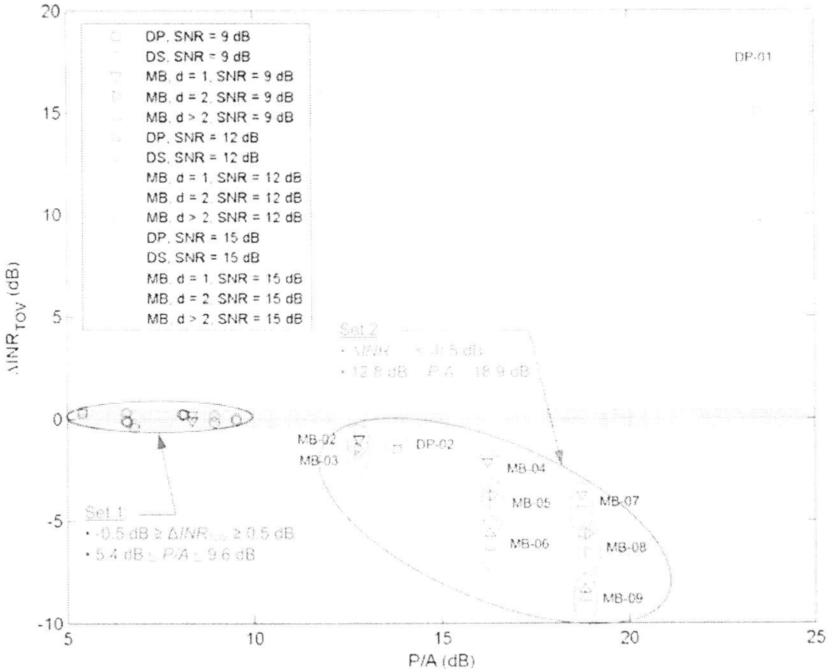

Figure 40. ΔINR_{TOV} versus P/A of UWB interference. Contours are drawn around signals within the same DTV susceptibility set.

In Section 4.1, it was mentioned that ΔINR_{TOV} varied slightly with *SNR* in Sets 1 and 2. This variation manifests itself as vertical displacement of ΔINR_{TOV} with *SNR* in figures 39 and 40. Displacement of *BD*, *BI*, ζ_{DTV}, and *P/A* with *SNR* are not applicable since these metrics were derived from RF measurements. Hence, horizontal displacement of *BD* and vertical displacement of *BI* in figure 38, horizontal displacement of ζ_{DTV} in figure 39, and horizontal displacement of *P/A* in figure 40 do not exist.

4.4. UWB Signal Emulation

As stated previously, the last objective of this study was to determine whether continuous and gated noise could be used to emulate DS and MB signals, respectively, in these and other susceptibility tests. This objective is important since gated noise is easier to implement and analyze than the UWB signals.

The main body of this chapter provided temporal, amplitude, and spectral analyses for each of the signals band-limited to B_{DTV}. With two exceptions, these analyses demonstrated that DS signals are similar to Gaussian noise. First, *P/A*s of the DS signals were 0 – 3 dB less than *P/A* of Gaussian noise. Consequently, *APD*s of the DS signals deviated slightly from Rayleigh. Second, the *PSD* of DS-06 sloped several dB over the DTV signal bandwidth in contrast to the flat spectrum of white Gaussian noise. This slope was attributed to the spectrum of the direct sequence codeword. Similarly, these analyses showed that gated noise was similar to MB signals with the exception that the *P/A* was 1 – 2 dB less than those of GN(MB).

The similarities were also present in the DTV susceptibility results, which were nearly identical for comparative signals. Hence, it appears reasonable to substitute continuous and gated noise signals for DS and MB signals, respectively, when replicating this experiment.

APD statistics as a function of frequency in Section B.4, however, suggest that GN(MB) signals cannot be used in place of MB signals for susceptibility tests on victim receivers with narrower bandwidths. These statistics show significant variation as a function of frequency when the signal is band-limited to 1.0 and 0.1 MHz. *APD* statistics versus frequency in the 1-MHz bandwidth case have a distinct periodic lobing pattern. *APD* statistics versus frequency in the 0.1-MHz bandwidth case have a distinct periodic spiking pattern centered at the pilot tone frequency. These features are not present in the GN(MB) signals and require further analysis.

5. Acknowledgments

The authors recognize John McCorkle of Freescale, Inc. for his leadership, insight, and insistence on the highest quality. We learned a great deal from working with John in all relevant technical areas. Dr. Paul Runkle of Signal Innovations provided a wealth of expertise in the field of digital signal processing. Given that signal processing was the crux of this project, Paul made significant contributions by ensuring the integrity and efficient execution of our simulations.

This investigation required the services of a number of ITS engineers with a broad range of expertise, some of whom are not included in the list of authors. The authors recognize Brent Bedford for his expertise in test and measurement equipment and techniques; Randy Hoffman, Richard Statz, and Steve Engelking for developing much of the low level software routines needed to automate the experiment via software control; Stephen Wolf for his expertise in video quality assessment; Frank Sanders for sharing knowledge learned in previous interference tests on C-band earth station receivers [9]; Andrew Rogers and Michael Mitchell for their attention to detail in hardware production and signal characterization measurements; and Dr. William Kissick for his support and leadership.

6. References

[1] "First Report and Order in the matter of ET Docket 98-153 (Revision of Part 15 of the Commission's Rules Regarding Ultra-Wideband Transmission Systems)," *FCC*. 02-48, adopted Feb. 14, 2002, released Apr. 22, 2002.

[2] W.A. Kissick, Ed., "The temporal and spectral characteristics of ultrawideband signals," *NTIA Report*. 01-383, Jan. 2001.

[3] J.R. Hoffman, M.G. Cotton, R.J. Achatz, R.N. Statz, and R.A. Dalke, "Measurements to determine potential interference to GPS receivers from ultrawideband transmission systems," *NTIA Report*. 01-384, Feb. 2001.

[4] J.R. Hoffman, M.G. Cotton, R.J. Achatz, and R.N. Statz, "Addendum to NTIA Report 01-384: Measurements to determine potential interference to GPS receivers from ultrawideband transmission systems," *NTIA Report.* 01-389, Sep. 2001.

[5] J.R. Hoffman, E.J. Haakinson, and Y. Lo, "Measurements to determine potential interference to public safety radio receivers from ultrawideband transmission systems," *NTIA Report.* 03-402, Apr. 2003.

[6] M. Cotton, R. Achatz, J. Wepman, and B. Bedford, "Interference potential of ultrawideband signals: Part 1 – Procedures to characterize ultrawideband emissions and measure interference susceptibility of C-band satellite digital television receivers," *NTIA Report.* TR-05-419, Feb. 2005.

[7] M. Cotton, R. Achatz, J. Wepman, and P. Runkle, "Interference potential of ultrawideband signals: Part 2 – Measurement of gated-noise interference to C-band satellite digital television receivers," *NTIA Report.* TR-05-429, Aug. 2005.

[8] J.C. Whitaker, "Digital television transmission systems," Ch. 13 in Standard Handbook of Video and Television Engineering, J.C. Whitaker, Ed., New York, NY: McGraw-Hill, 2000.

[9] F. Sanders, "Measurements of pulsed co-channel interference in a 4-GHz digital Earth station receiver," *NTIA Report.* 02-393, May 2002.

APPENDIX A: THEORETICAL PSD OF DP AND DS SIGNALS

A general expression for the power spectral density (PSD) of ultrawideband (UWB) signals using a fixed time-base dither was provided in Section 3 of [2]. This section provides the derivation of the expression and uses it to evaluate the PSD of DP and DS signals.

UWB signals that modulate pulses to convey binary data can be expressed as

$$u(t) = \sum_{n=-\infty}^{\infty} \alpha_{0n} y_0(t - nT - \beta_n) - \alpha_{1n} y_1(t - nT - \beta_n) = \sum_{n=-\infty}^{\infty} \sum_{k=0}^{1} \alpha_{kn} y_k(t - nT - \beta_n) \qquad \text{(A-1)}$$

where y_k is the channel symbol, e.g., y_0 represents a zero and y_1 represents a one, T is the nominal symbol period, n is the symbol period index, and β_n is a random variable that defines the dithering as described by the dither function $q(\beta)$. α_{kn} chooses the appropriate symbol for each information bit according to

$$\alpha_{kn} = \begin{cases} 1 - a_n & (k = 0) \\ a_n & (k = 1) \end{cases} \quad \text{and} \quad a_n = \begin{cases} 0 & (\text{with probability } \xi_0) \\ 1 & (\text{with probability } \xi_1 = 1 - \xi_0) \end{cases}$$

This formulation can be altered to map any two distinct information bit values, e.g., {1, -1}, with no change in the final PSD expressions. Note that a_n and β_n are independent and identically distributed (iid).

The PSD is the Fourier transform of the autocorrelation function, i.e., $R_{uu}(t_1,t_2) = \mathrm{E}\{u(t_1)u^*(t_2)\}$, where the expectation operator, defined as $\mathrm{E}\{X\} = \int xq(x)dx$ is distributive and for independent random variables the expectation of the product is equal to the product of the individual expectations. The complex conjugate can be dropped because u(t) is a real function, and the autocorrelation function is expressed as

$$R_{uu}(t_1,t_2) = \mathrm{E}\{u(t_1)u(t_2)\} = \mathrm{E}\left\{\sum_{n,m}\sum_{k,l}\alpha_{kn}\alpha_{lm}v_k(t_1 - nT - \beta_n)v_l(t_2 - mT - \beta_m)\right\}$$

Note that summations involving n and/or m imply a range $\pm\infty$, summations involving k and/or l imply a range from 0 to 1, and integrals with unspecified limits imply the limits $\pm\infty$.

Evaluation of $R_{uu}(t_1, t_2)$ involves summation of the function

$$\gamma_{nm}(t_1,t_2) = \sum_{k,l}\mathrm{E}\{\alpha_{kn}\alpha_{lm}\}\mathrm{E}\{v_k(t_1 - nT - \beta_n)v_l(t_2 - mT - \beta_m)\}$$

where the expectation was moved inside the summation operators and split into the product of expectations of the independent random variables. The expectation of the product of the channel symbols is difficult to simplify in a meaningful way. To avoid this difficulty, the following variable is introduced

$$\hat{\gamma}_{nm}(t_1,t_2) = \sum_{k,l}\mathrm{E}\{\alpha_{kn}\}\mathrm{E}\{\alpha_{lm}\}\mathrm{E}\{v_k(t_1 - nT - \beta_n)\}\mathrm{E}\{v_l(t_2 - mT - \beta_m)\}$$

$\sum_{n\neq m}\gamma_{nm}(t_1,t_2) = \sum_{n\neq m}\hat{\gamma}_{nm}(t_1,t_2)$ because the symbols are independent in different symbol periods. However, $\sum_{n}\gamma_{nn}(t_1,t_2) \neq \sum_{n}\hat{\gamma}_{nn}(t_1,t_2)$ because the symbols are not independent in the same symbol period. Since both quantities are easily evaluated along the diagonal, $\sum_{n}\hat{\gamma}_{nn}(t_1,t_2)$ is subtracted and $\sum_{n}\hat{\gamma}_{nn}(t_1,t_2)$ is added to formulate the expression in a way that can be evaluated analytically, i.e.,

$$R_{uu}(t_1,t_2) = \sum_{n,m}\gamma_{nm}(t_1,t_2) = \sum_{n,m}\hat{\gamma}_{nm}(t_1,t_2) + \sum_{n}\left[\gamma_{nn}(t_1,t_2) - \hat{\gamma}_{nn}(t_1,t_2)\right] \quad \text{(A-2)}$$

As will be demonstrated, this approach has the added benefit of distinguishing between the discrete and continuous spectra of the resulting PSD.

Consider the first term of equation (A-2), $\sum_{n,m} \gamma_{nm}(t_1, t_2)$. E{$\alpha_{kn}$} is evaluated as

$$E\{\alpha_{1n}\} = E\{a_n\} = (0)\xi_0 + (1)\xi_1 = \xi_1$$

$$E\{\alpha_{0n}\} = E\{1 - a_n\} = E\{1\} - E\{a_n\} = \xi_0 \quad .$$

Hence, E{α_{kn}} = ξ_k is independent of n and can be moved outside $\sum_n$, yielding

$$\sum_{n,m} \gamma_{nm}(t_1, t_2) = \left[\sum_k \xi_k \sum_n E\{y_k(t_1 - nT - \beta_n)\}\right]\left[\sum_l \xi_l \sum_m E\{y_l(t_2 - mT - \beta_m)\}\right] \quad .$$

Expanding the expectation into integral form and introducing $\phi_k(t) = \sum_n y_k(t - nT)$ gives

$$\sum_n E\{y_k(t - nT - \beta_n)\} = \sum_n \int y_k(t - nT - \beta)q(\beta)d\beta = \int \phi_k(t - \beta)q(\beta)d\beta = (\phi_k * q)(t) \quad .$$

where * is the convolutional operator. Defining the function gives

$$\sum_{n,m} \gamma_{nm}(t_1, t_2) = [(\psi * q)(t_1)][(\psi * q)(t_2)] \quad . \tag{A-3}$$

where the periodic function $\psi(t)$ can be expressed as a Fourier series

$$\psi(t) = \sum_m \Psi_m e^{j2\pi mt/T} \quad , \qquad \Psi_m = \frac{1}{T}\int_{-T/2}^{T/2} \psi(t) e^{-j2\pi mt/T} dt \quad .$$

This allows for the convolution in equation (A-3) to be expressed as

$$(\psi * q)(t) = \int \sum_m \Psi_m e^{j2\pi m(t-\beta)/T} q(\beta)d\beta = \sum_m \Psi_m e^{j2\pi mt/T} \int q(\beta) e^{-j2\pi m\beta/T} d\beta$$

$$= \sum_m \Psi_m Q\left(\frac{m}{T}\right) e^{j2\pi mt/T} \quad .$$

where Q is the Fourier transform of q. The Fourier coefficients in terms of ξ_k and y_k are

$$\Psi_m = \frac{1}{T}\sum_k \xi_k \int y_k(t')e^{-j2\pi mt'/T}dt' = \frac{1}{T}\sum_k \xi_k Y_k\left(\frac{m}{T}\right) .$$

where change of variable, $t' = t - nT$, was used to simplify and Y_k is the Fourier transform of y_k. Substitution into equation (A-3) and changing the independent variables to $\tau = t_2 - t_1$ and $t = t_2$ yields

$$\sum_{n,m}\gamma_{nm}(\tau,t) = \frac{1}{T^2}\sum_{n,m}\left[\sum_k \xi_k Y_k\left(\frac{n}{T}\right)\right]\left[\sum_k \xi_k Y_k\left(\frac{m}{T}\right)\right]Q\left(\frac{n}{T}\right)Q\left(\frac{m}{T}\right)e^{j2\pi[(m-n)t-m\tau]/T} .$$

(A-4)

The next term in equation (A-2) to be considered is

$$\sum_n \gamma(t_1,t_2) = \sum_{k,l}\xi_k\xi_l \sum_n \left(\int y_k(t_1 - nT - \beta)q(\beta)d\beta\right)\left(\int y_l(t_2 - nT - \beta)q(\beta)d\beta\right)$$
$$= \sum_{k,l}\xi_k\xi_l \iint Y_k(f_1)Y_l(f_2)Q(f_1)Q(f_2)\left[\sum_n e^{-j2\pi(f_1-f_2)nT}\right]e^{j2\pi(f_1t_1-f_2t_2)}df_1df_2 .$$

where $\int y_k(t - nT - \beta)q(\beta)d\beta = \int Y_k(f)Q(f)e^{-j2\pi fnT}e^{j2\pi ft}df$.

The term in square brackets may be expressed as the sum of delta functions using the identity

$$\sum_n e^{-j2\pi fnT} = \frac{1}{T}\sum_n \delta\left(f - \frac{n}{T}\right) .$$

(A-5)

Simple evaluation of the f_2 integral, changing the independent variables to $\tau = t_2 - t_1$ and $t = t_2$, and consolidating the k and l terms yields

$$\sum_n \gamma_m(t_1,t_2) = \frac{1}{T}\sum_n e^{j2\pi nt/T}\int\left[\sum_k \xi_k Y_k(f)\right]\left[\sum_l \xi_l Y_l\left(\frac{n}{T} - f\right)\right]Q(f)Q\left(\frac{n}{T} - f\right)e^{-j2\pi f\tau}df$$
$$= \frac{1}{T}\sum_n e^{j2\pi nt/T}\left[\left(\sum_k \xi_k y_k(\tau)\right) * \left(\sum_l \xi_l y_l(-\tau)e^{j2\pi n\tau/T}\right) * \left[q(\tau) * \left(q(-\tau)e^{j2\pi n\tau/T}\right)\right]\right] .$$

(A-6)

where $\int Y_k\left(\frac{n}{T}-f\right)e^{j2\pi f\tau}df = y_k(-\tau)e^{j2\pi n\tau/T}$.

The last term to be evaluated in equation (A-2) is

$$\sum_n r_{nn}(t_1,t_2) = \sum_n \sum_{k,l} \mathrm{E}\{\alpha_{kn}\alpha_{ln}\}\mathrm{E}\{y_k(t_1 - nT - \beta_n)y_l(t_2 - nT - \beta_n)\}$$

E $\{\alpha_{kn}, \alpha_{ln}\}$ is independent of *n*, that is,

Hence, E $\{\alpha_{kn}, \alpha_{ln}\} = \xi_k\delta_{kl}$, where δ_{kl} is the Kronecker delta function. Expanding the expectation into integral form, taking the double Fourier transform, and consolidating n and β terms yields

$$\sum_n r_{nn}(t_1,t_2) = \sum_k \xi_k \sum_n \int y_k(t_1 - nT - \beta)y_k(t_2 - nT - \beta)q(\beta)d\beta$$

$$= \sum_k \xi_k \iint Y_k(f_1)Y_k(f_2)\left[\sum_n e^{-j2\pi(f_1-f_2)nT}\right]\left[\int q(\beta)e^{-j2\pi(f_1-f_2)\beta}d\beta\right]e^{j2\pi(f_1t_1-f_2t_2)}df_1df_2$$

Utilizing equation (A-5) to evaluate the f_2 integral, changing independent variables to $\tau = t_2 - t_1$ and t = t_2, rearranging terms, and recognizing the Fourier transform with respect to τ yields

$$\sum_n r_{nn}(\tau,t) = \frac{1}{T}\sum_n e^{j2\pi nt/T}Q\left(\frac{n}{T}\right)\int\sum_k \xi_k Y_k(f)Y_k\left(\frac{n}{T}-f\right)e^{-j2\pi f\tau}df$$

$$= \frac{1}{T}\sum_n e^{j2\pi nt/T}Q\left(\frac{n}{T}\right)\sum_k \xi_k y_k(\tau) * \left(y_k(-\tau)e^{j2\pi n\tau/T}\right) \quad .$$

(A-7)

Equations (A-2), (A-4), (A-6), and (A-7) combine to give an expression for the autocorrection function in terms of τ and t. The statistics for $R_{uu}(\tau, t)$ are periodic with period T. Such processes are commonly referred to as cyclostationary. For these processes, it is useful to calculate the average over all possible observation times within a period. The time average over one period denoted as follows

$$\langle R_{uu}(\tau)\rangle = \frac{1}{T}\int_0^T R_{uu}(\tau,t)dt \quad .$$

Evaluating this integral involves the integral identity

$$\frac{1}{T}\int_0^T e^{j2\pi k\tau/T}\,d\tau = \begin{cases} 1 & (k=0) \\ 0 & (\text{otherwise}) \end{cases} \quad .$$

where k is a integer. Hence, the time average of equation (A-4) is non-zero only for $n = -m$, and

the time average of equations (A-6) and (A-7) are non-zero only for n = 0. The time-average autocorrelation function reduces to

$$\langle R_{uu}(\tau)\rangle = \frac{1}{T^2}\sum_n\left|\sum_k \xi_k Y_k\left(\frac{n}{T}\right)\right|^2\left|Q\left(\frac{n}{T}\right)\right|^2 e^{j2\pi n\tau/T}$$
$$+\frac{1}{T}\left\{\sum_k \xi_k y_k(\tau)*y_k(-\tau)-\left[\left(\sum_k \xi_k y_k(\tau)\right)*\left(\sum_k \xi_k y_k(-\tau)\right)\right]*\left[q(\tau)*q(-\tau)\right]\right\} \quad .$$

where $Q(0)$ is the integrated probability function $q(B)$ equal to 1.

Finally, taking the Fourier transform of $\langle R_{uu}(\tau)\rangle$ yields

$$\langle S_{uu}(f)\rangle = \mathrm{FT}\{\langle R_{uu}(\tau)\rangle\} = \frac{1}{T^2}\left|\sum_k \xi_k Y_k(f)\right|^2 |Q(f)|^2 \sum_n \delta\left(f-\frac{n}{T}\right)$$
$$-\frac{1}{T}\left\{\sum_k \xi_k |Y_k(f)|^2 - \left|\sum_k \xi_k Y_k(f)\right|^2 |Q(f)|^2\right\} \quad . \tag{A-8}$$

which is comprised of discrete and continuous components that depend on the symbol spectrum and the Fourier transform of the density function used to randomize the signal.

A.1. DP-UWB

For DP signals, q(β) is uniformly distributed (and more importantly symmetric); hence, the signal can be expressed according to the general expression (A-1). The DP channel symbol y_1 is a time advanced version of y_0 defined as

$$y_1\left(\tau+\frac{\beta_{max}T}{2}\right) = y_0(\tau) \equiv p_{DP}(\tau) \quad .$$

where β_{max} is the dither fraction and $p_{DP}(t)$ is the pulse shape. Assuming equiprobable DP information bits, i.e., $\xi_k = ½$, and knowing that time shift translates to a phase shift in the frequency domain results in

$$\sum_k \xi_k |Y_k(f)|^2 = \frac{1}{2}\left[\left|\int p_{DP}(t)e^{-j2\pi ft}dt\right|^2 + \left|e^{j\pi f\beta_{max}T}\int p_{DP}(t)e^{-j2\pi ft}dt\right|^2\right] = |P_{DP}(f)|^2$$

$$\left|\sum_k \xi_k Y_k(f)\right|^2 = \frac{1}{4}|P_{DP}(f)|^2\left|1+e^{j\pi f\beta_{max}T}\right|^2 = \frac{1}{2}|P_{DP}(f)|^2\left[1+\cos(\pi f\beta_{max}T)\right] .$$

Substitution in equation (A-8) yields

$$\langle S_{xx}(f)\rangle = \frac{1}{2T^2}\left|P_{DP}(f)Q(f)\right|^2\left[1+\cos(\pi f\beta_{max}T)\right]\sum_n \delta\left(f-\frac{n}{T}\right)$$
$$+\frac{1}{T}|P_{DP}(f)|^2\left\{1-\frac{|Q(f)|^2\left[1-\cos(\pi f\beta_{max}T)\right]}{2}\right\}$$

Figures A-1 – A-4 illustrate the power spectral densities of DP-01 – DP-04, respectively. The magnitudes of the spectra were normalized to the peak of the continuous spectrum. For DP-01 – DP-04, q(β) is uniformly distributed over half of the pulse period. Consequently, $\beta_{max} = 0.5$ and $Q(f) = \mathrm{sinc}(\pi\beta_{max}Tf)$. This sinc function has nulls at frequencies equal to 2k/T (k = ±1, ±2, ...), which zeros out every other spectral line and causes the spacing between discrete spectral lines to be 2/T. In these plots, the vertical shaded regions in the top plots correspond to the DTV channel displayed in the bottom plots. Notice that in all cases the spectral lines were either below the continuous spectrum or outside the DTV channel.

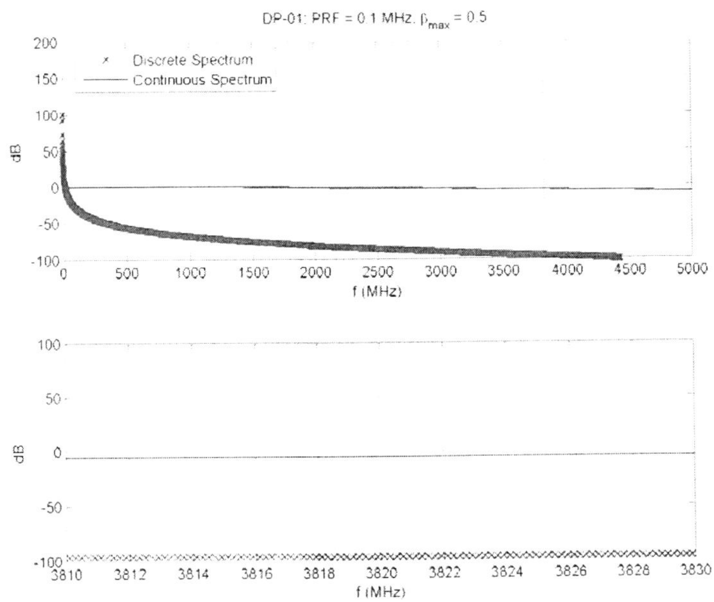

Figure A-1. Theoretical discrete and continuous spectra of DP-01.

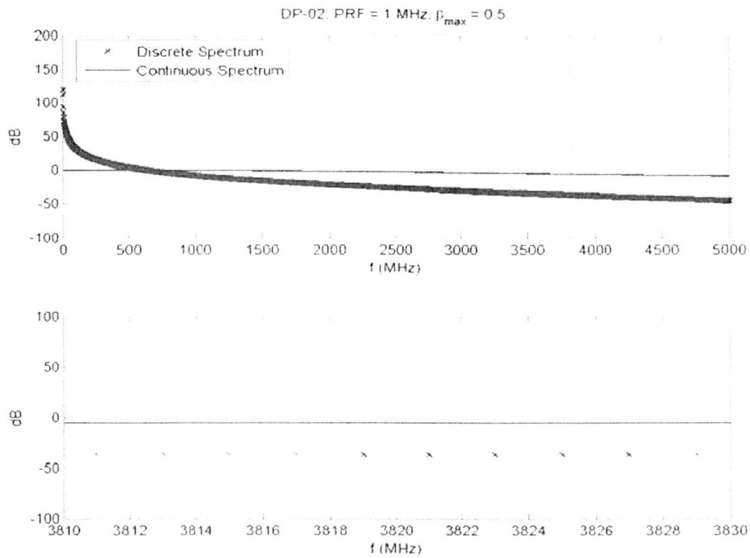

Figure A-2. Theoretical discrete and continuous spectra of DP-02.

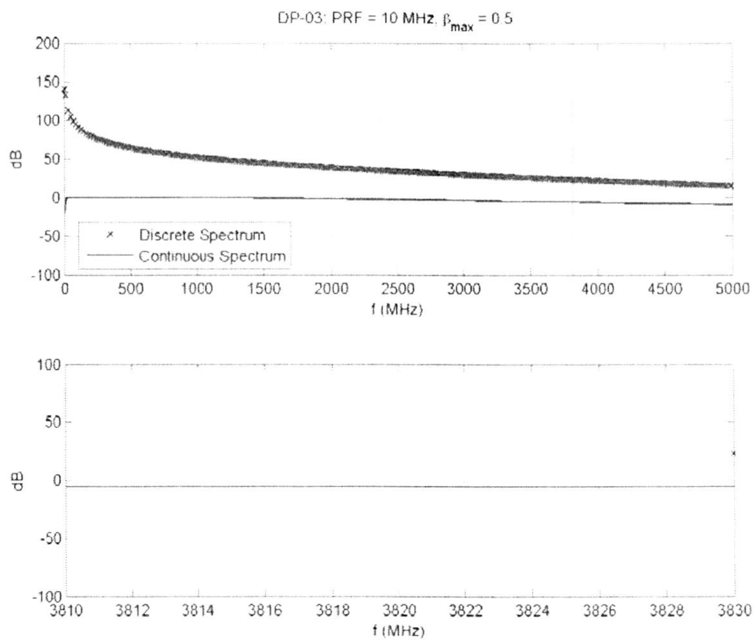

Figure A-3. Theoretical discrete and continuous spectra of DP-03.

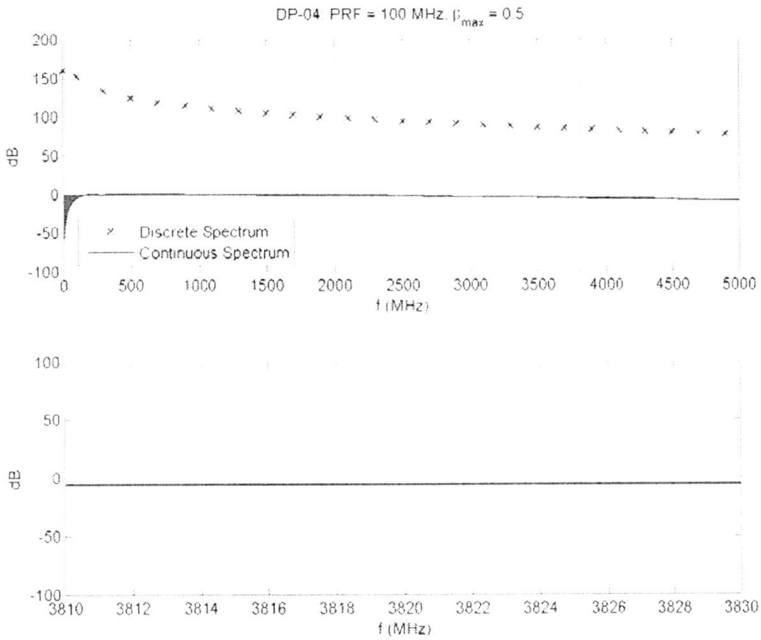

Figure A-4. Theoretical discrete and continuous spectra of DP-04.

A.2. DS-UWB

DS channel symbols are not dithered, i.e., Q(f) = FT{q(β)} = FT{δ(0)} = 1. They are defined as the convolution of a codeword and the DS pulse shape (p_{DS}), i.e.,

$$-y_1(t) = y_0(t) \equiv \sum_{l=0}^{L-1} c_l\, p_{DS}(t - lT_{chip}) \quad ,$$

where L is the codeword length with values of {1, 3, 6, 12, or 24}, c_l is the lth codeword chip with values {-1, 0, 1}, and T_{chip} is the chip period. Assuming DS information bit values are equiprobable, the negative relation between y_1 and y_0 makes

$$\left|\sum_k p_k Y_k(f)\right|^2 = \frac{1}{2}\int \left[y_0(t) - (-y_0(t))\right] e^{-j2\pi ft}\, dt = 0 \quad .$$

and causes the discrete component of the DS-UWB PSD to be zero. The other sum in equation (A-7) evaluates to

$$\sum_k \xi_k \left|Y_k(f)\right|^2 = \left|\sum_{l=0}^{L-1} c_l \int p_{DS}(t - lT_{chip}) e^{-j2\pi ft} dt\right|^2$$

$$= \left|P_{DS}(f)\right|^2 \left|\sum_{l=0}^{L-1} c_l e^{-j2\pi f l T_{chip}}\right|^2$$

$$= \left|P_{DS}(f)\right|^2 \left|C(f)\right|^2 \quad .$$

and substitution into equation (A-8) yields

$$\langle S_{uu}(f)\rangle = \frac{1}{T} \left|P_{DS}(f)\right|^2 \left|C(f)\right|^2 \quad .$$

Figures A-5 and A-6 illustrate |C(f)| for DS-04 and DS-06, respectively. In these plots, the magnitudes of the spectra were normalized to their peak, and the vertical shaded regions in the top plots correspond to the DTV channel displayed in the bottom plots. Notice the resemblance between the |C(f)| and the measured PSD of DS-06 illustrated in Section 2.3.3. This agreement is because $|P_{DS}(f)|$ is flat over the frequency range 3.7 – 4.2 GHz as illustrated in figure 3 of Part 1 [6]. Not shown in this appendix are frequency flat sparse codes for DS-01, DS-02, DS-03, and DS-05, which produce the flat measured PSDs shown in Section 2.3.3.

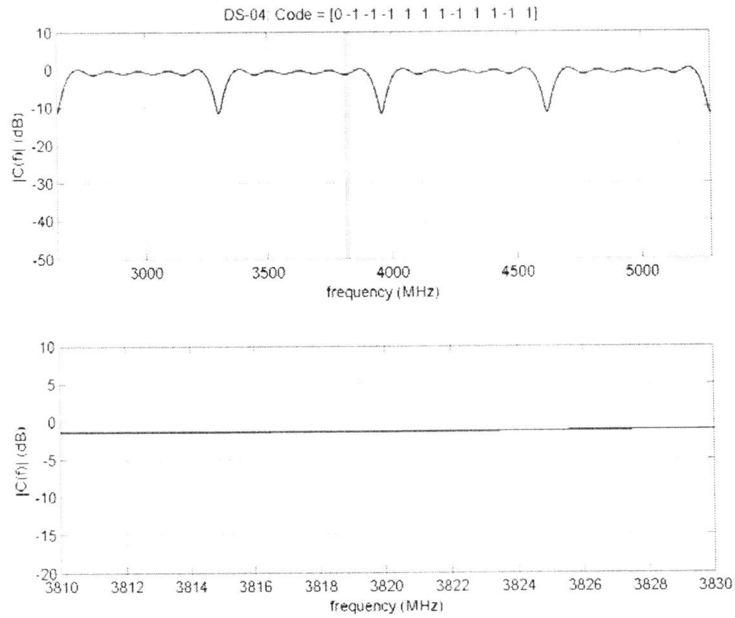

Figure A-5. |C(f)| of the DS-04 code word.

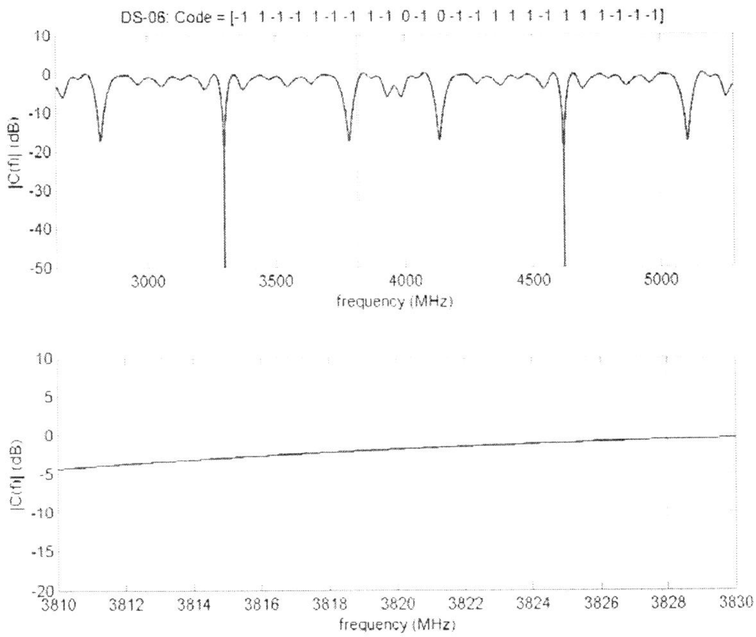

Figure A-6. |C(f)| of the DS-06 code word.

APPENDIX B. CHARACTERIZATION MEASUREMENTS

This appendix provides a thorough description of the measurement procedures and processing used to characterize the ultrawideband (UWB) signals. It provides an extensive set of temporal, amplitude, and spectral analysis results that support Sections 2.3 and 3.3.

B.1. Measurement Setup

The UWB signals were measured under optimal conditions at the radio frequency (RF) output of the vector signal generator (VSG) as described in figure B-1. Measurements at RF were made with the vector signal analyzer (VSA) and digital oscilloscope (DO).

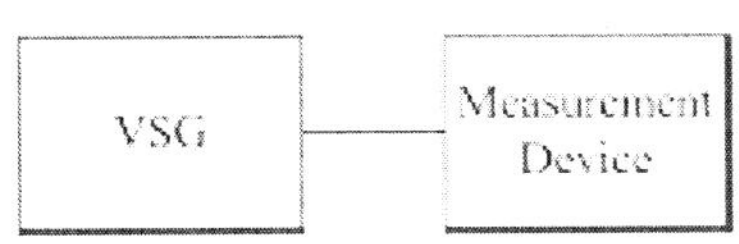

Figure B-1. Setup for RF signal characterization measurements.

Prototype DS and MB devices were made available for these tests; however, we chose to generate simulated signals with a VSG to have more control over modulation parameters and test conditions. Prototype DS and MB signals were characterized to verify corresponding

VSG signals. Characteristics of the prototype DS signals matched those of the VSG DS signals. General characteristics of the prototype MB signals resembled those of the VSG MB signals. However, differences attributed to local-oscillator (LO) feed-through in the prototype MB signals were observed. Prototype-generated signal characterization measurements are not included in this chapter.

The VSG was set up with the procedures described in Appendix B of Part 1 [6]. UWB waveforms were created via software simulation, down-conversion to baseband, and decimation to the sampling frequency of the VSG. In order to ensure optimal use of the analog-to-digital converters in the VSG, DP and DS signals were normalized according to their maximum values shown in table B-1. In contrast, MB-02 – MB-09 were normalized by the maximum value of MB-01 from which they were derived. In the composite plots of Section 2.3, discrepancies due to different normalization factors were removed so the relative average powers are proportional to T_{pulse}. Plots in this appendix, however, were not adjusted according to the individual normalization factors.

VSA instrument settings are summarized in table B-2. As discussed in Appendix E of Part 1, VSG amplitudes were adjusted to ensure that the dynamic range of the VSA was optimally utilized. Table B-1 provides the VSG amplitudes used for RF signal characterization of DP and DS signals. The VSG amplitude for MB signals was -28.62 dBm.

Table B-1. Maximum Waveform Values and VSG Amplitudes for RF VSA Measurements

Type	Index	Maximum	VSG Amplitude (dBm)
DP	01	0.014	-31.61
	02	0.014	
	03	0.018	
	04	0.053	
DS	01	0.048	-29.86
	02	0.025	-30.71
	03	0.016	-31.62
	04	0.032	-33.31
	05	0.009	-33.61
	06	0.029	-33.23

Table B-2. RF VSA Measurement Settings

Post-Processing Bandwidths (MHz)	Span (MHz)	Number of Samples	Center Frequency (MHz)	Input Range (dBm)
36, 19.51, 10	36	> 360,000	3820	-30
1	4.5	> 450,000		
0.1	0.5625	> 562,500		

DO measurements at the maximum sample rate, i.e., 20 GSps, provided real, time-domain representations of RF signals. DO measurements were made with the VSG output level set to a higher value, i.e., 0 dBm, to compensate for low DO sensitivity at the maximum sample rate.

UWB interference signals were also measured at the IF output of the low-noise block downconverter (LNB) with the setup described in figure 20 of Part 1 with the VSG output level corresponding to INR_{TOV}. The simplified block diagram of the hardware components that affect the IF signal characterization measurements is shown in figure B-2. Bandpass filter BPF1 has a 1-dB bandwidth equal to 40 MHz.

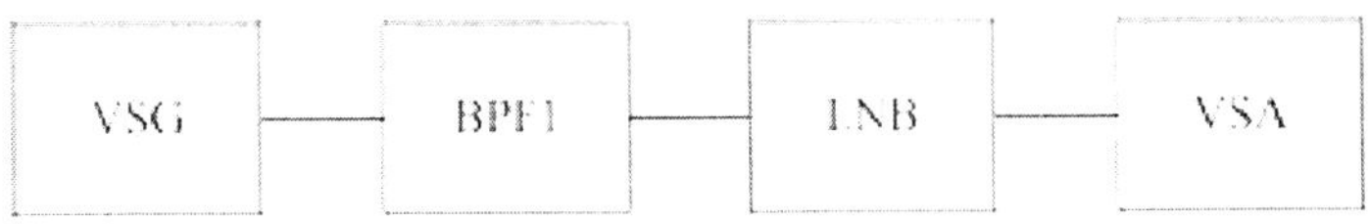

Figure B-2. Simplified setup for IF signal characterization measurements.

Measurements at IF were made with the VSA only. VSA settings for IF measurements were the same as those used for RF measurements described in table B-2, except the center frequency was set to 1330 MHz.

B.2. Temporal Analysis

Time or crossing statistics are provided via the pulse duration distribution (PDD) and the pulse interval distribution (PID). PDD is presented in terms of the percentage of pulses that exceed various time durations in seconds. It provides statistics on how long a signal crosses and stays above a specified threshold. PID gives the probability distribution for the times a signal falls and stays below a threshold. PDD and PID are plotted on Rayleigh graph where the ordinate is plotted on a log scale. Poisson-distributed crossings with exponentially-distributed PD and PI values would plot as a straight line on this graph.

Figure B-3 is an illustration of PD statistics from VSA measurements of the gated-noise signal, GN-10, i.e., τ_{on} = 1 μs and 50% duty cycle, when the crossing threshold is set to -70 dBm. There are three PD categories: non-interrupted on-times in black, interrupted on-times in green, and anomalous spikes in red. Non-interrupted pulse durations are identified by the horizontal portion at low percentiles. Nearly 40% of PDs are approximately 1.13 μs corresponding to a τ_{on} slightly broadened by band-limiting.

Figure B-4 is an illustration of PI estimates from VSA measurements of GN-10 when the crossing threshold is set to -70 dBm. Interestingly, the interrupted off-time category is absent because there are negligible disruptions from higher amplitudes. The resulting PID is nearly a step function. Similar to the PD analysis, non-interrupted PI estimates are identified by the horizontal portion at low percentiles.

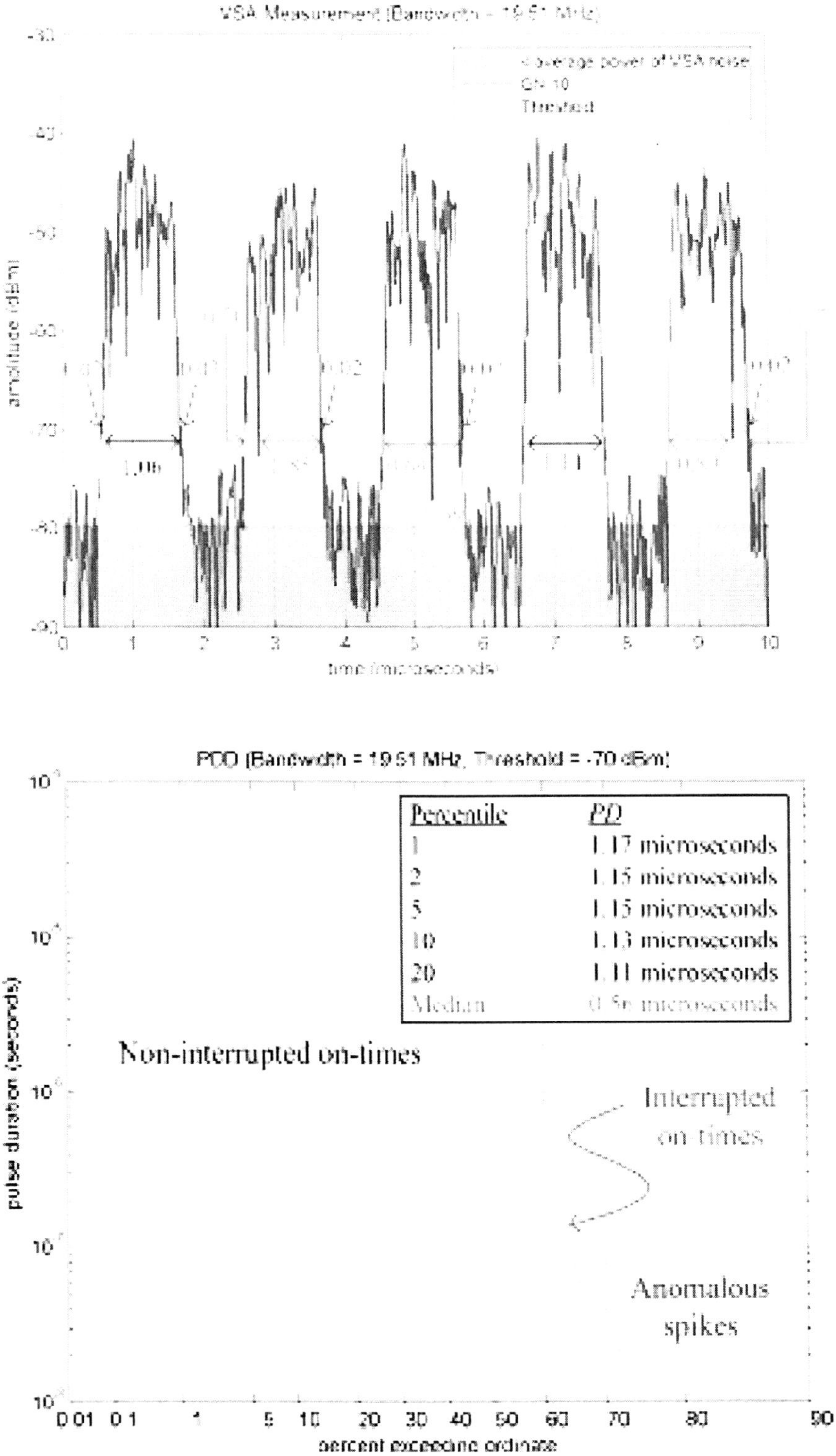

Figure B-3. Pulse duration statistics of gated noise with 50% duty cycle and 1-μs on-time.

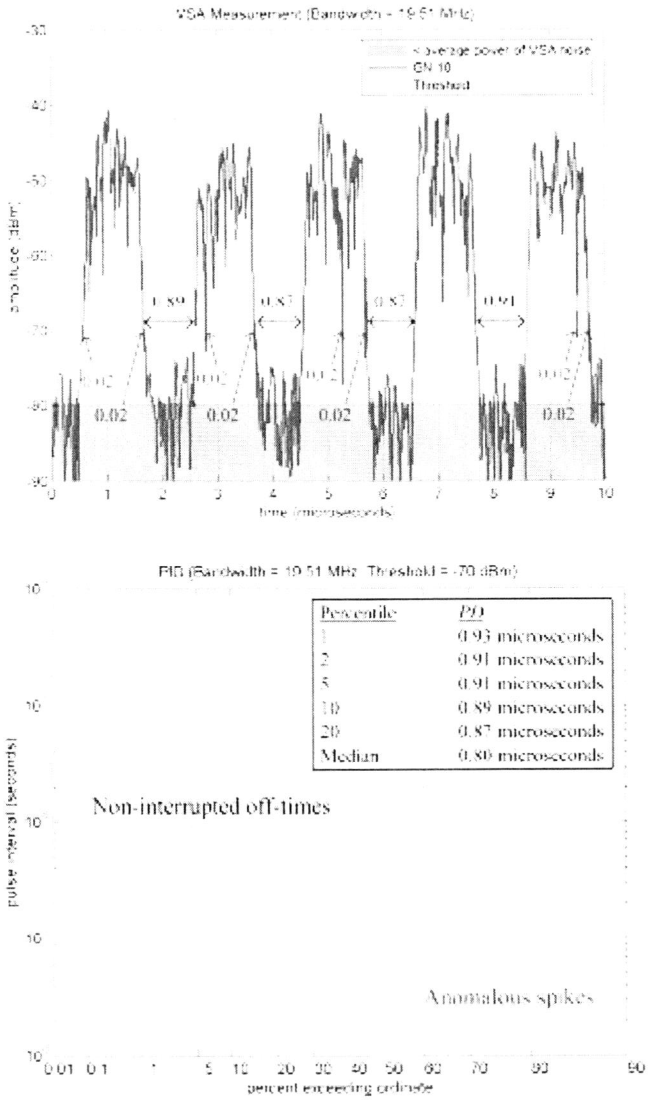

Figure B-4. Pulse interval statistics of gated noise with 50% duty cycle and 1-μs on-time.

Figure B-5 is an illustration of PI estimates from VSA measurements of complex Gaussian noise, GN-01, when the crossing threshold is set to -70 dBm. For the given measurement procedure, this choice for the crossing threshold is too low to quantify meaningful time statistics of continuous signals. As illustrated, GN-01 only triggered time statistic estimates at anomalous noise spikes. Therefore, it is useful to provide curves for a range of crossing thresholds.

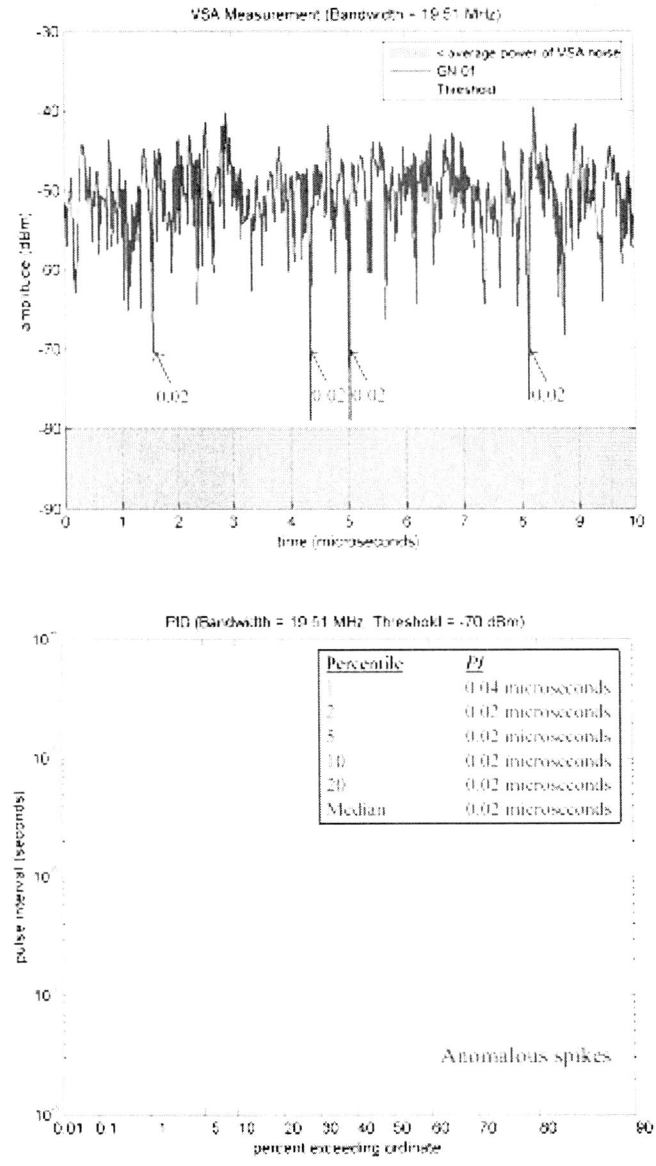

Figure B-5. Pulse interval statistics of complex Gaussian noise.

Figures B-6 – B-24 provide PDD and PID for the UWB signals when the crossing threshold was set to mean amplitude, -50 dBm, -60 dBm, and -70 dBm. Tenth percentile PD and PI estimates when the crossing threshold was set to -70 dBm were chosen as meaningful metrics to quantify the on- and off-times of band-limited signals and were specified in the main text as burst duration (BD) and burst interval (BI), respectively.

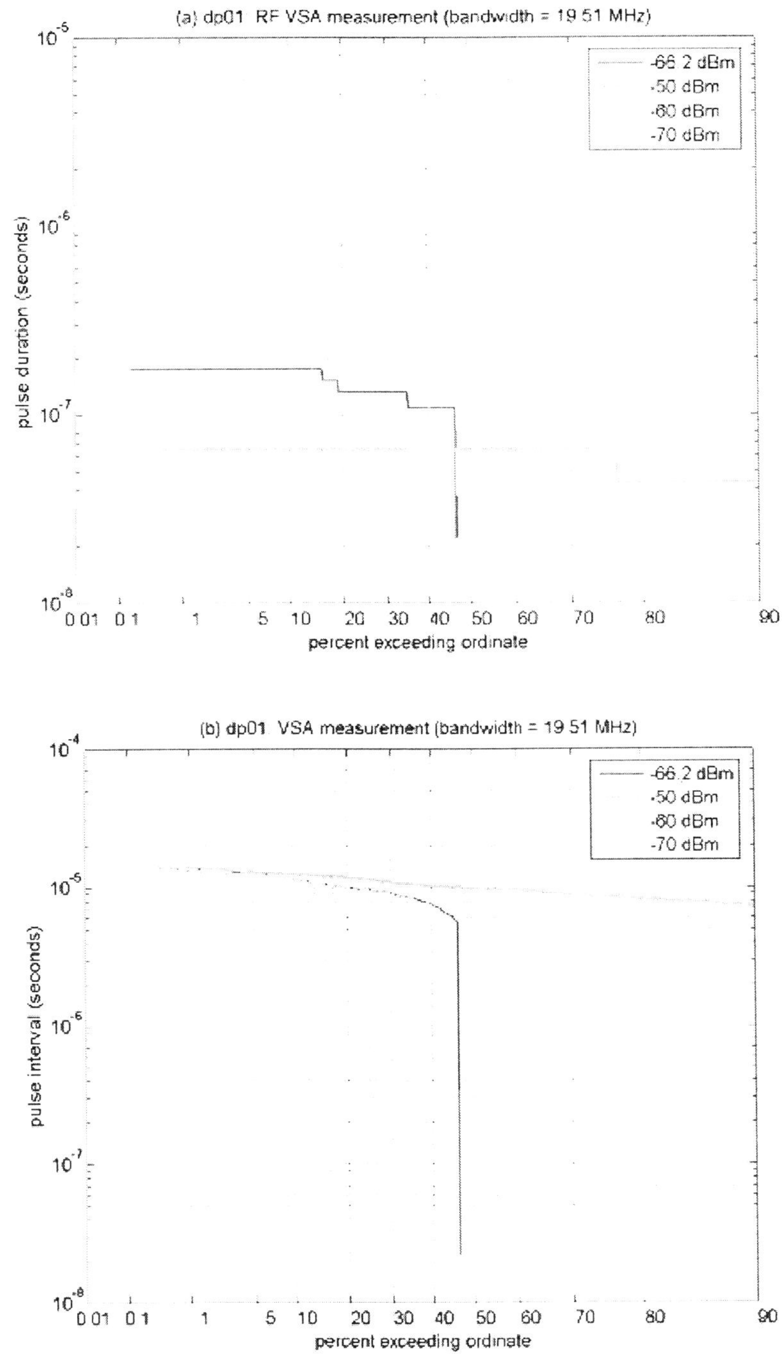

Figure B-6. RF temporal analyses of DP-01 (T_{pulse} = 10,000 ns, w = 0.094 ns, f_D = 0.5).

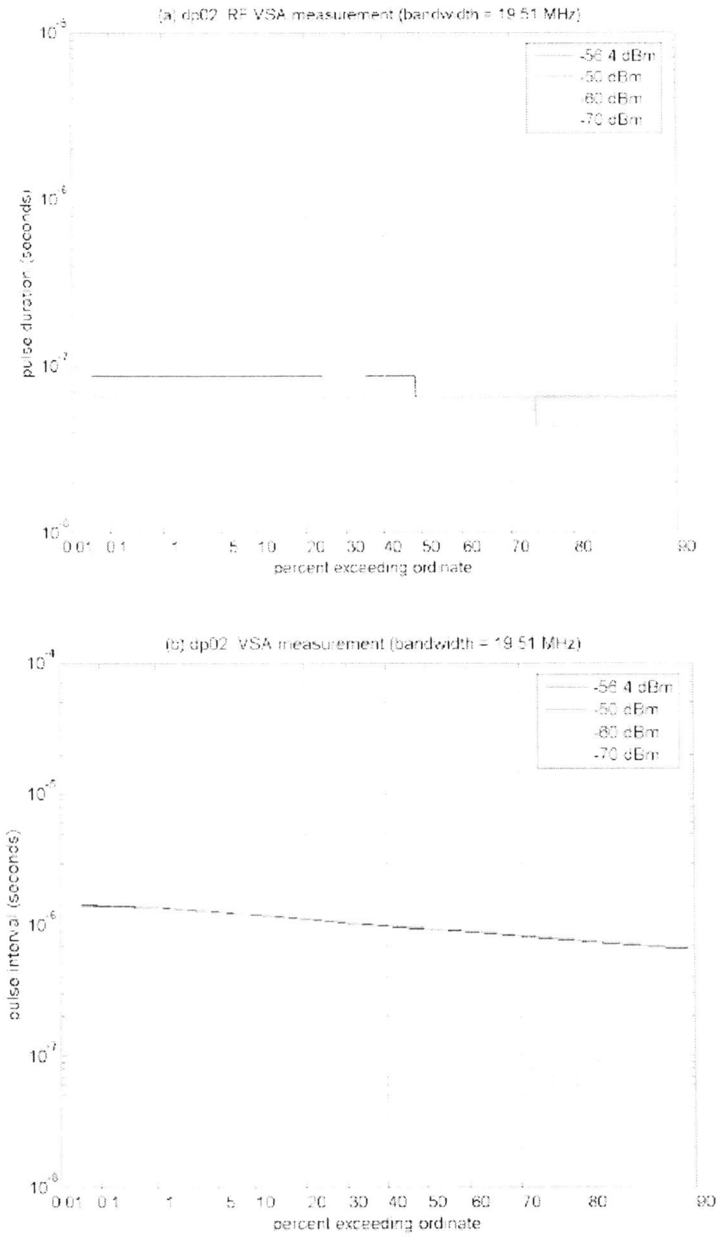

Figure B-7. RF temporal analyses of DP-02 (T_{pulse} = 1000 ns, w = 0.094 ns, f_D = 0.5).

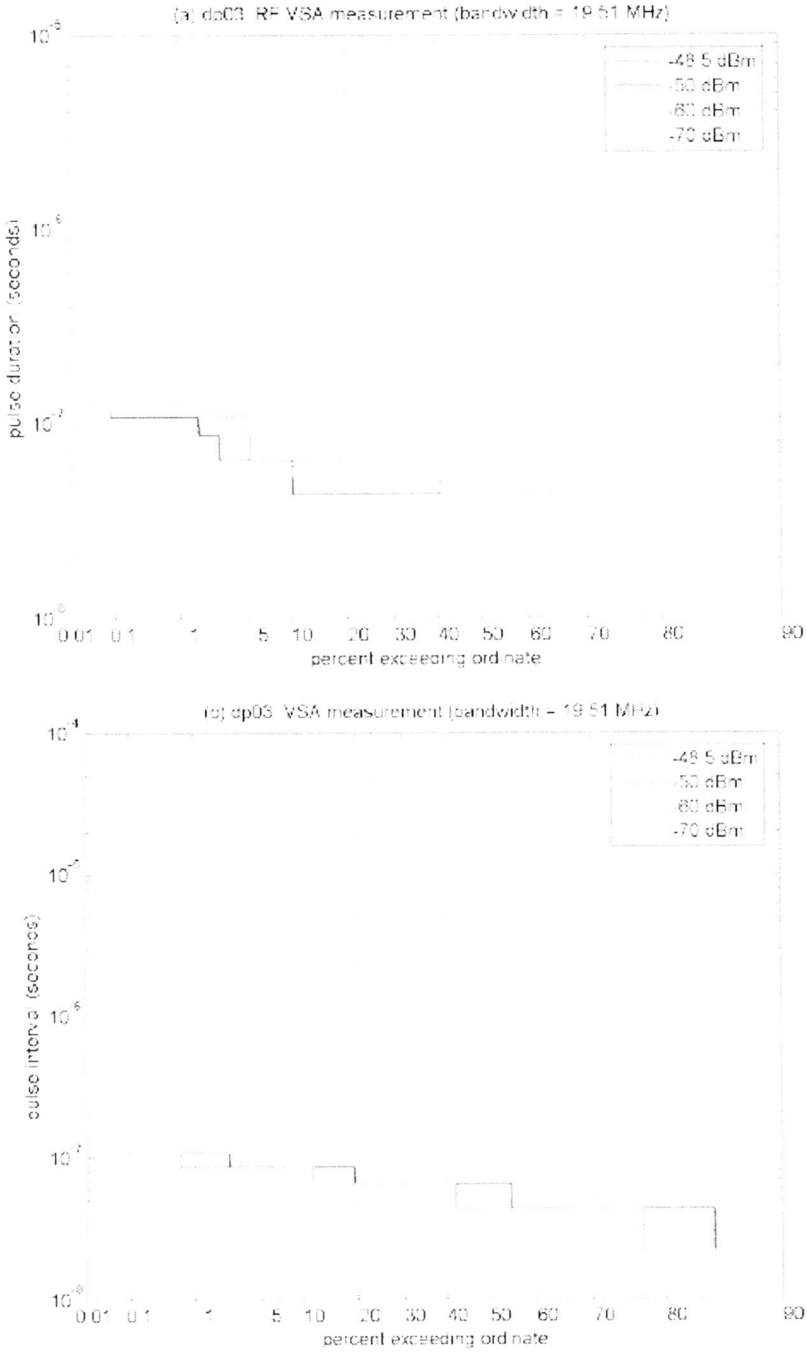

Figure B-8. RF temporal analyses of DP-03 (T_{pulse} = 100 ns, w = 0.094 ns, f_D = 0.5).

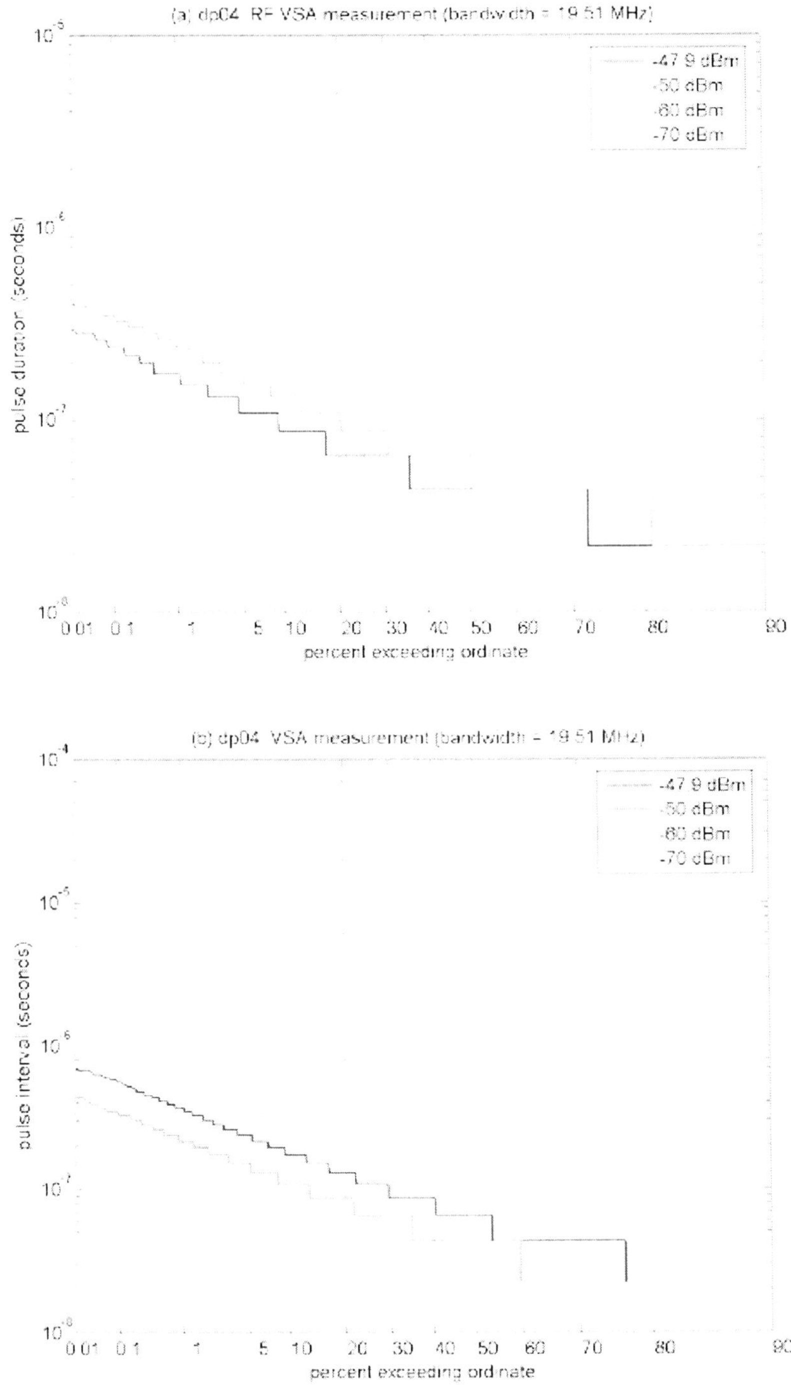

Figure B-9. RF temporal analyses of DP-04 (T_{pulse} = 10 ns, w = 0.094 ns, f_D = 0.5).

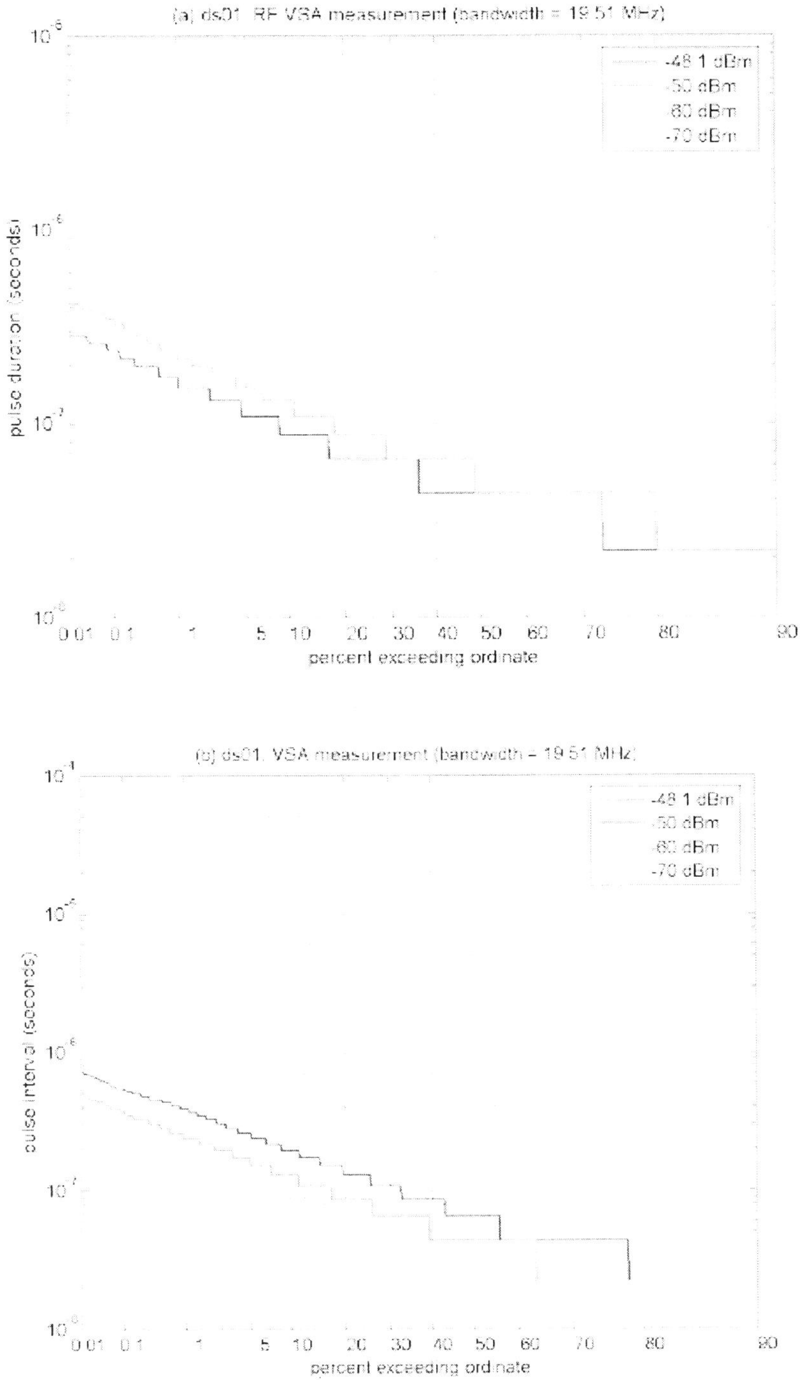

Figure B-10. RF temporal analyses of DS-01 (L = 1).

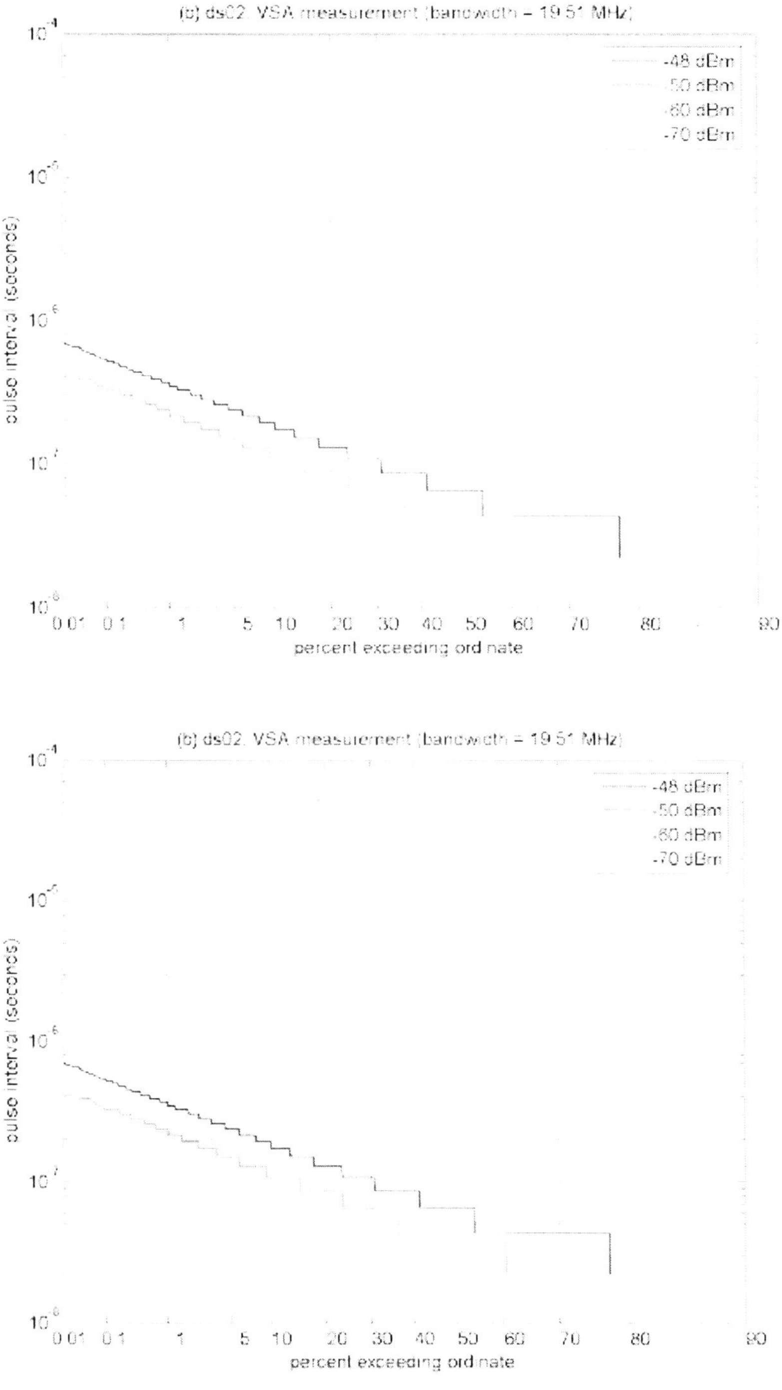

Figure B-11. RF temporal analyses of DS-02 (L = 3, sparse code).

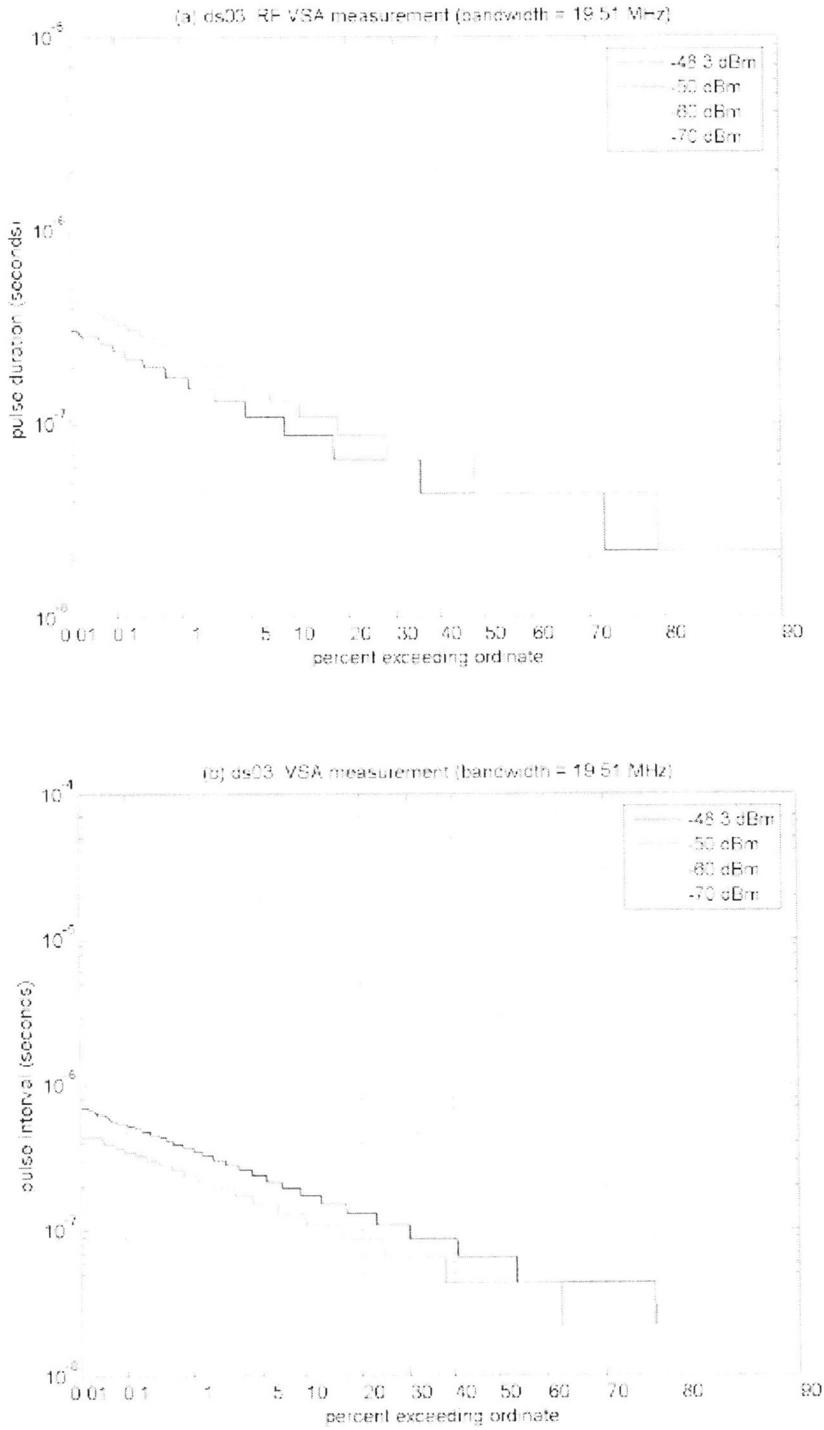

Figure B-12. RF temporal analyses of DS-03 (L = 6, sparse code).

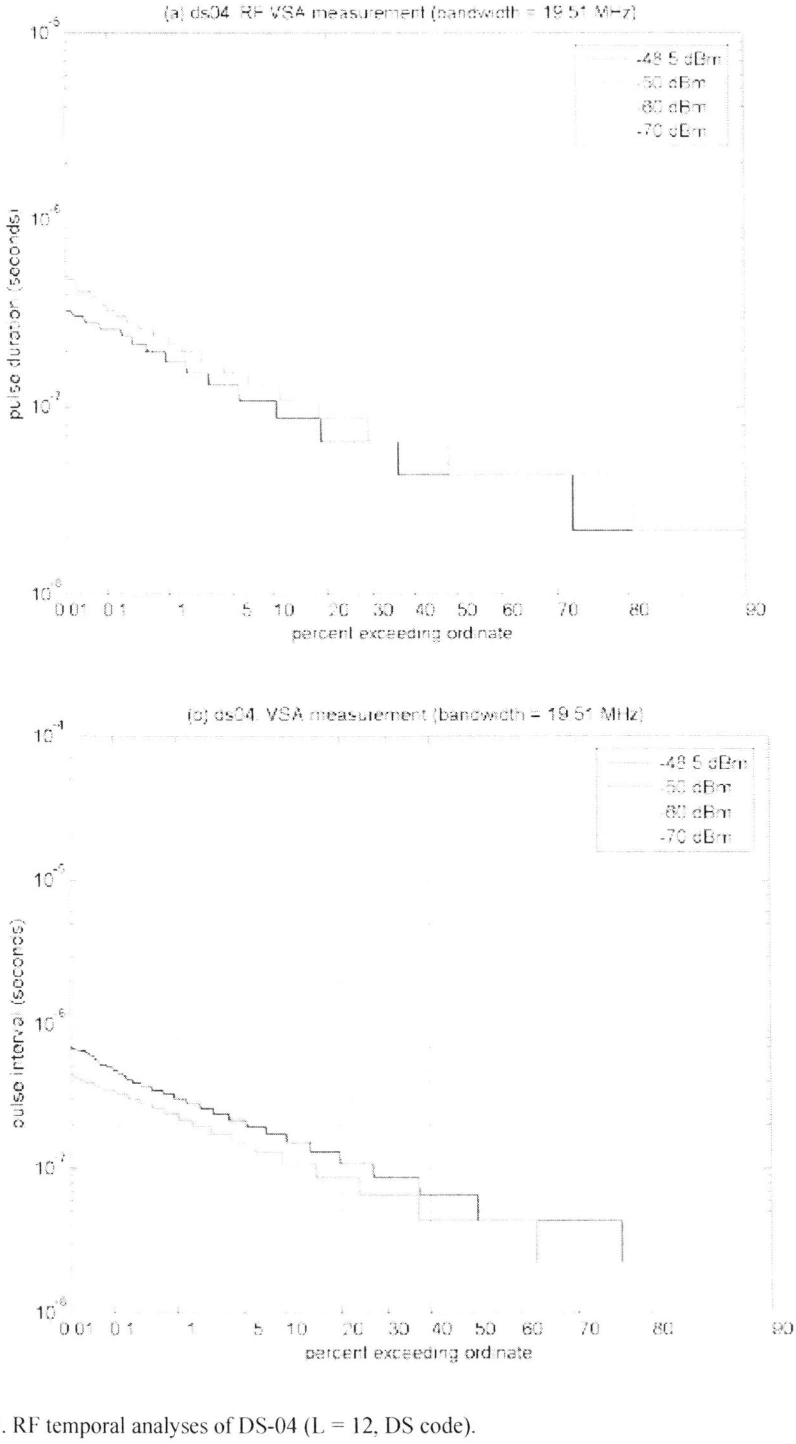

Figure B-13. RF temporal analyses of DS-04 (L = 12, DS code).

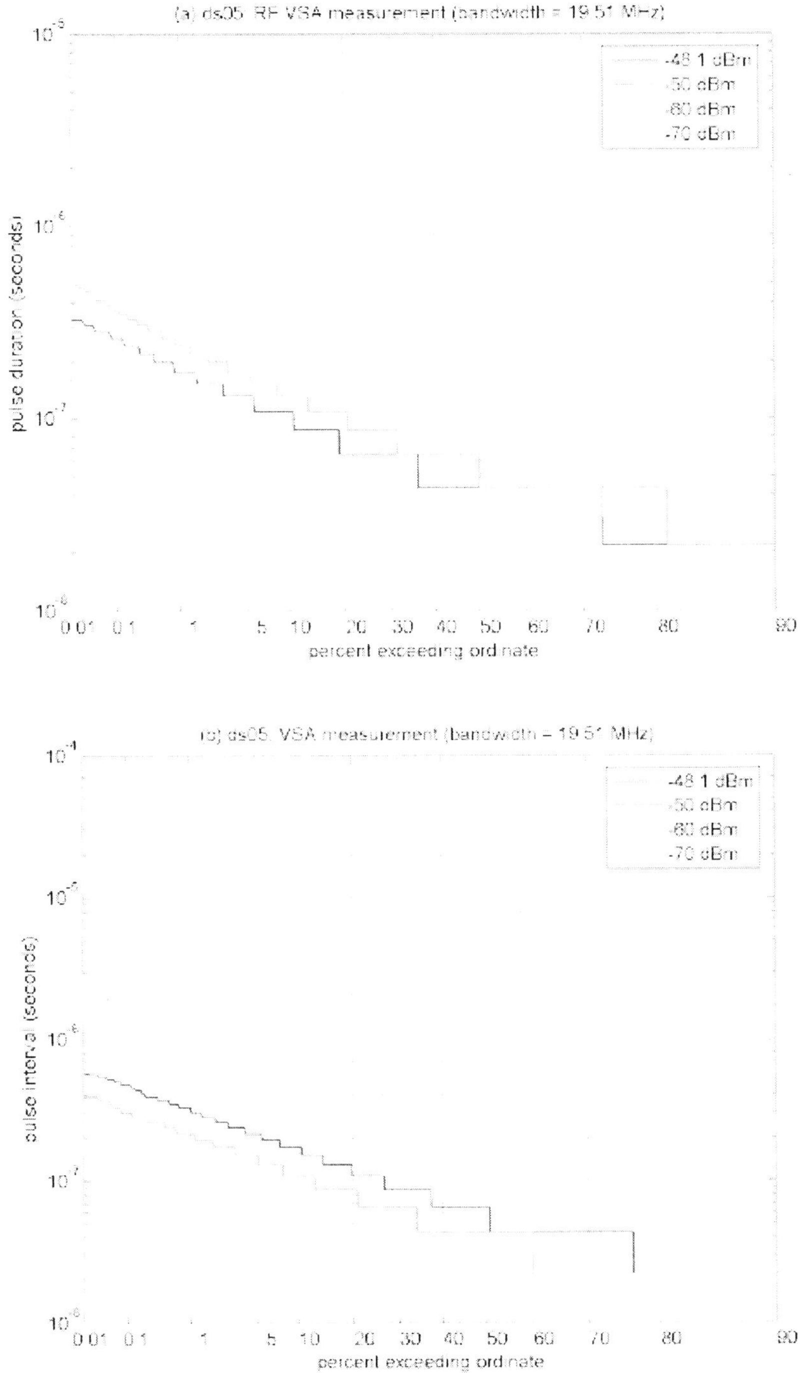

Figure B-14. RF temporal analyses of DS-05 (L = 12, sparse code).

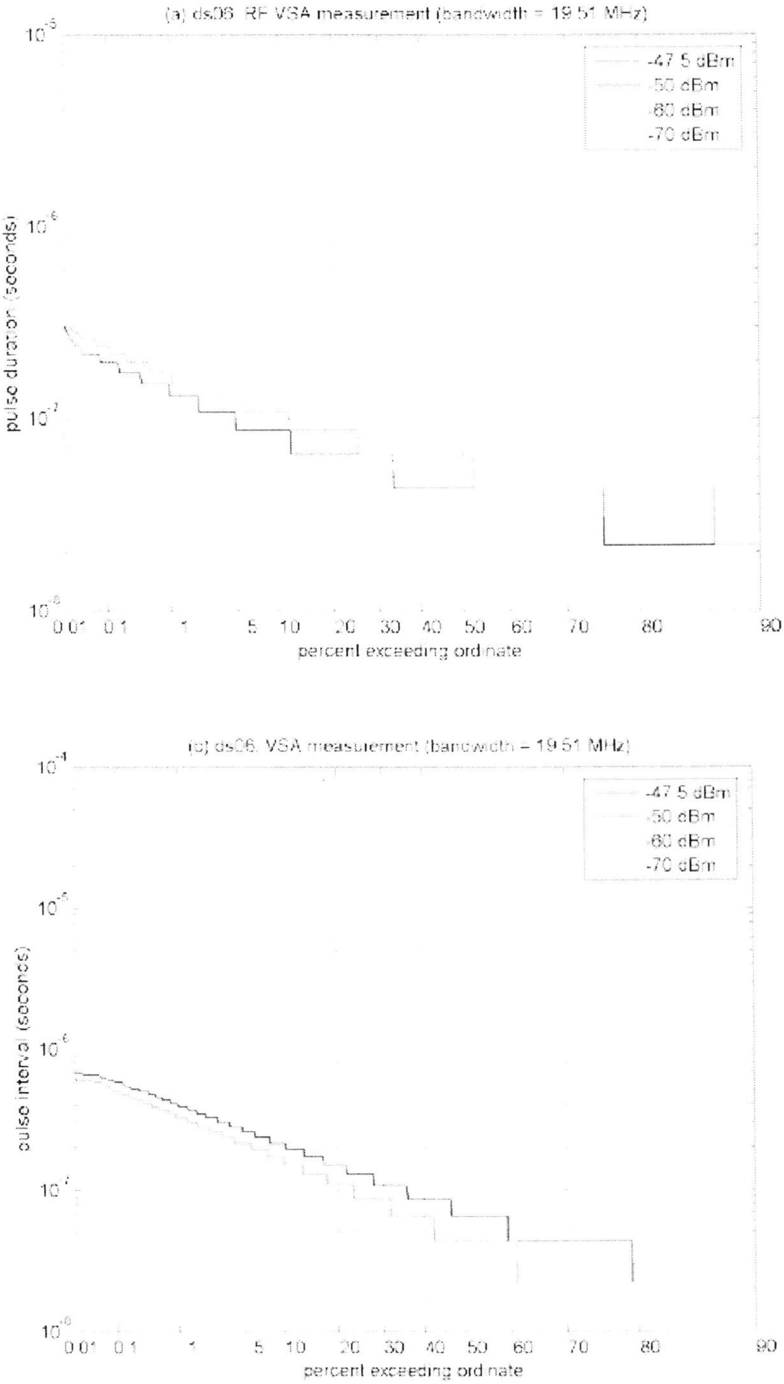

Figure B-15. RF temporal analyses of DS-06 (L = 24, DS code).

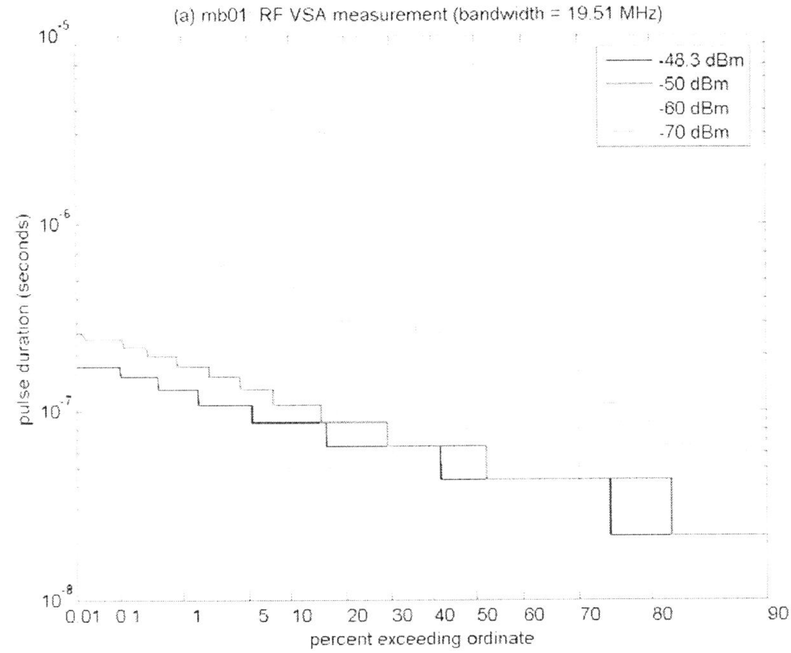

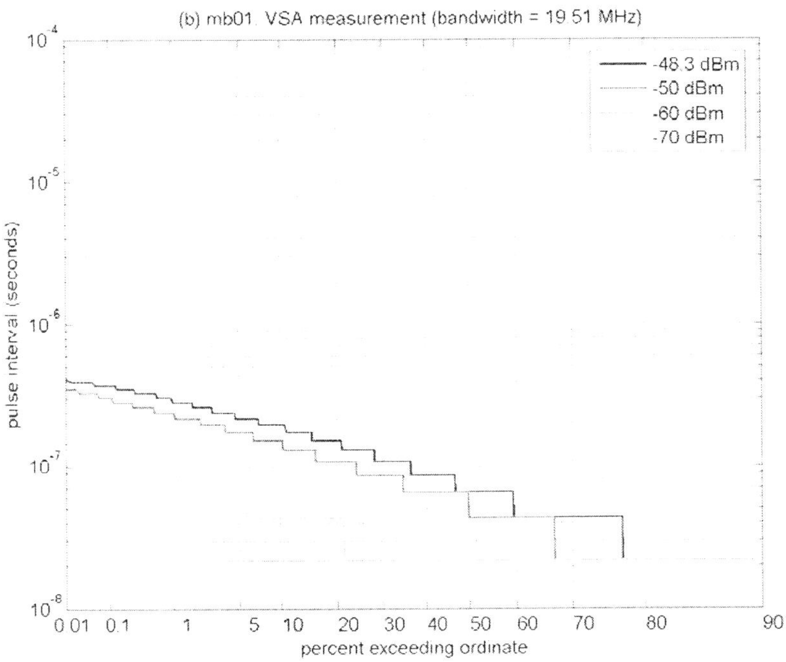

Figure B-16. RF temporal analyses of MB-01 (b = 1).

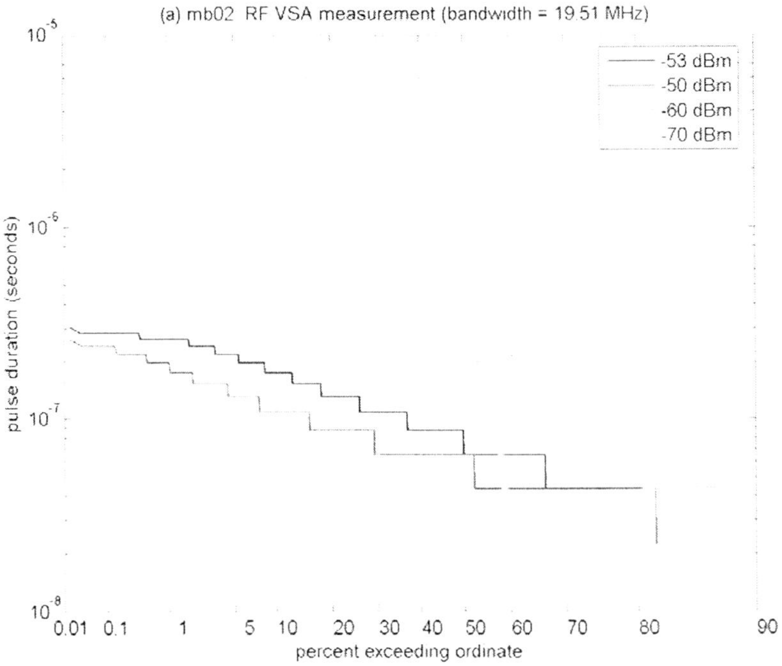

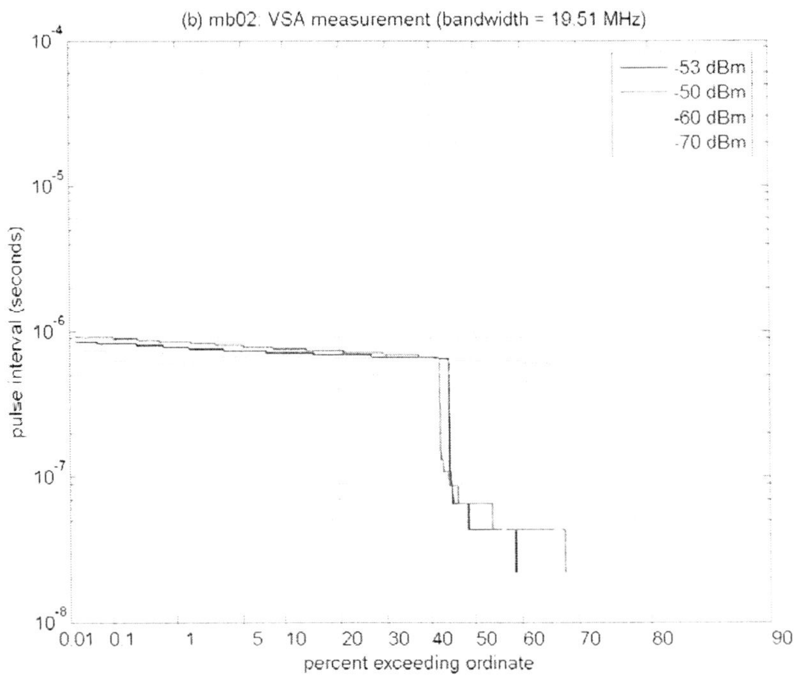

Figure B-17. RF temporal analyses of MB-02 (b = 3, d = 1).

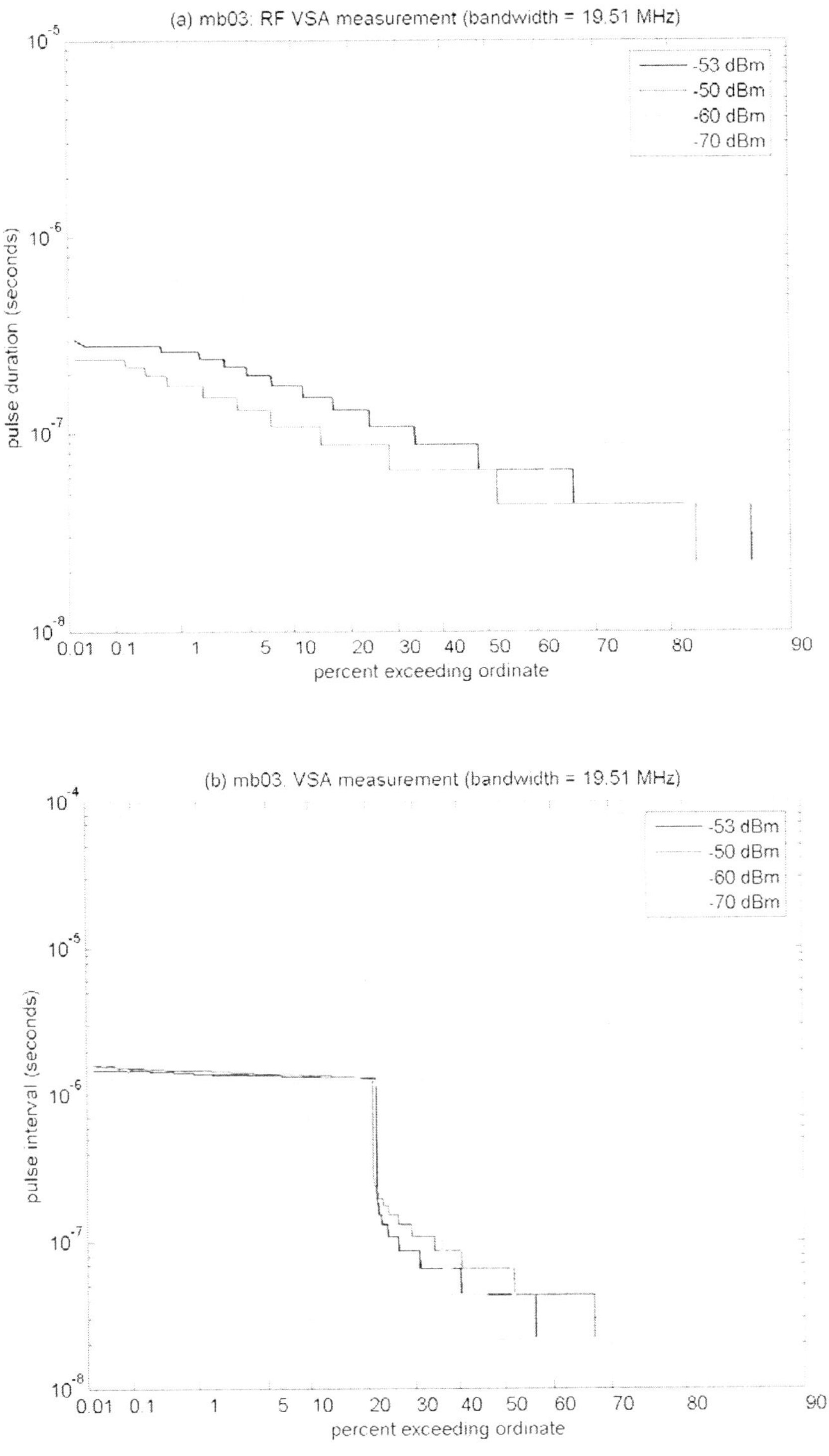

Figure B-18. RF temporal analyses of MB-03 (b = 3, d = 2).

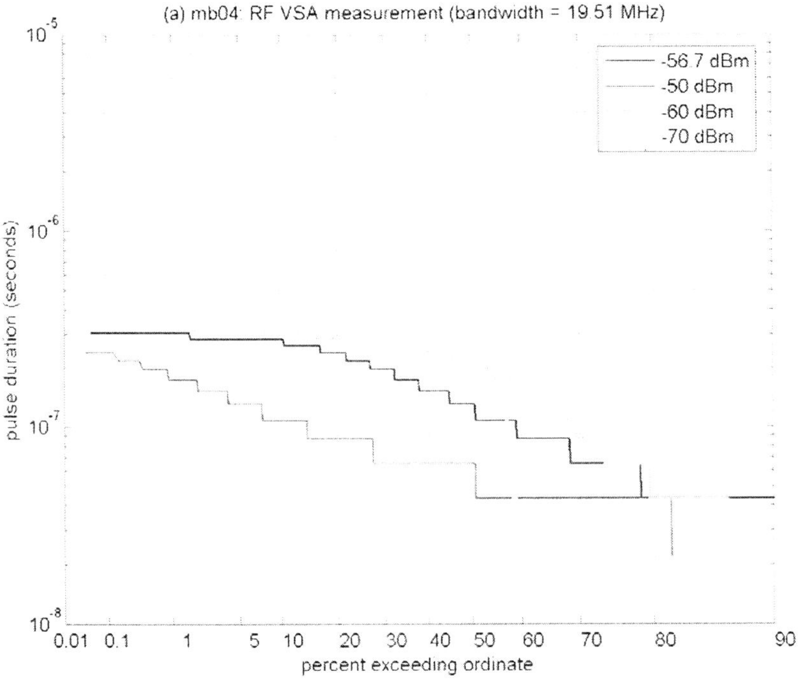

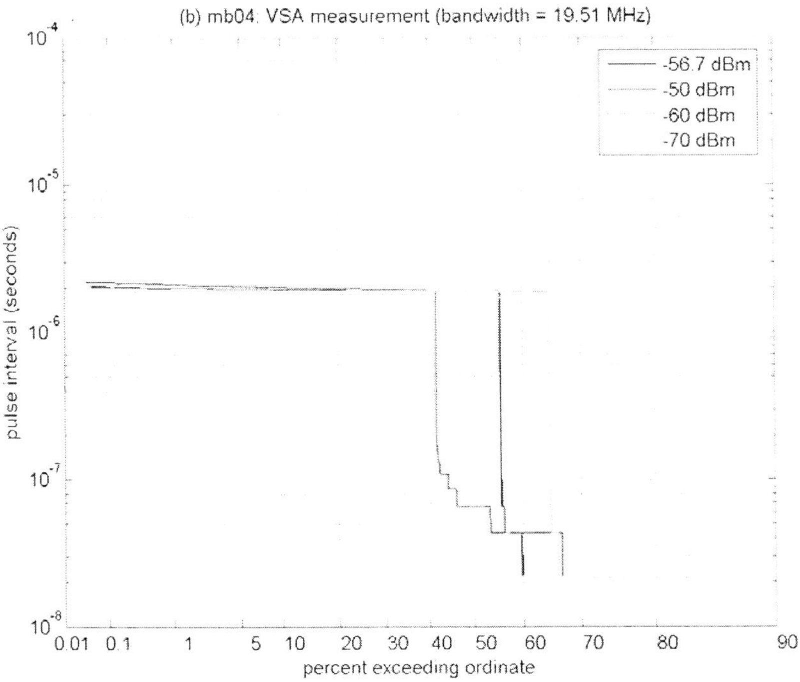

Figure B-19. RF temporal analyses of MB-04 (b = 7, d = 1).

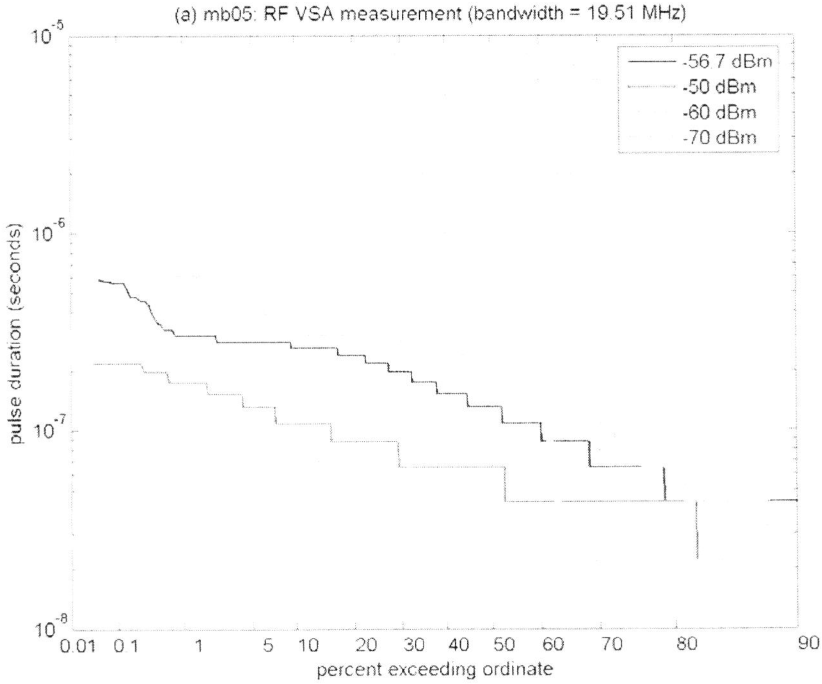

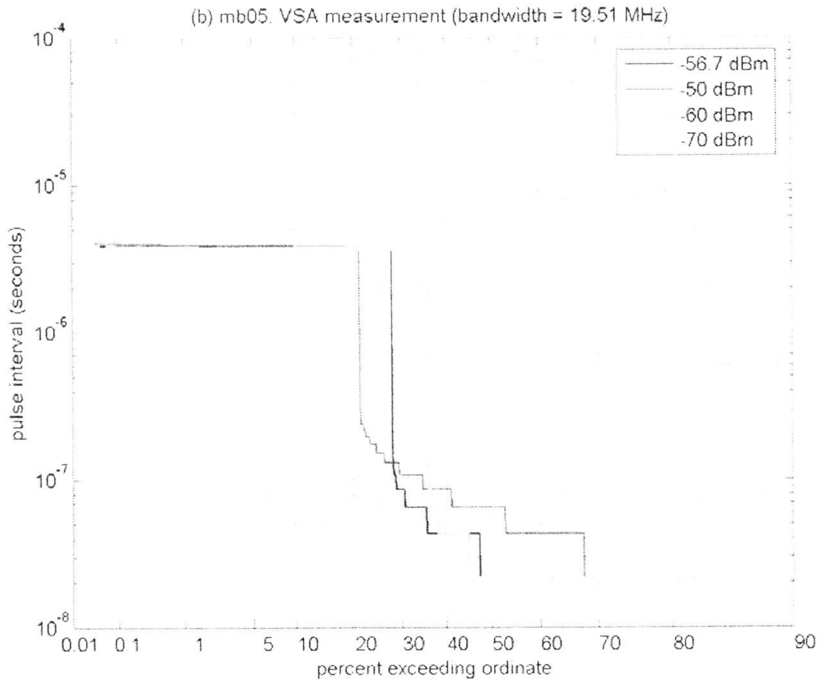

Figure B-20. RF temporal analyses of MB-05 (b = 7, d = 2).

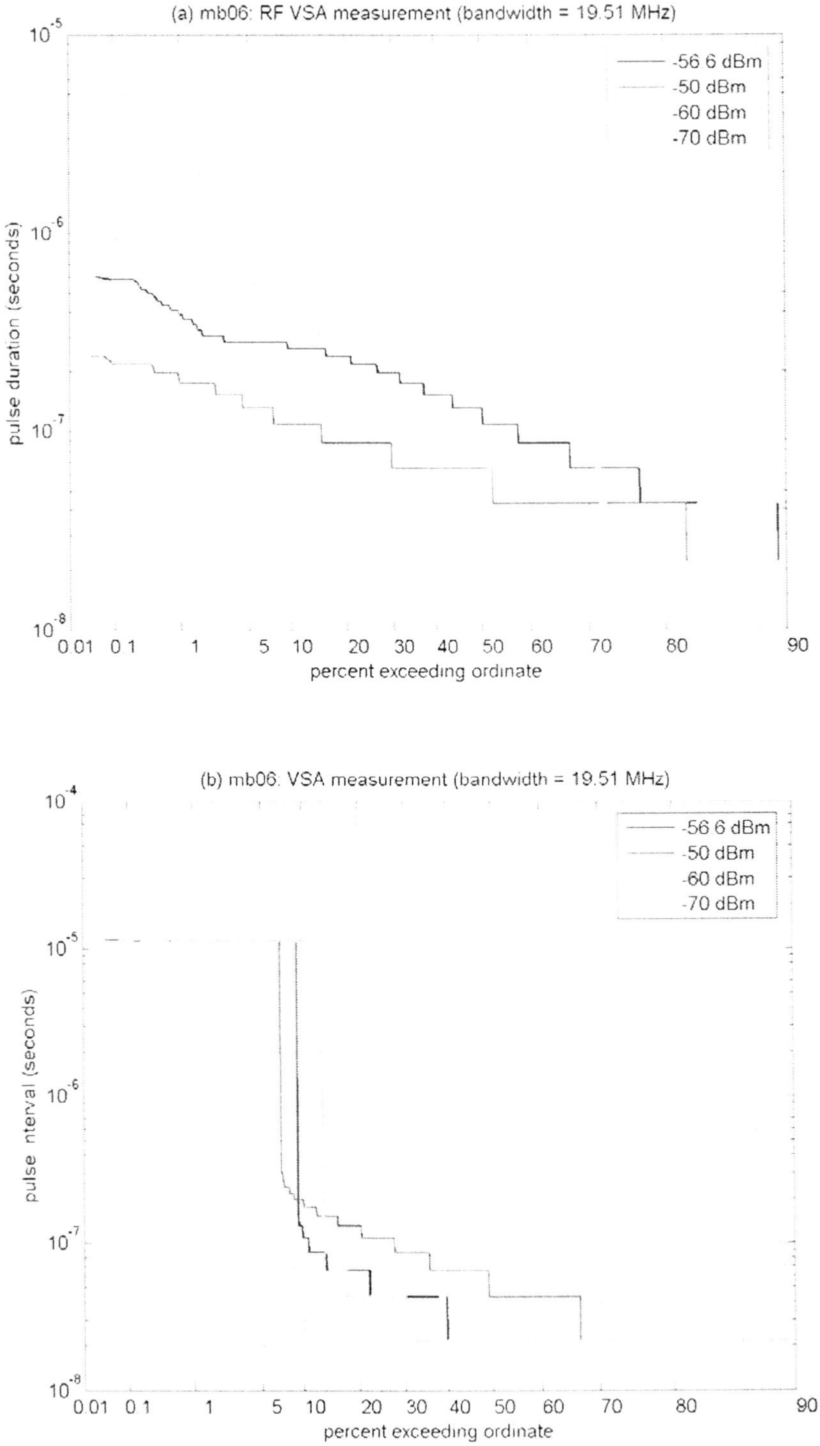

Figure B-21. RF temporal analyses of MB-06 (b = 7, d = 6).

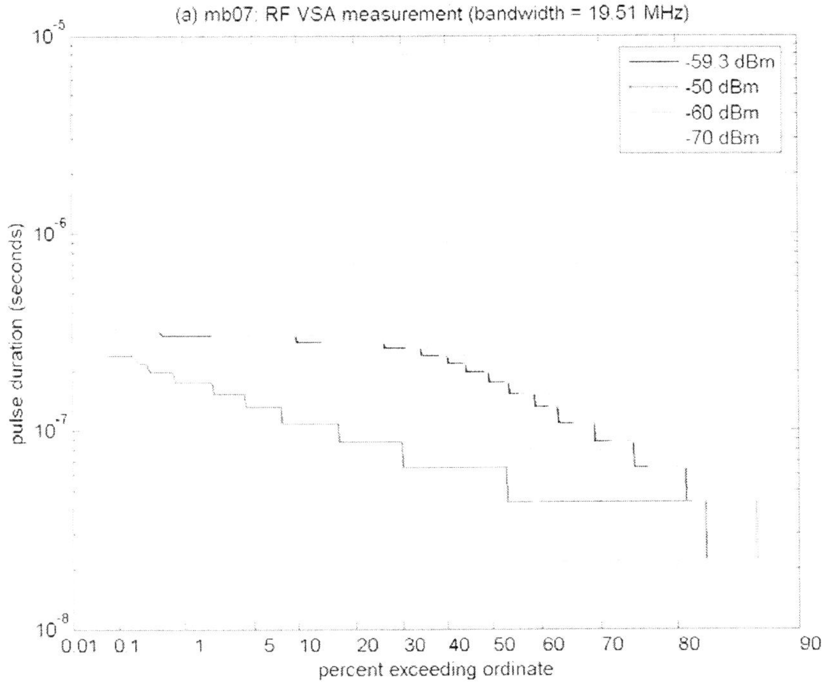

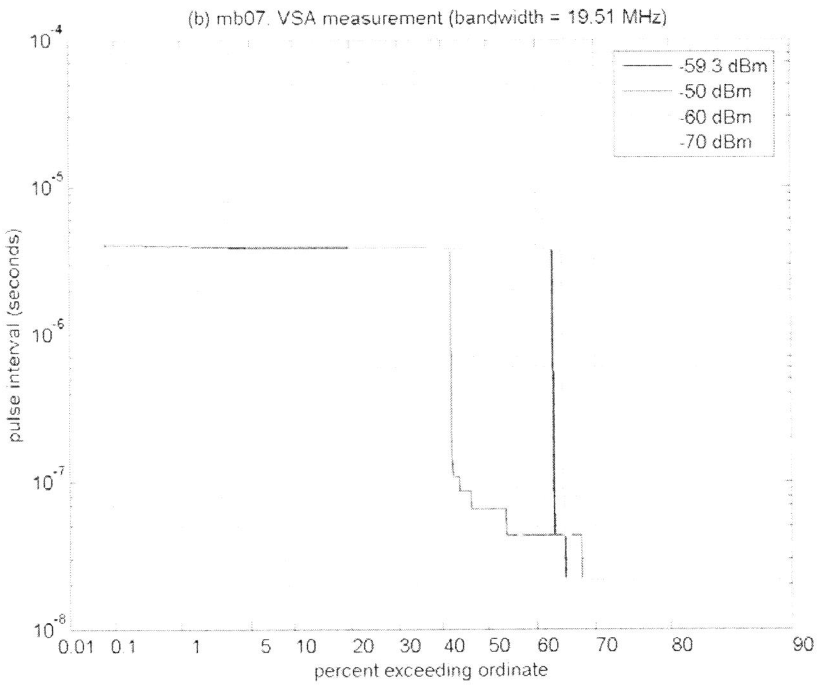

Figure B-22. RF temporal analyses of MB-07 (b = 13, d = 1).

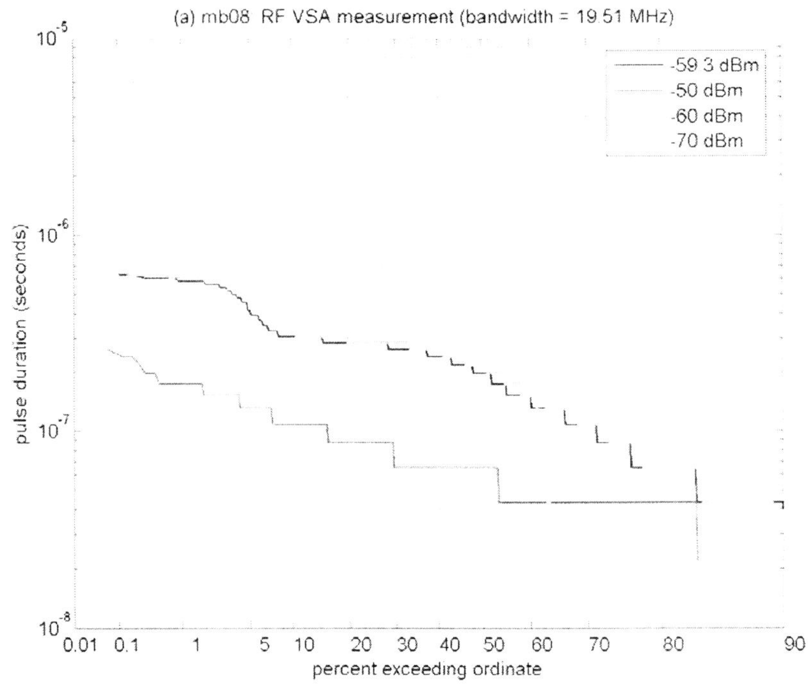

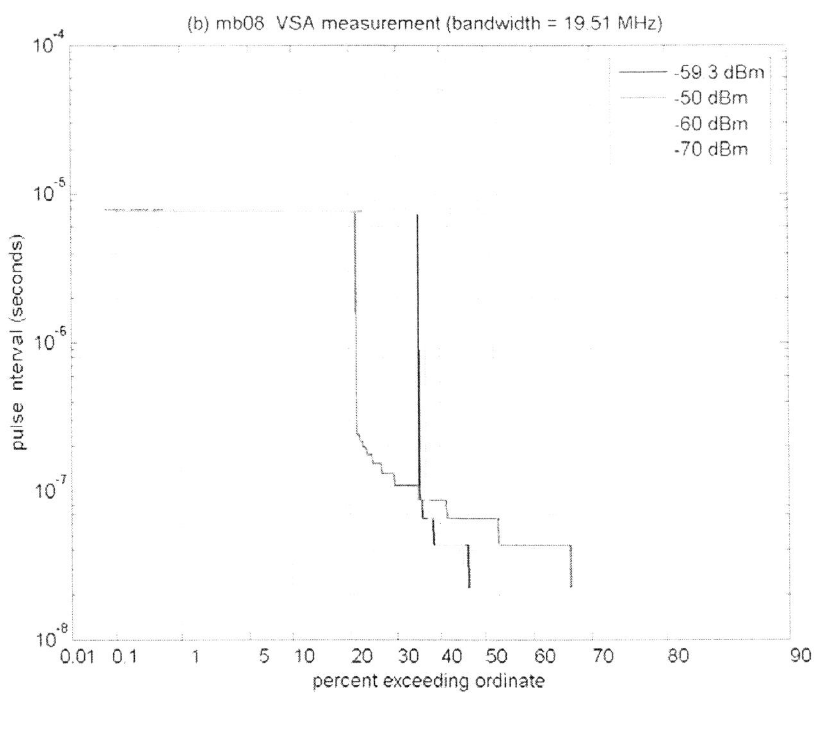

Figure B-23. RF temporal analyses of MB-08 (b = 13, d = 2).

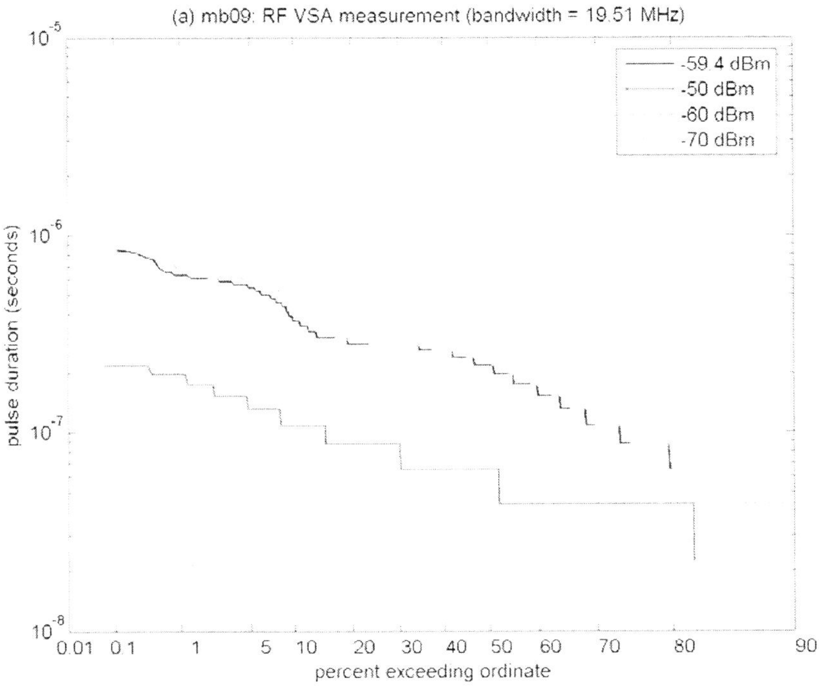

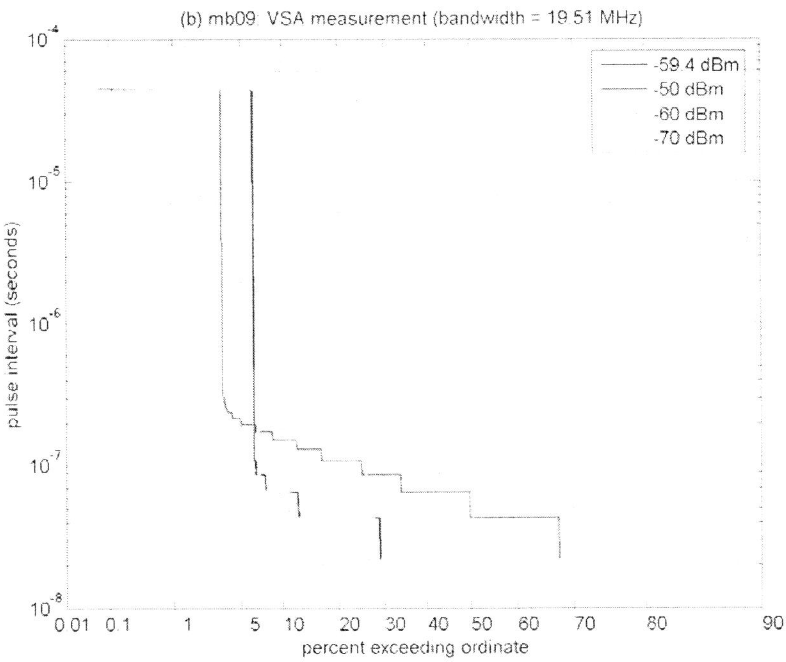

Figure B-24. RF temporal analyses of MB-09 (b = 13, d = 12).

B.3. Amplitude Analysis

This subsection provides measured time series and amplitude probability distributions (APDs) of the UWB signals. Odd numbered plots in figures B-25 – B-62 provide RF measurement results organized as:

a. Real time series measured with the DO and amplitude time series measured with the VSA and band-limited to B_{DTV} = 19.51 MHz.
b. APDs measured with the VSA and band-limited to 36, B_{DTV}, 10, 1, and 0.1 MHz.

Even numbered plots in figures B-25 – B-62 provide IF measurement results organized as:

a. Amplitude time series measured with the VSA and band-limited to B_{DTV}.
b. APDs measured with the VSA and band-limited to B_{DTV}.

Table B-3 provides peak-to-average (P/A) values for the UWB signals measured at RF and band-limited to various bandwidths in post-measurement processing. Table B-4 provides P/A values for the UWB signals measured at IF under TOV operational conditions and band-limited to B_{DTV} in post-measurement processing.

Table B-3. Measured RF P/A (dB) of UWB Signals Band-Limited to Different Bandwidths‡

Type	Index	36.0 MHz	B_{DTV}	10.0 MHz	1.0 MHz	0.1 MHz
DP	01	25.4	23.5	20.6	11.5	6.8
	02	15.7	13.8	11.0	6.1	9.4
	03	7.8	5.4	6.0	9.0	10.1
	04	7.7	8.1	9.0	9.4	9.4
DS	01	9.3	9.5	9.5	9.5	9.5
	02	8.8	9.0	9.5	9.8	9.6
	03	7.8	8.1	8.8	9.9	9.7
	04	6.8	6.6	8.0	9.6	9.6
	05	6.8	6.6	8.0	9.8	9.8
	06	6.6	6.8	8.3	9.5	9.8
MB	01	9.6	8.3	6.9	7.5	9.2
	02	13.5	12.8	11.1	5.6	8.6
	03	13.8	12.8	11.3	7.4	8.7
	04	16.9	16.2	14.7	6.4	8.1
	05	17.1	16.4	14.8	10.8	8.2
	06	17.2	16.3	15.0	14.8	10.0
	07	19.4	18.7	17.3	8.6	7.4
	08	19.4	18.9	17.3	13.3	7.6
	09	19.4	18.8	17.4	17.7	14.9

‡ Shaded cells highlight Rayleigh-amplitude statistics.

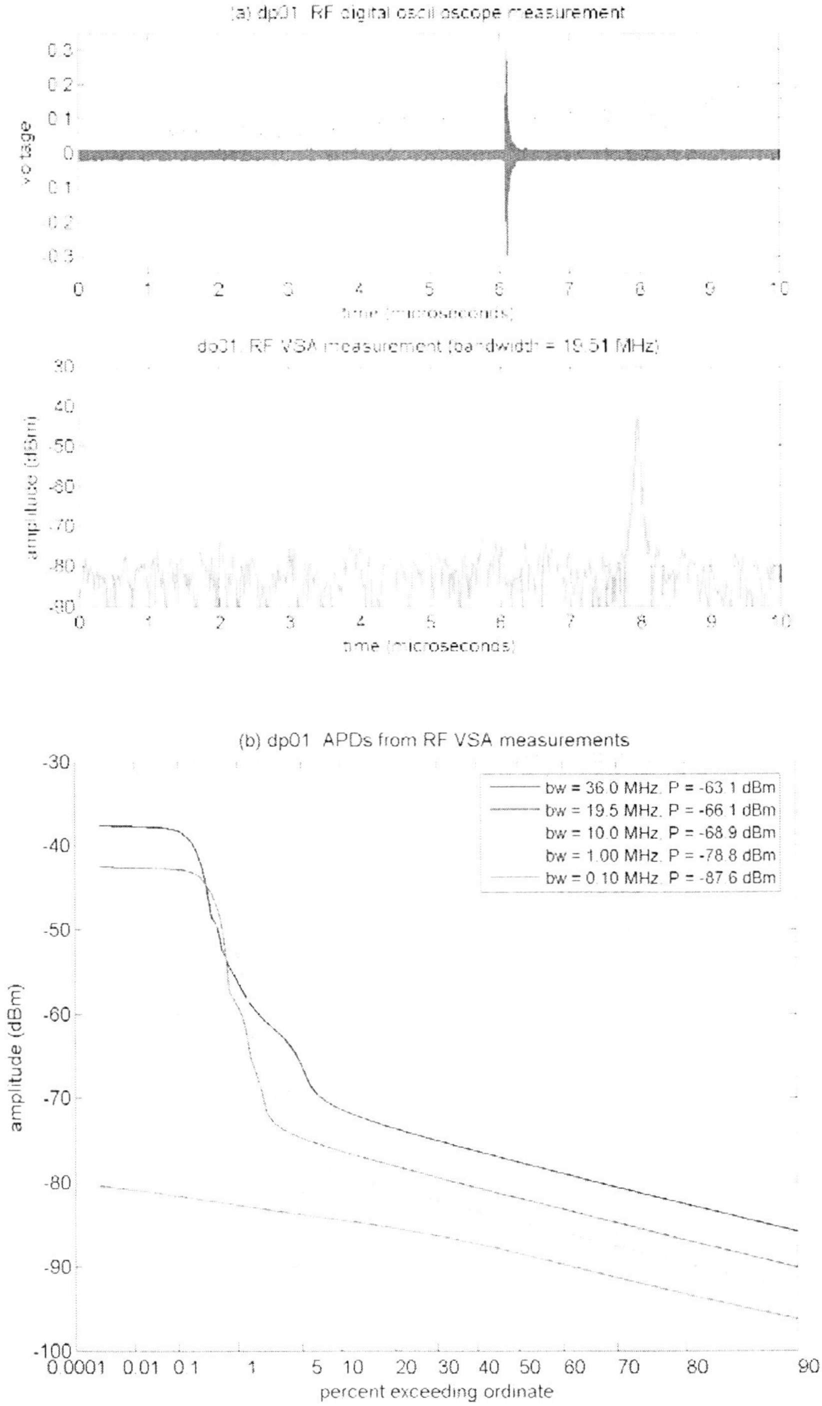

Figure B-25. RF amplitude analyses of DP-01 (T_{pulse} = 10,000 ns, w = 0.094 ns, f_D = 0.5).

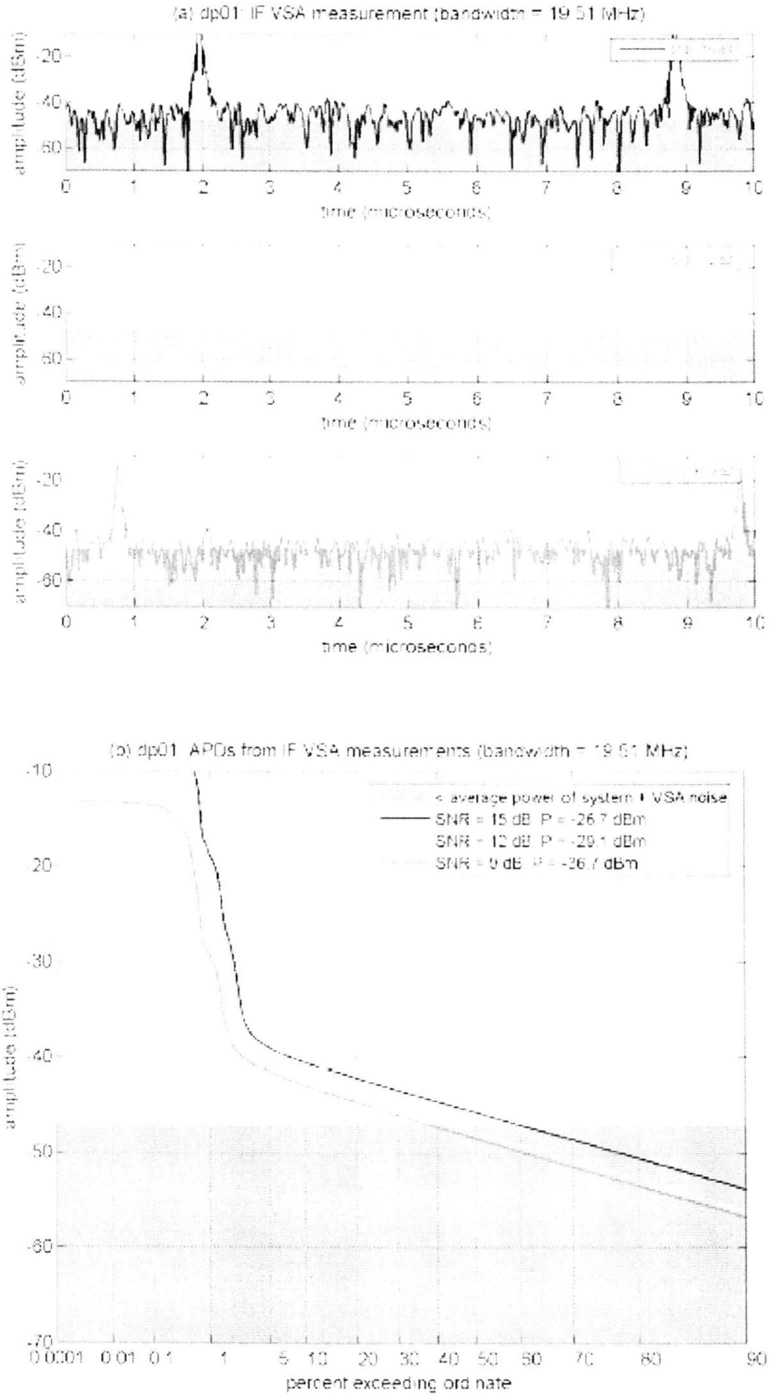

Figure B-26. IF amplitude analyses of DP-01 at INR_{TOV}.

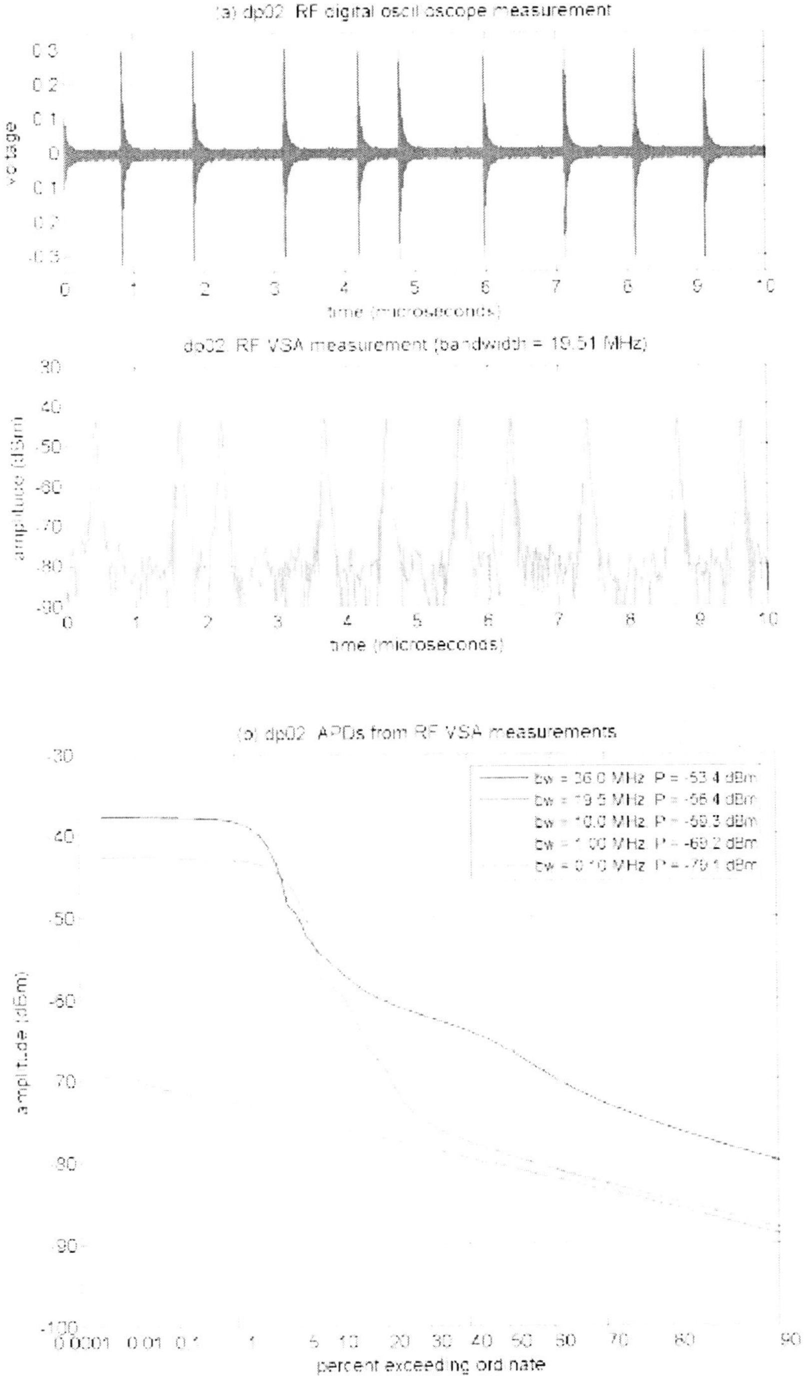

Figure B-27. RF amplitude analyses of DP-02 (T_{pulse} = 1000 ns, w = 0.094 ns, f_D = 0.5).

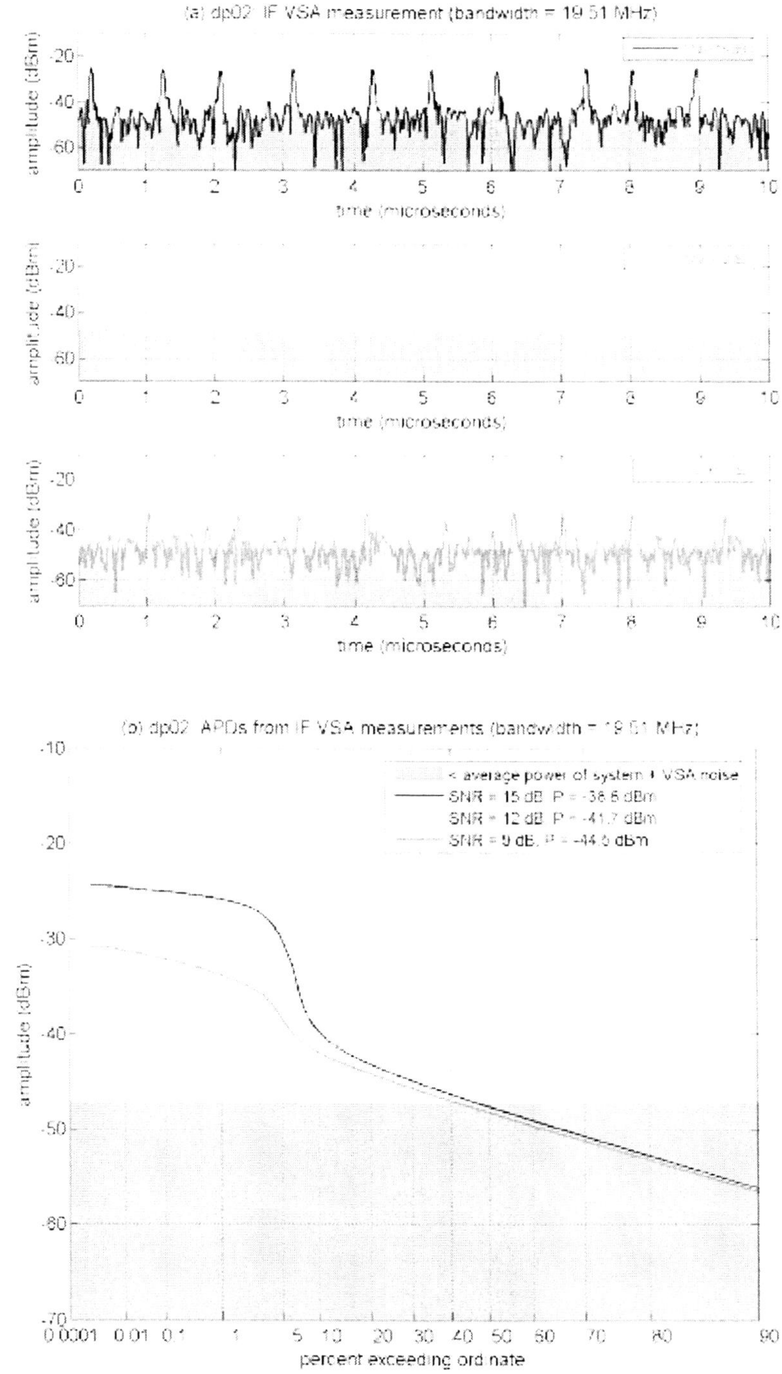

Figure B-28. IF amplitude analyses of DP-02 at INR_{TOV}.

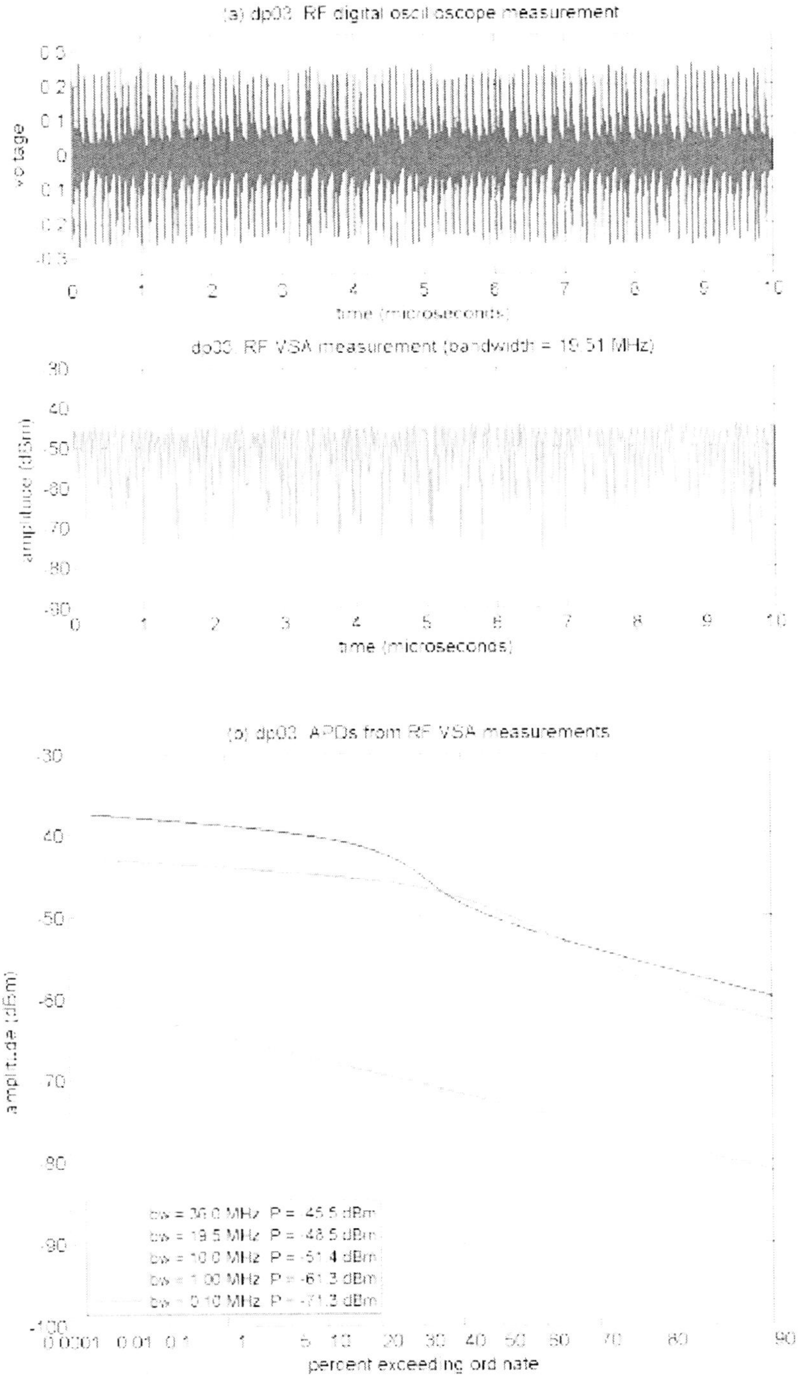

Figure B-29. RF amplitude analyses of DP-03 (T_{pulse} = 100 ns, w = 0.094 ns, f_D = 0.5).

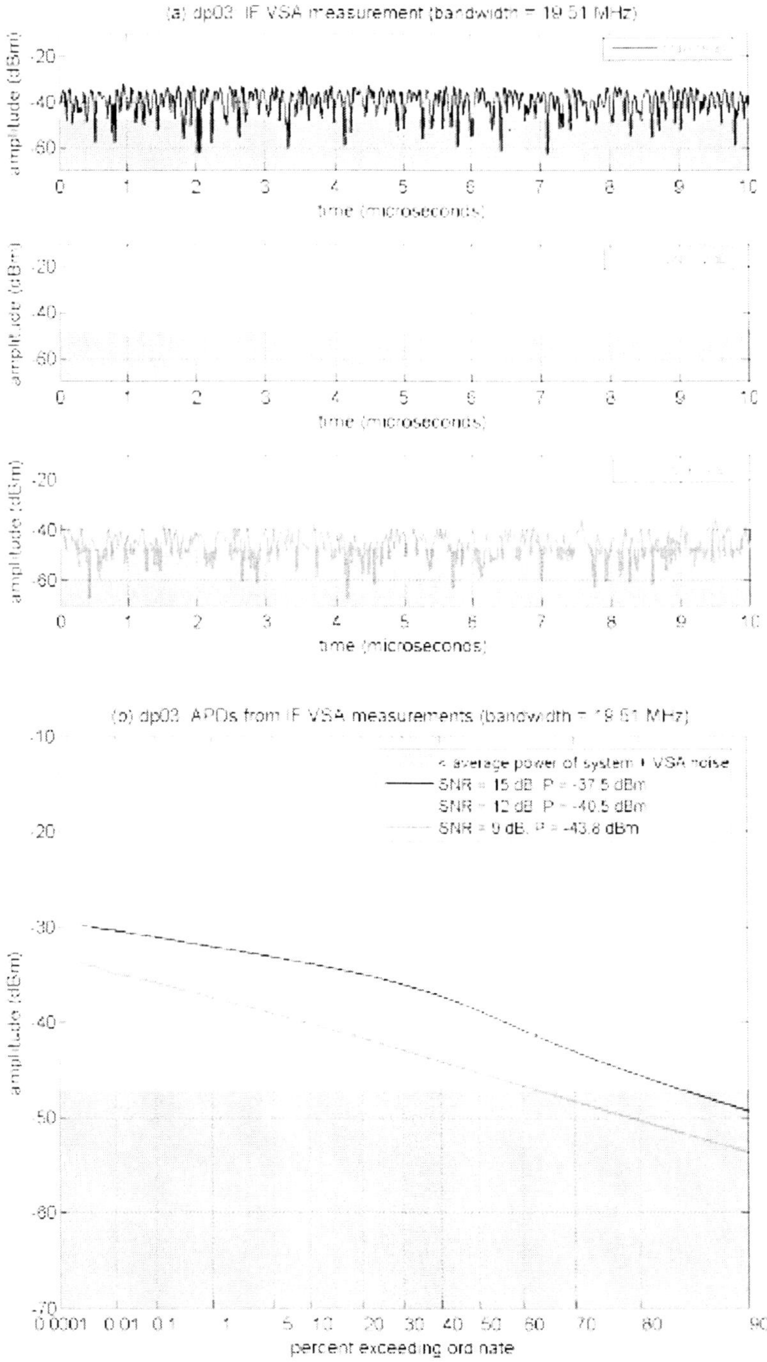

Figure B-30. IF amplitude analyses of DP-03 at INR_{TOV}.

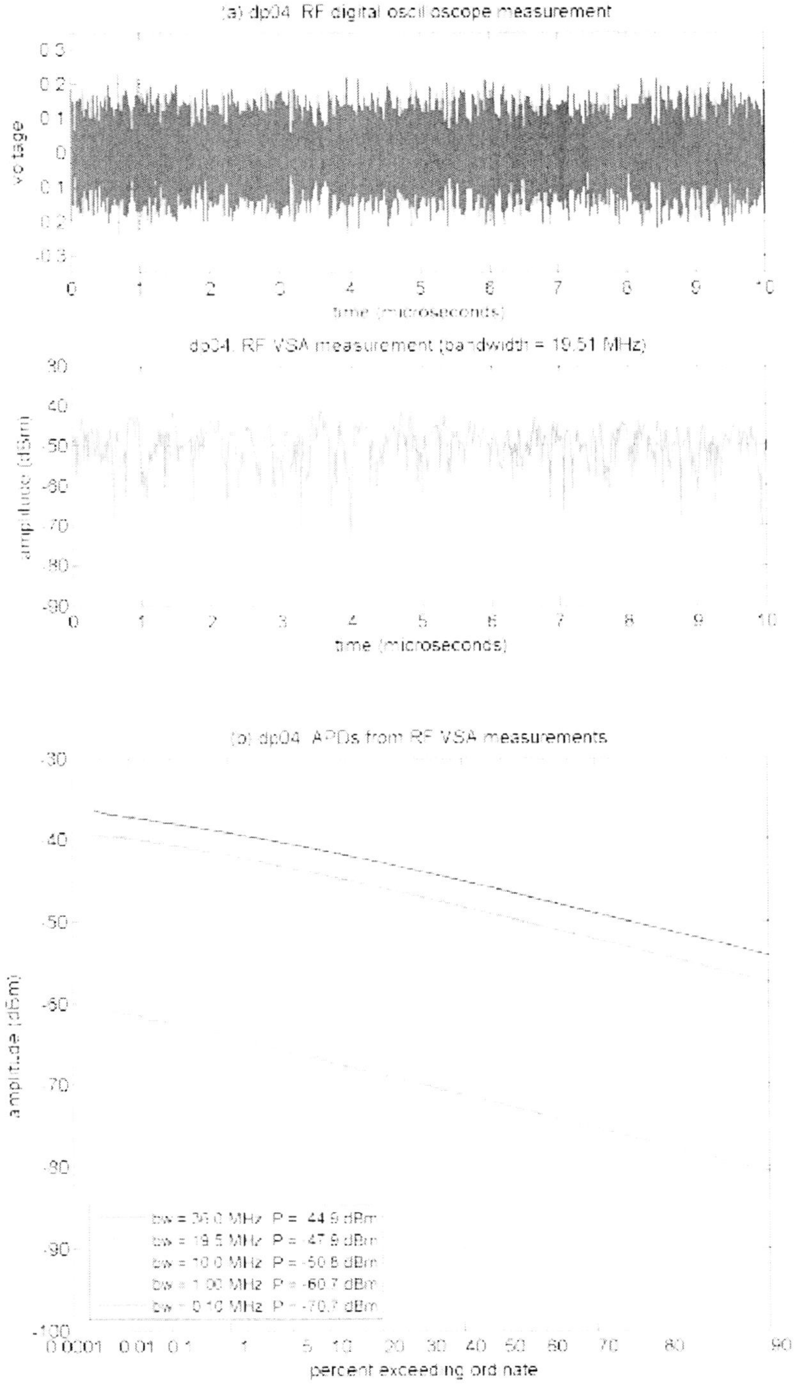

Figure B-31. RF amplitude analyses of DP-04 (T_{pulse} = 10 ns, w = 0.094 ns, f_D = 0.5).

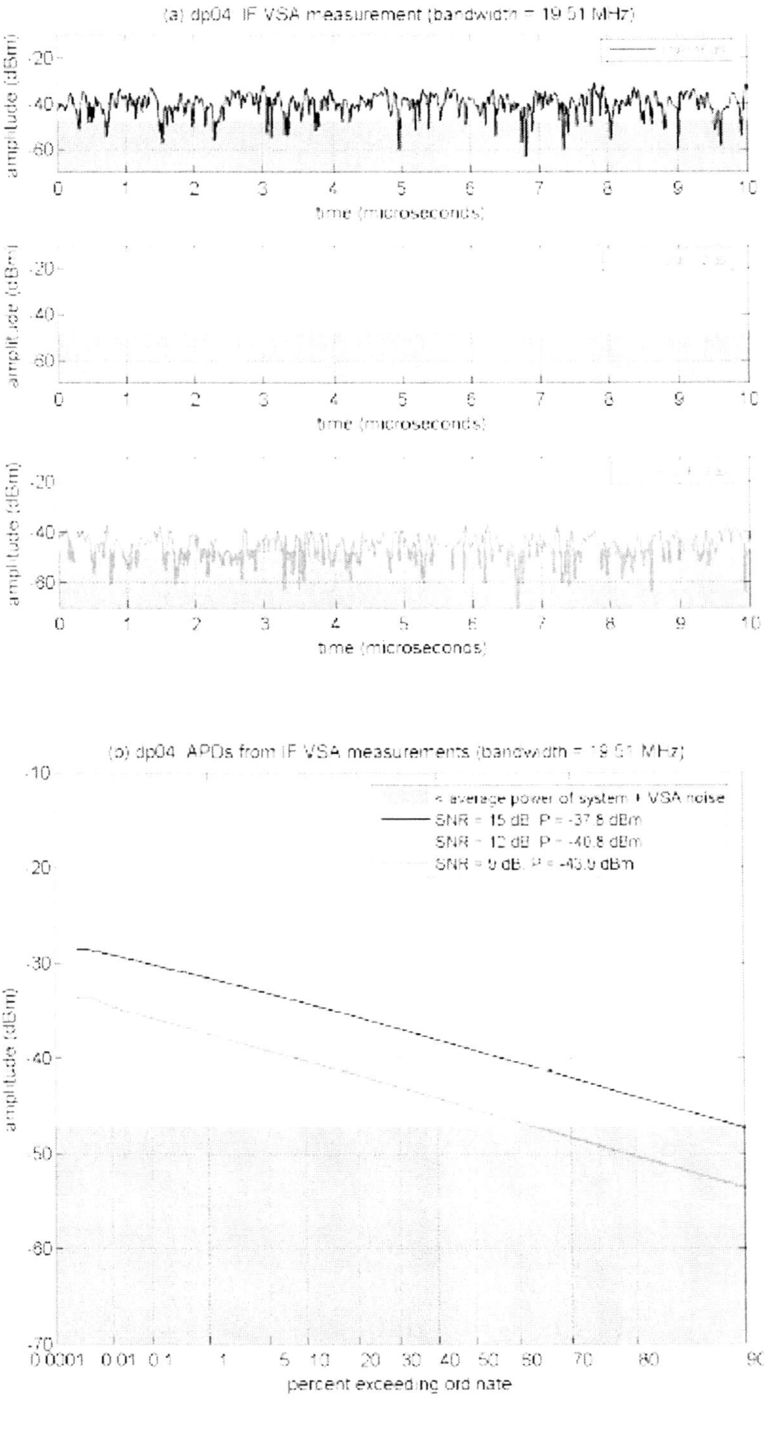

Figure B-32. IF amplitude analyses of DP-04 at INR_{TOV}.

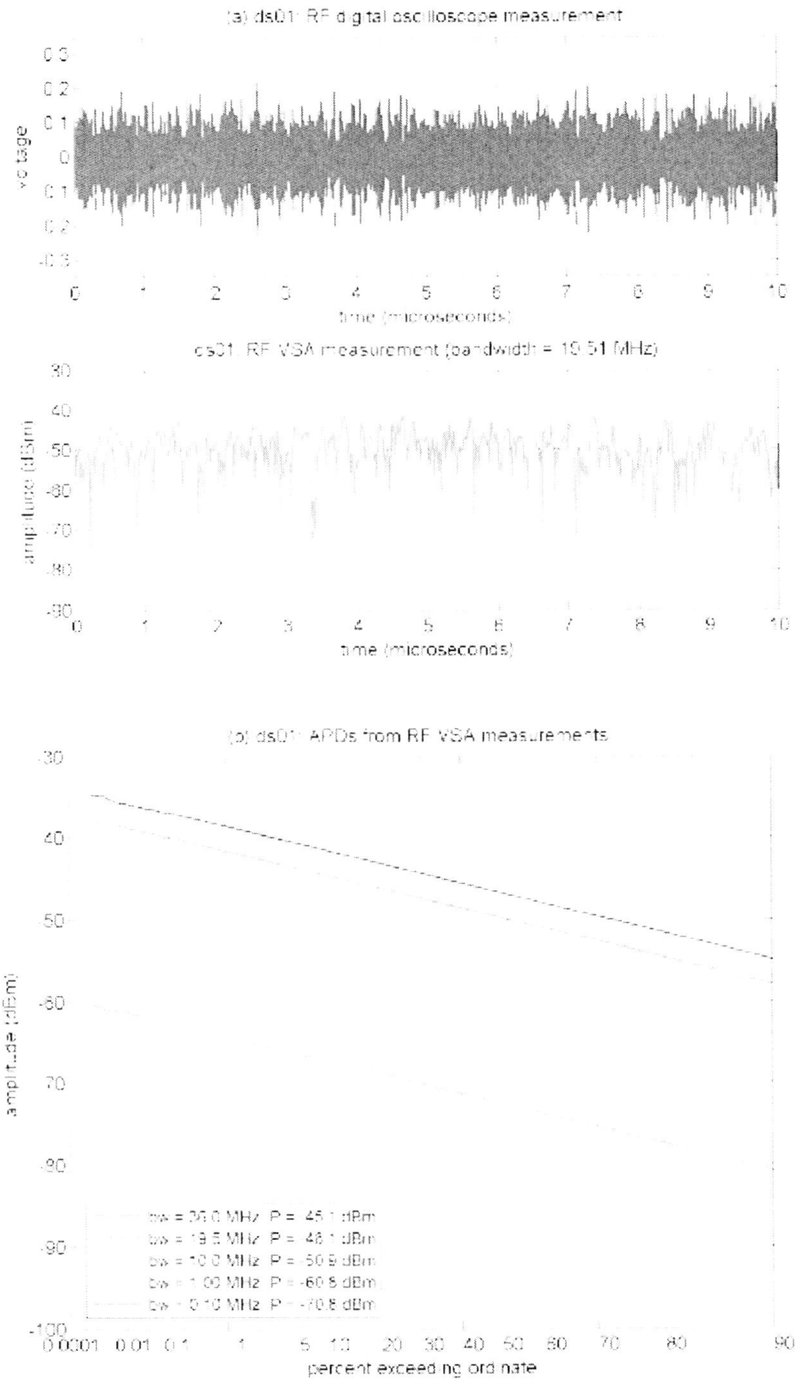

Figure B-33. RF amplitude analyses of DS-01 (L = 1).

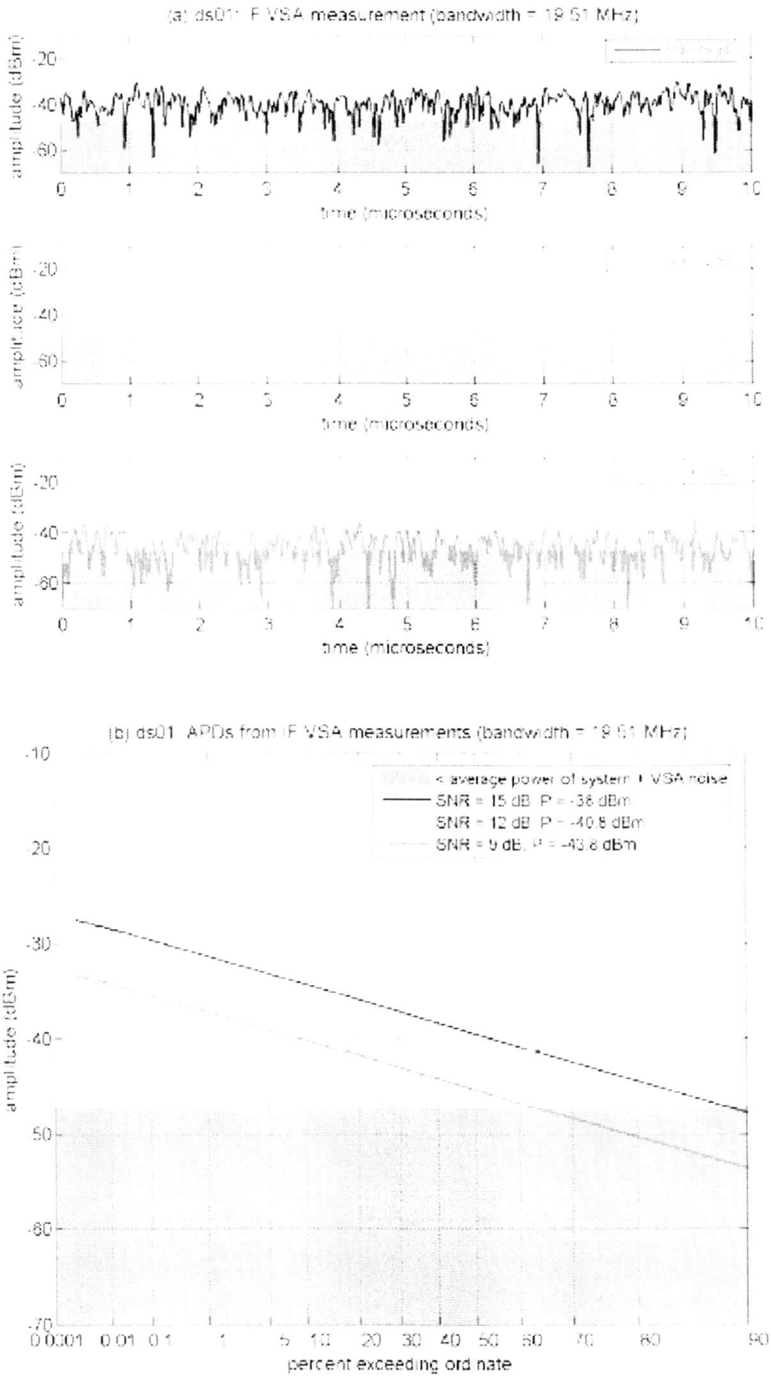

Figure B-34. IF amplitude analyses of DS-01 at INR_{TOV}.

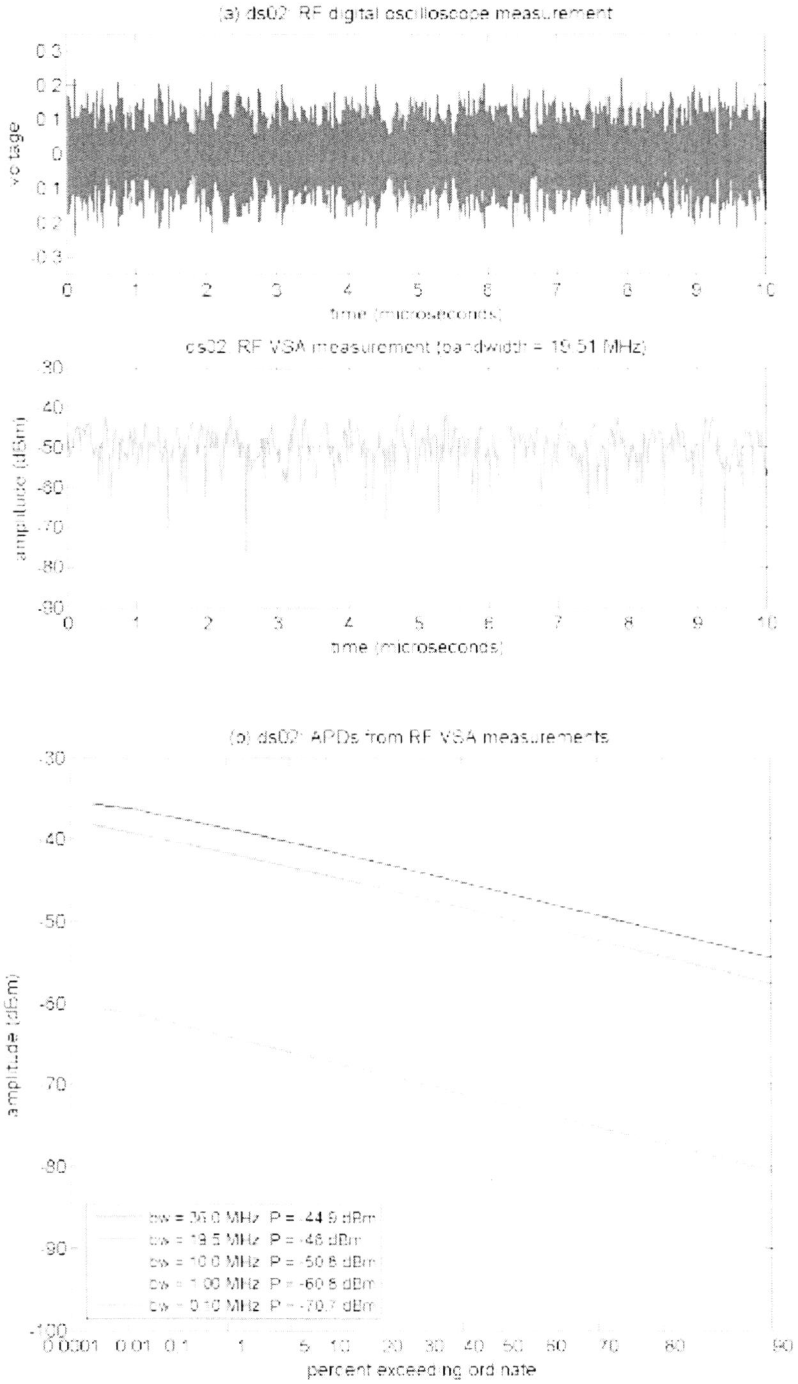

Figure B-35. RF amplitude analyses of DS-02 (L = 3, sparse code).

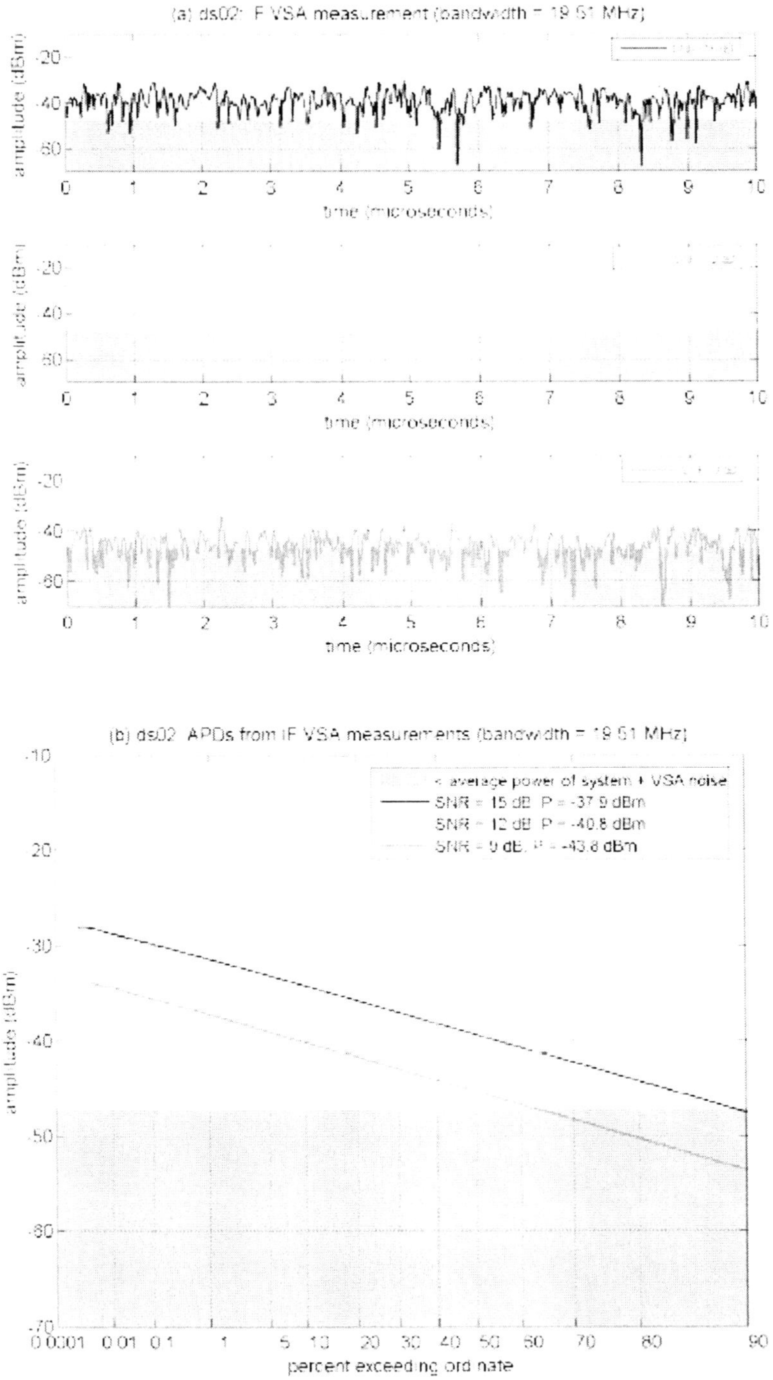

Figure B-36. IF amplitude analyses of DS-02 at INR_{TOV}.

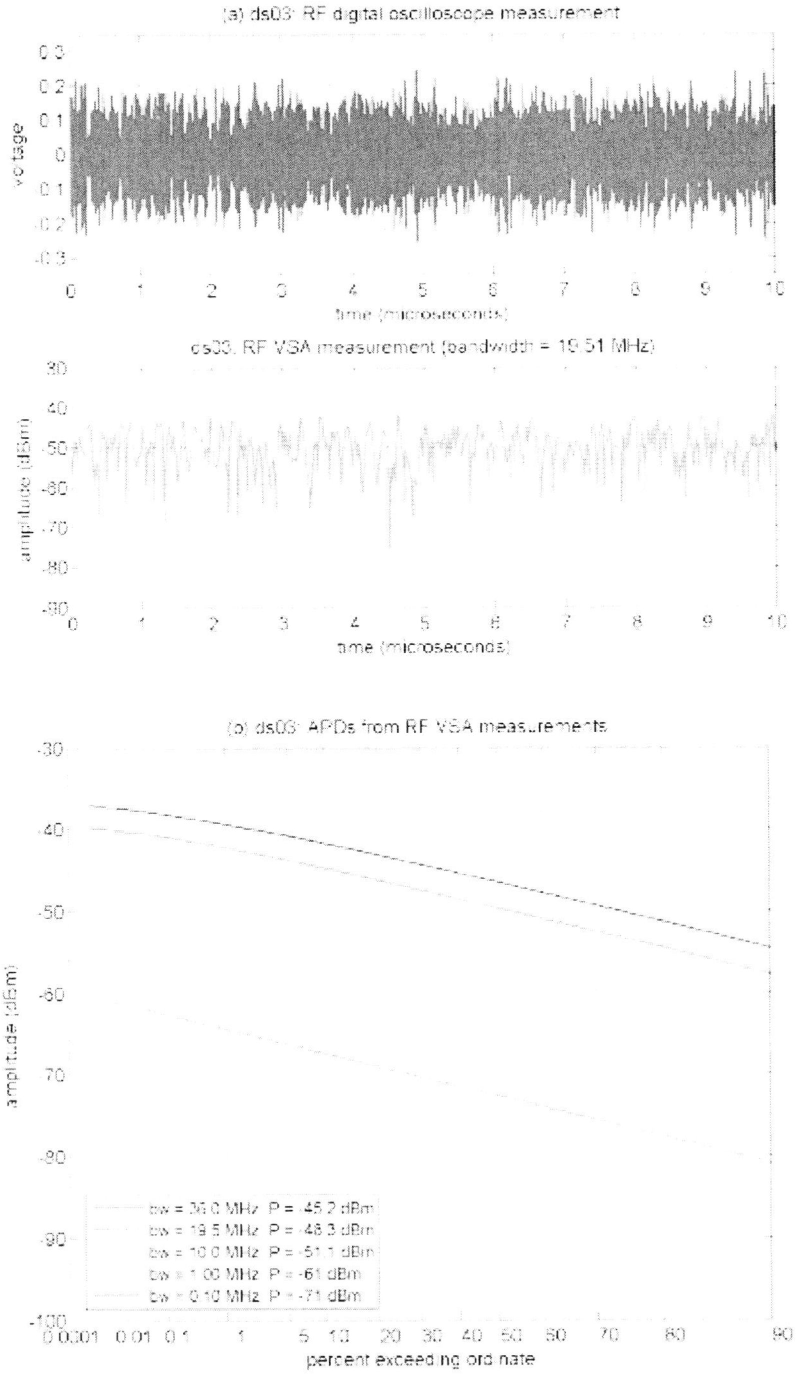

Figure B-37. RF amplitude analyses of DS-03 (L = 6, sparse code).

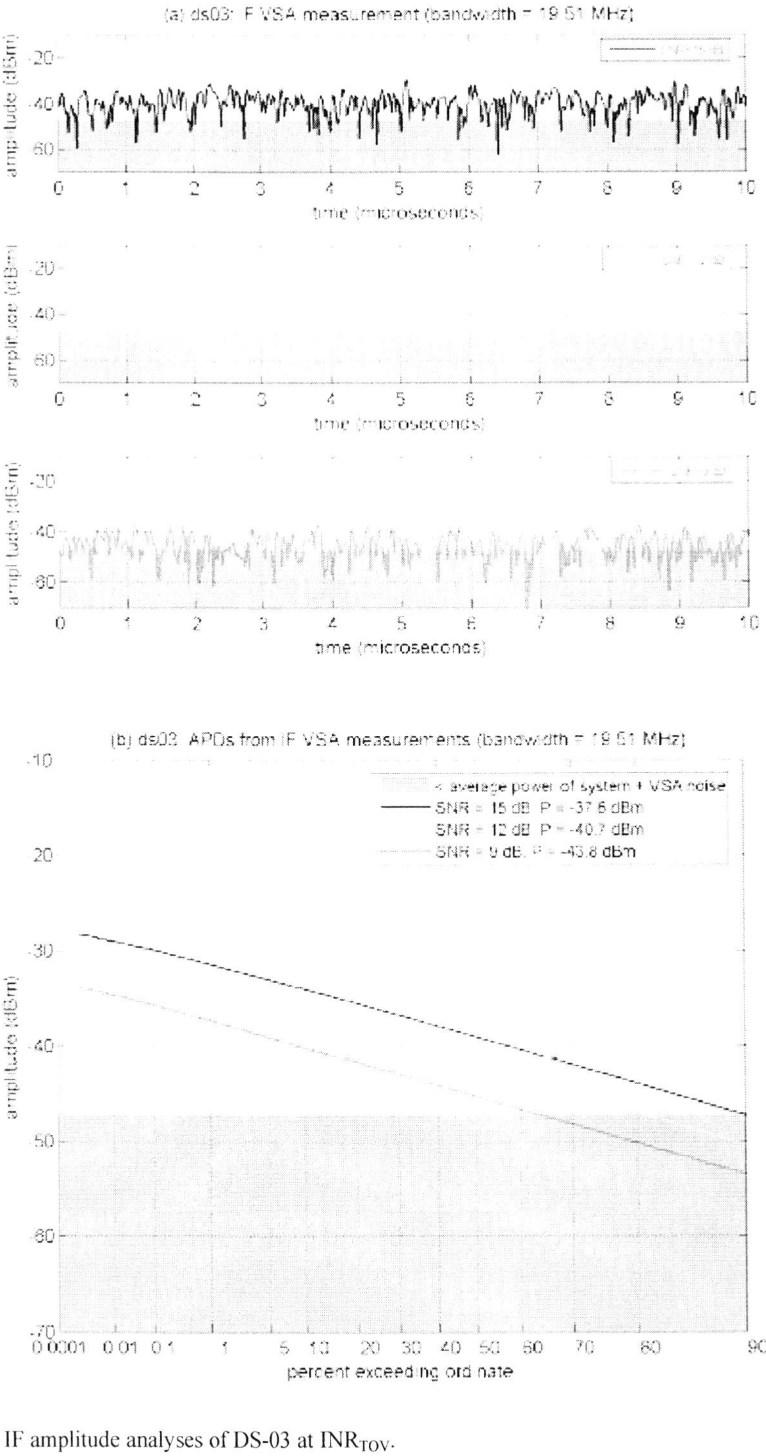

Figure B-38. IF amplitude analyses of DS-03 at INR_{TOV}.

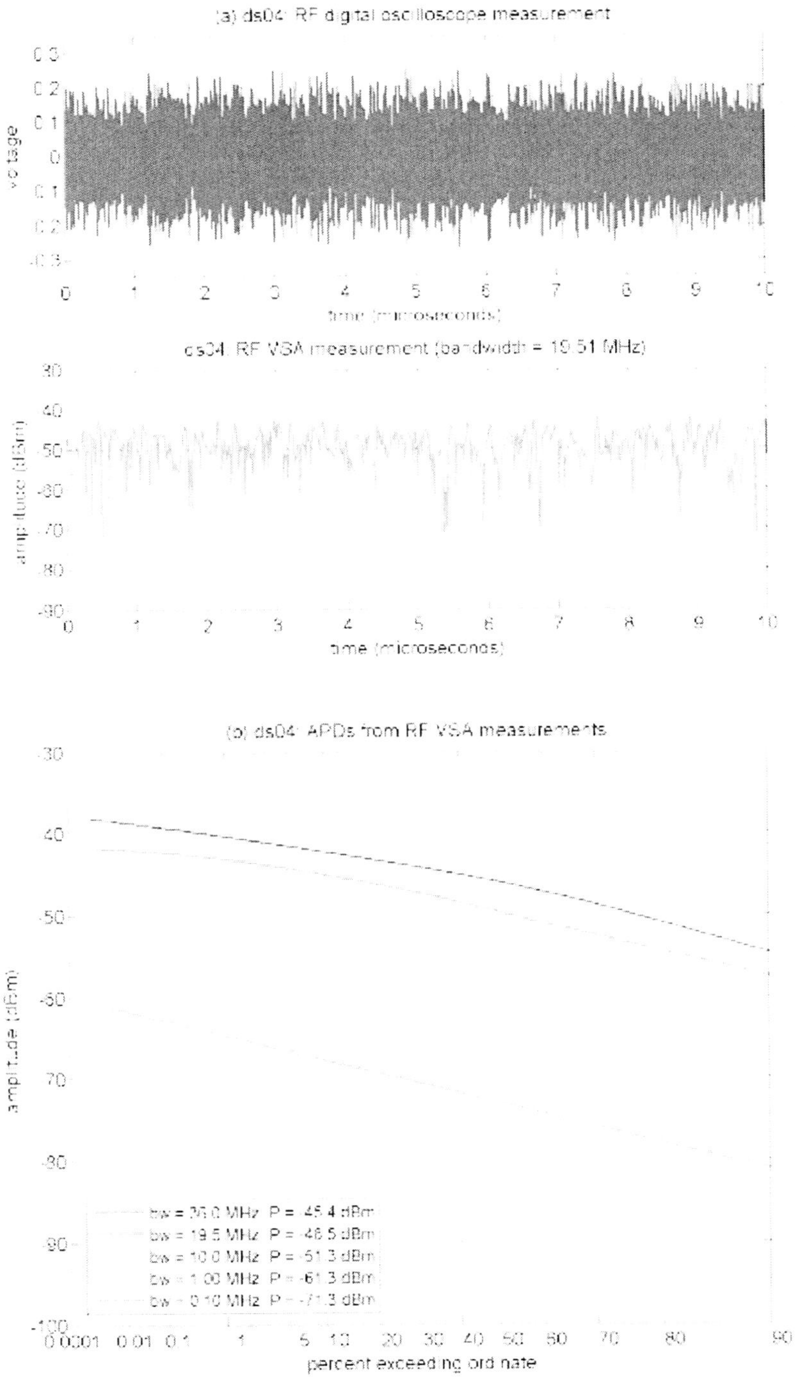

Figure B-39. RF amplitude analyses of DS-04 (L = 12, DS code).

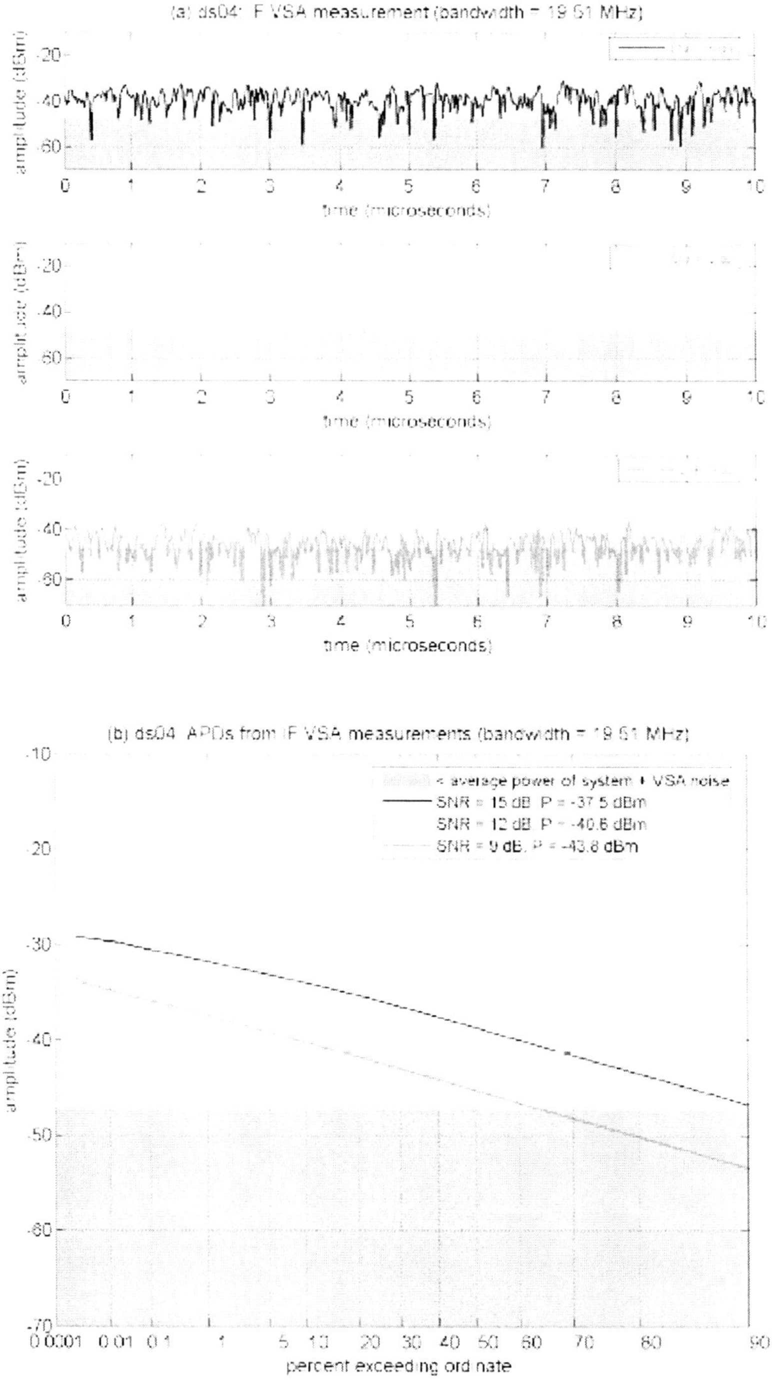

Figure B-40. IF amplitude analyses of DS-04 at INR_{TOV}.

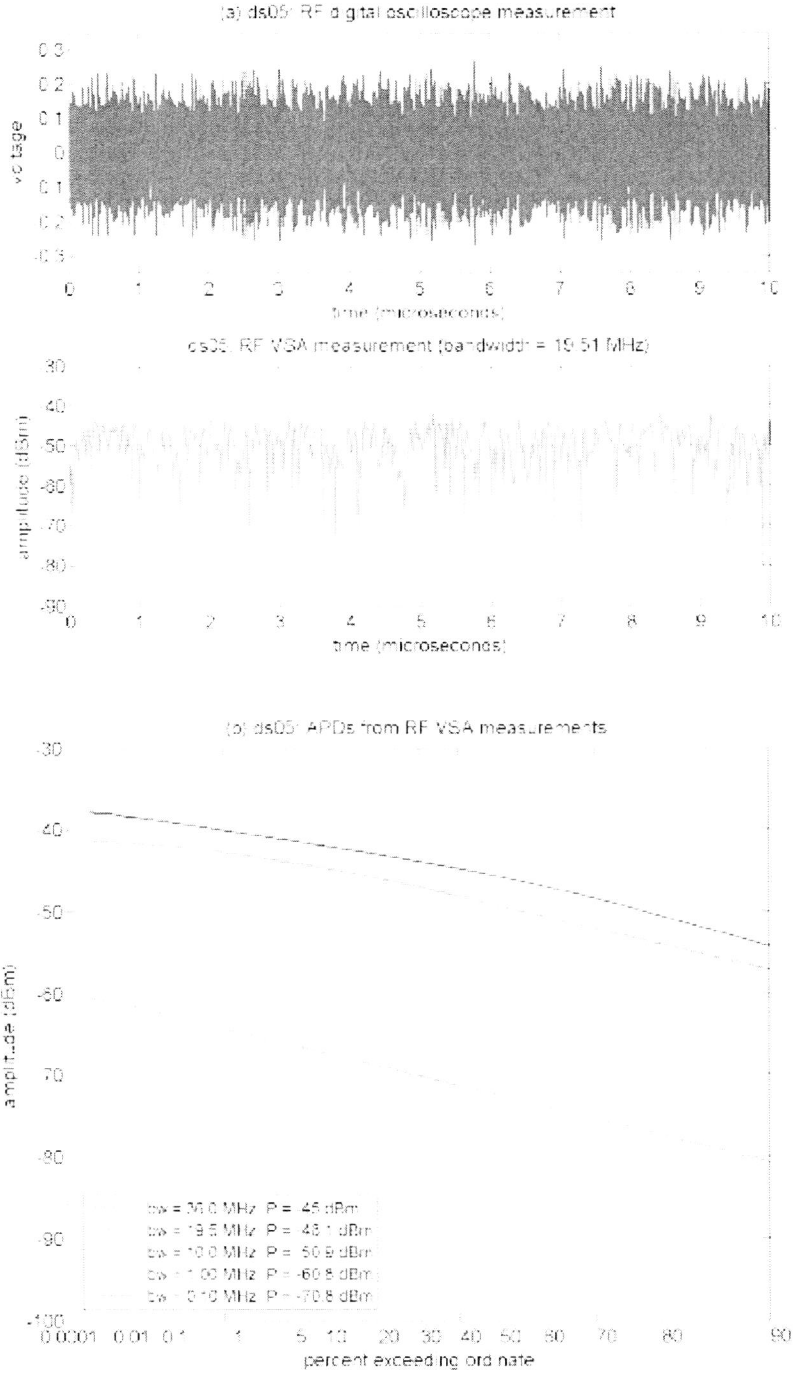

Figure B-41. RF amplitude analyses of DS-05 (L = 12, sparse code).

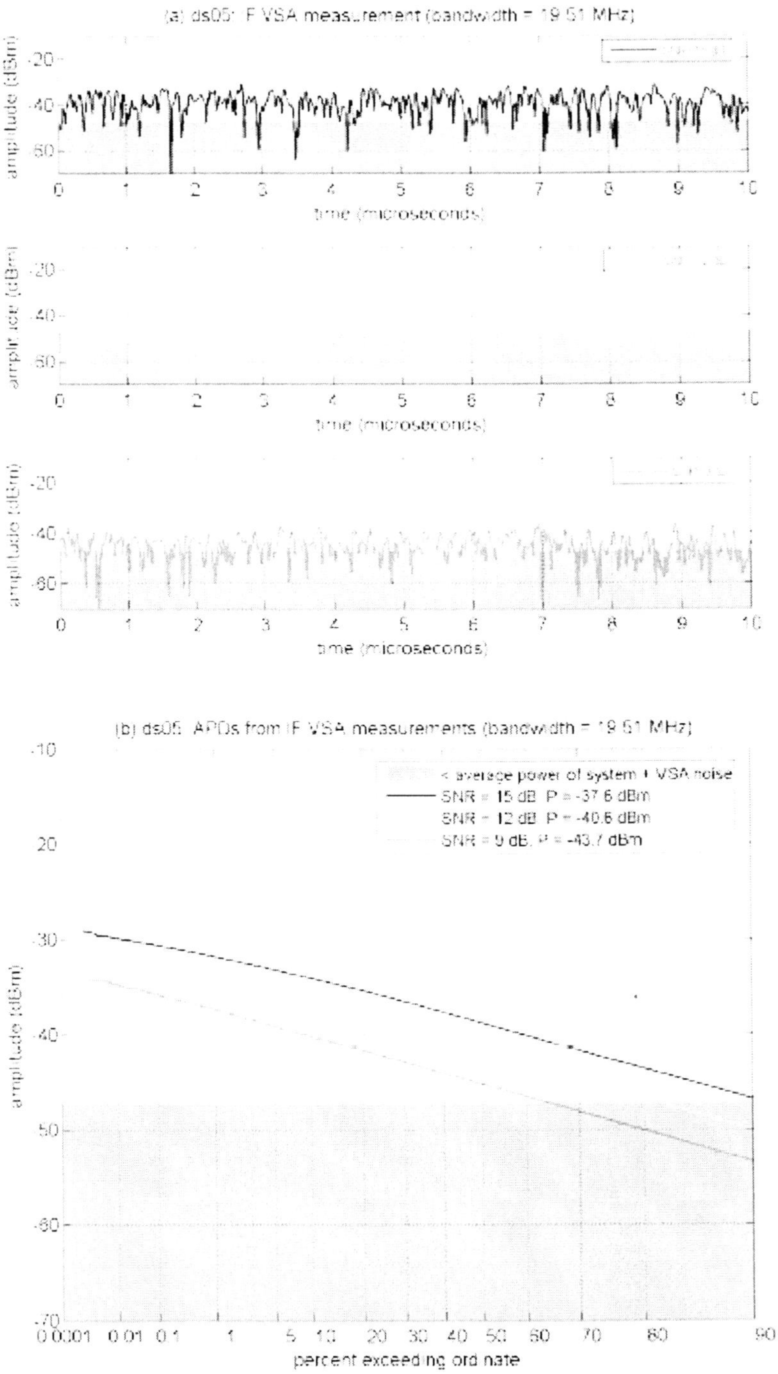

Figure B-42. IF amplitude analyses of DS-05 at INR_{TOV}.

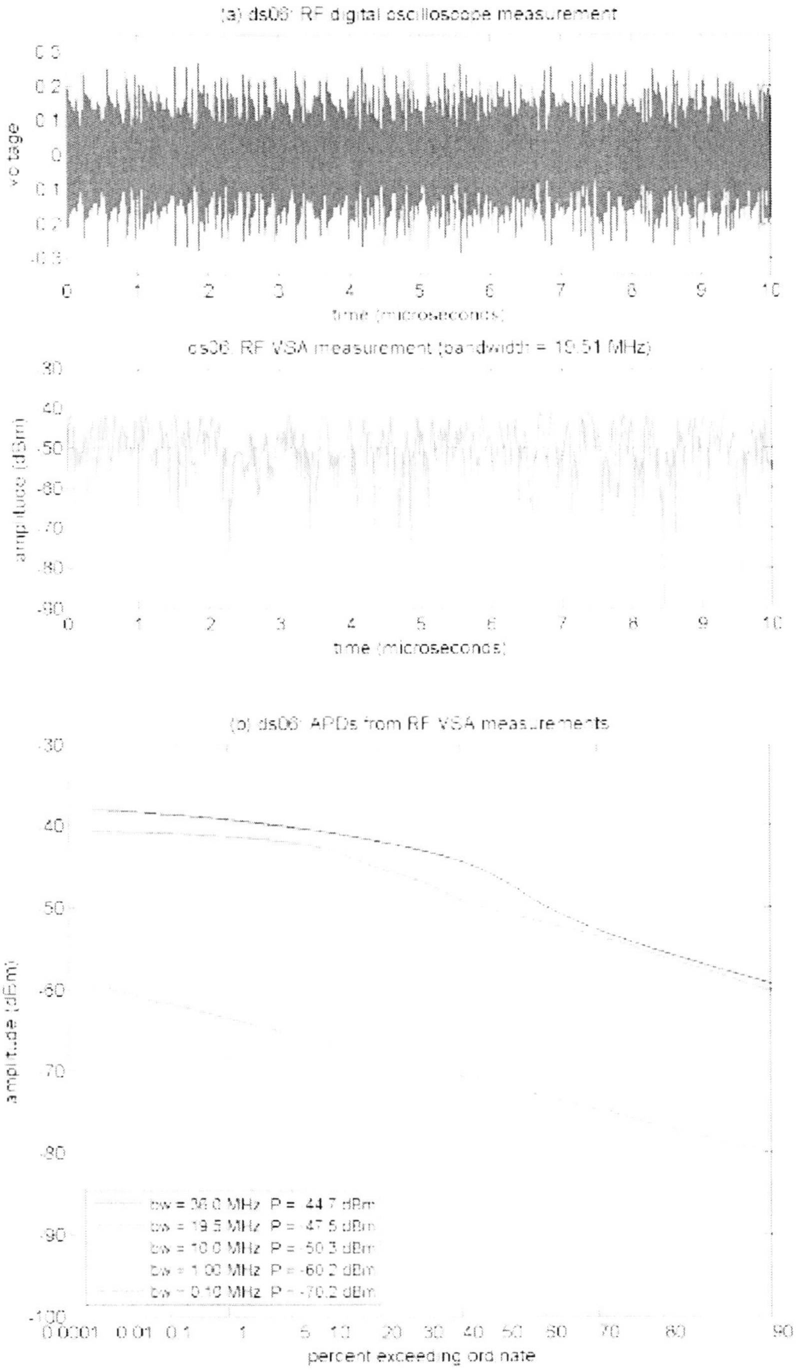

Figure B-43. RF amplitude analyses of DS-06 (L = 24, DS code).

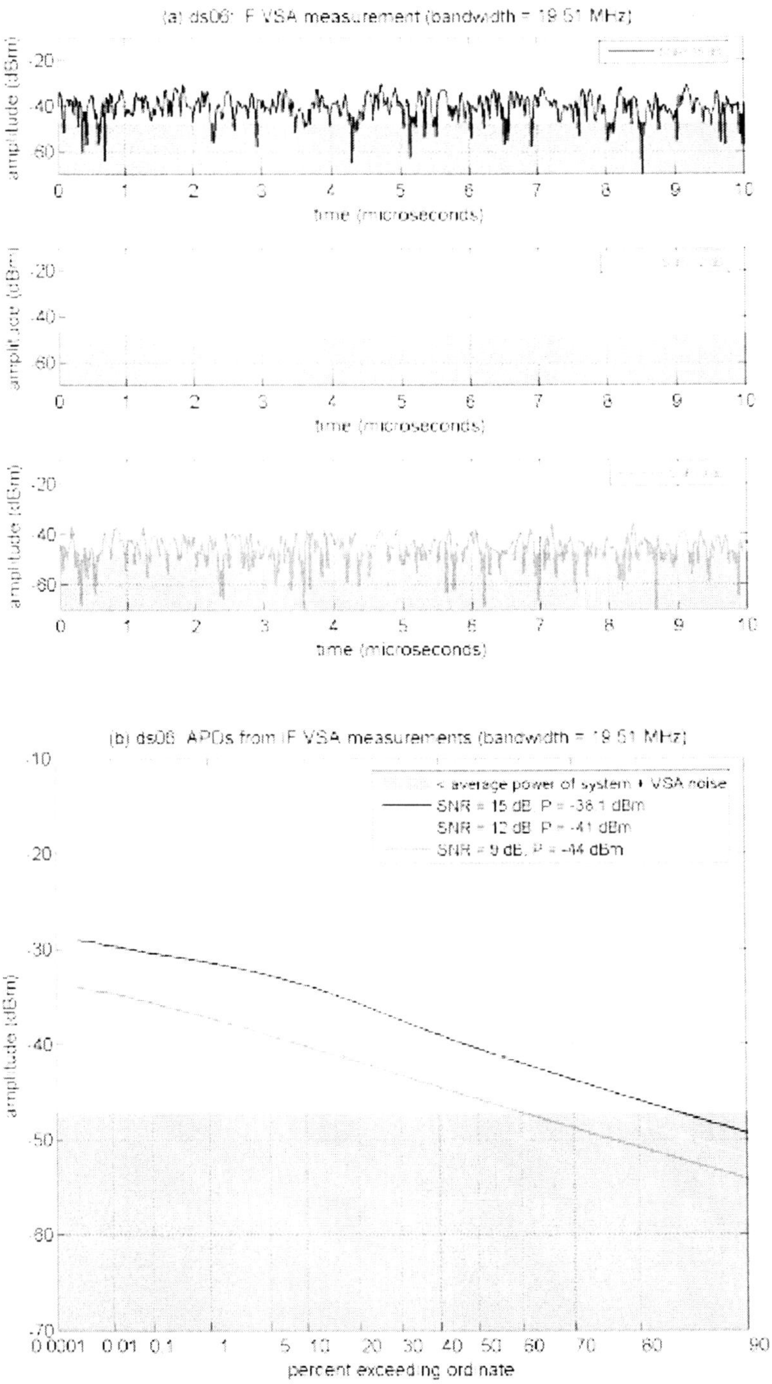

Figure B-44. IF amplitude analyses of DS-06 at INR_{TOV}.

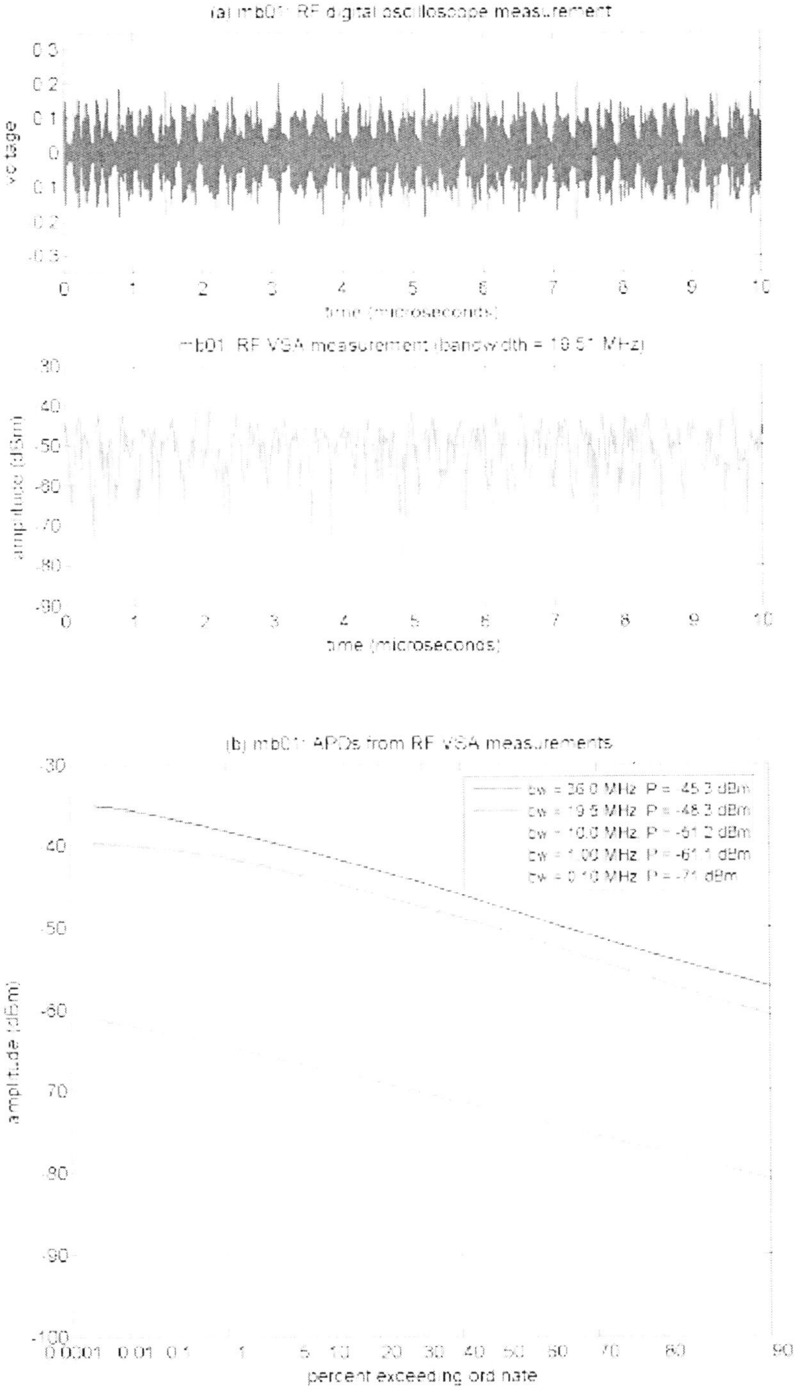

Figure B-45. RF amplitude analyses of MB-01 (b = 1).

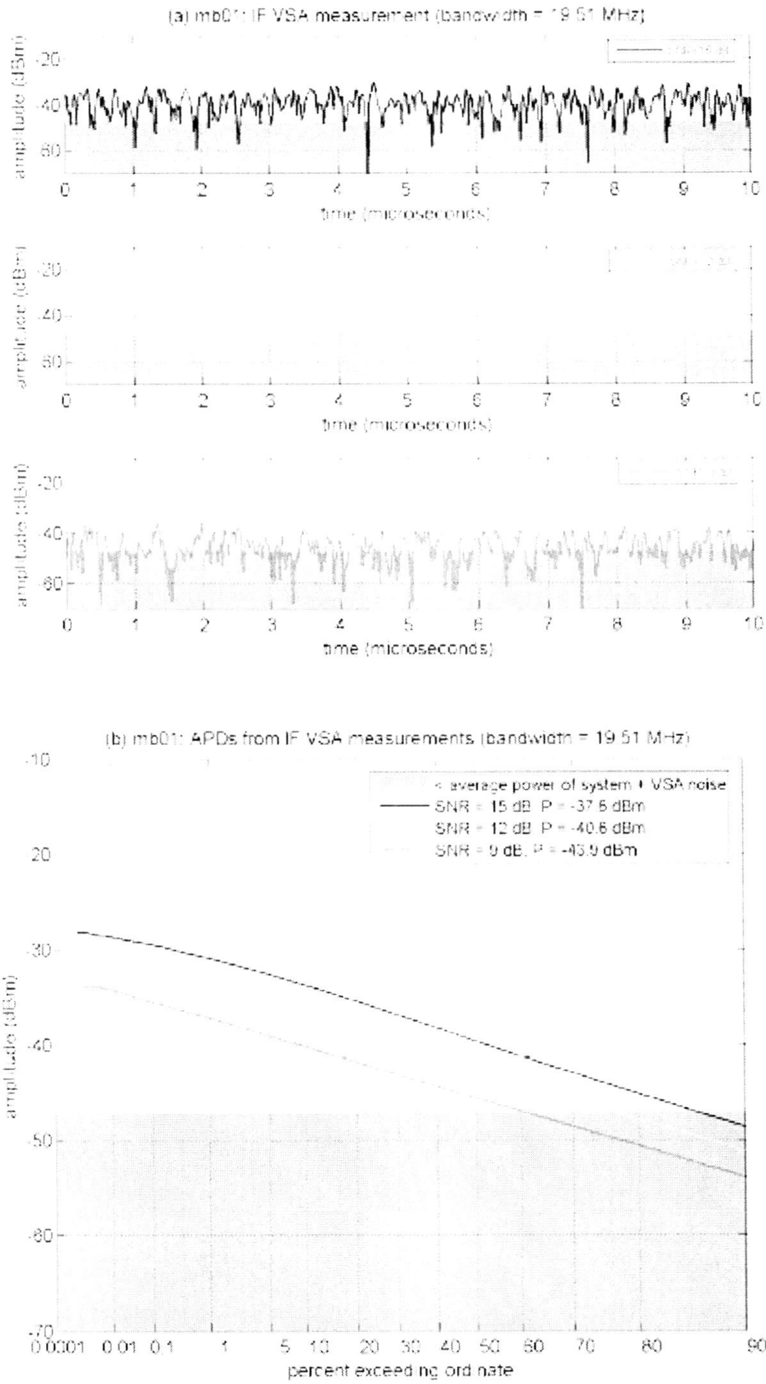

Figure B-46. IF amplitude analyses of MB-01 at INR_{TOV}.

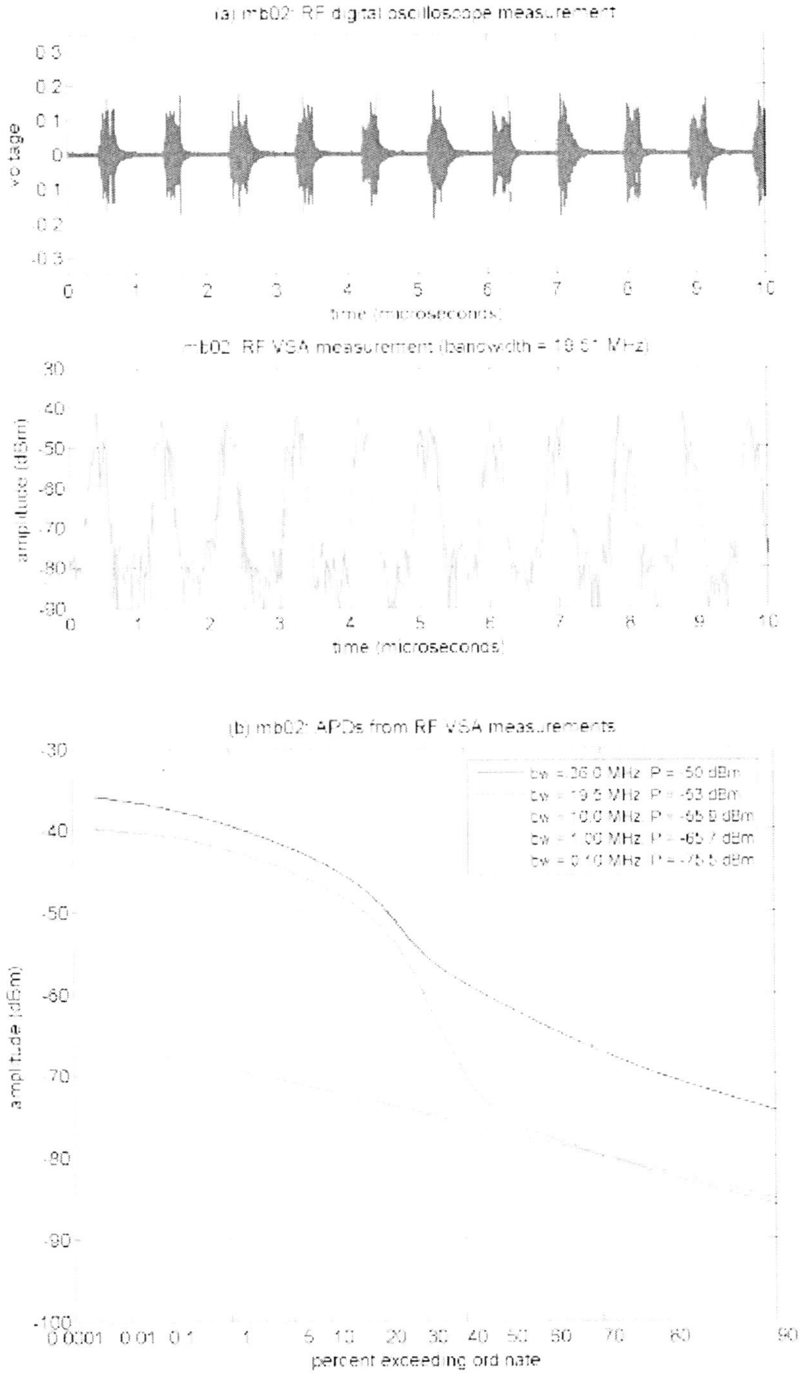

Figure B-47. RF amplitude analyses of MB-02 (b = 3, d = 1).

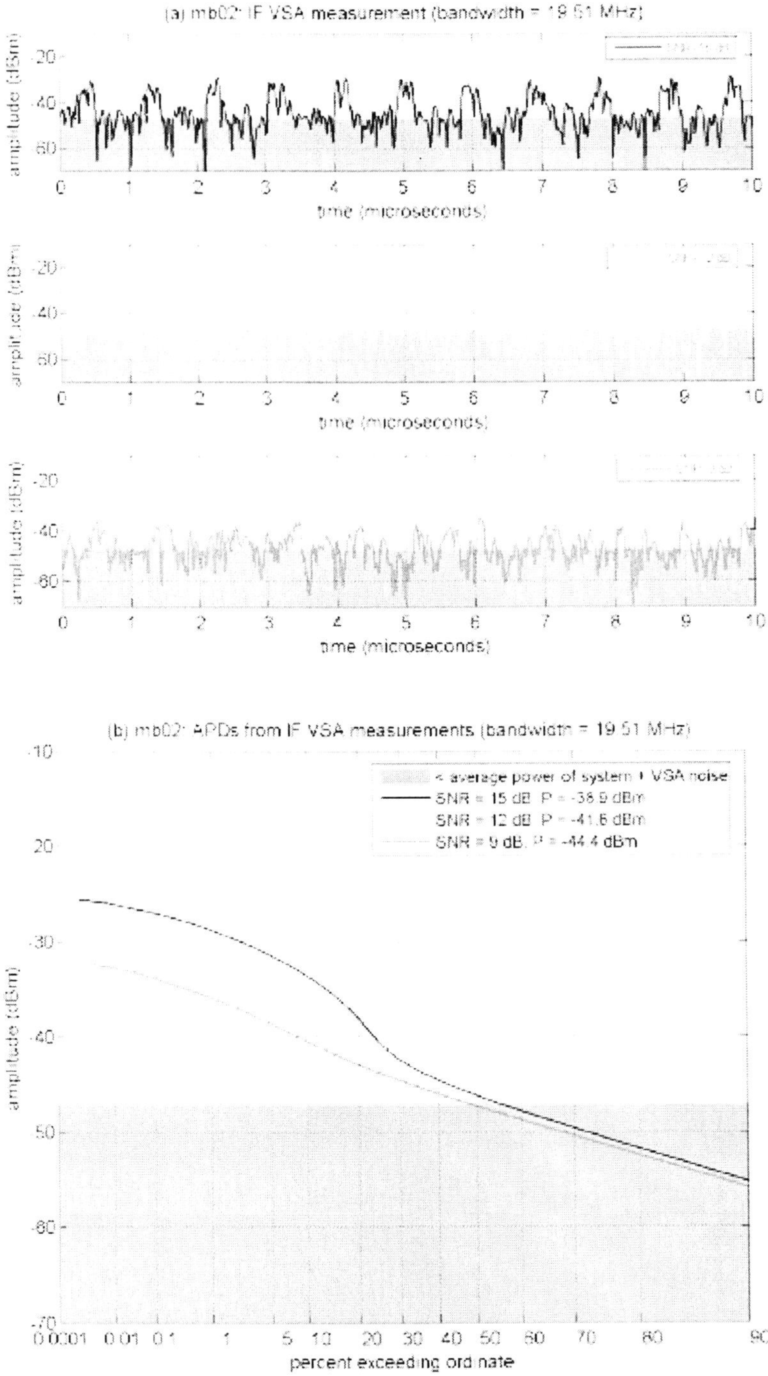

Figure B-48. IF amplitude analyses of MB-02 at INR_{TOV}.

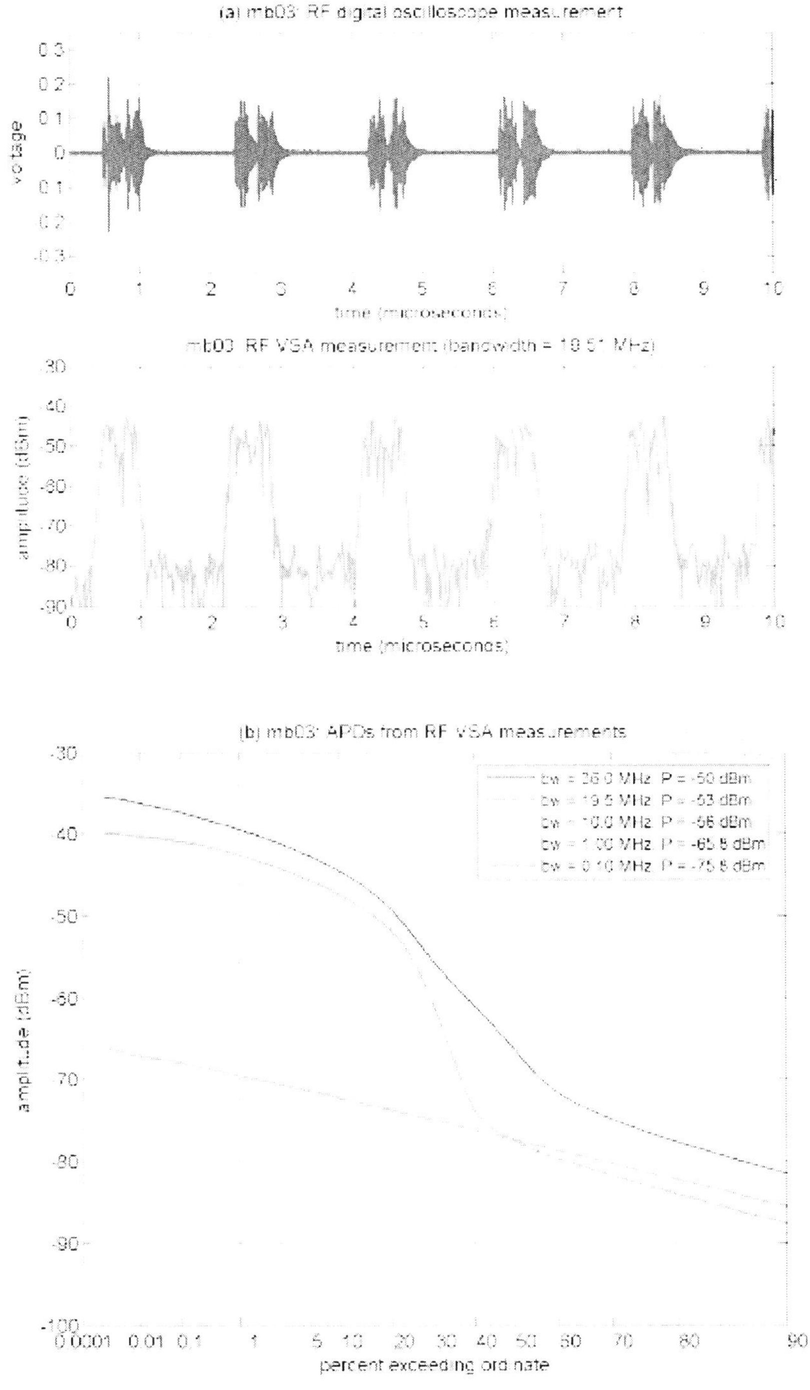

Figure B-49. RF amplitude analyses of MB-03 (b = 3, d = 2).

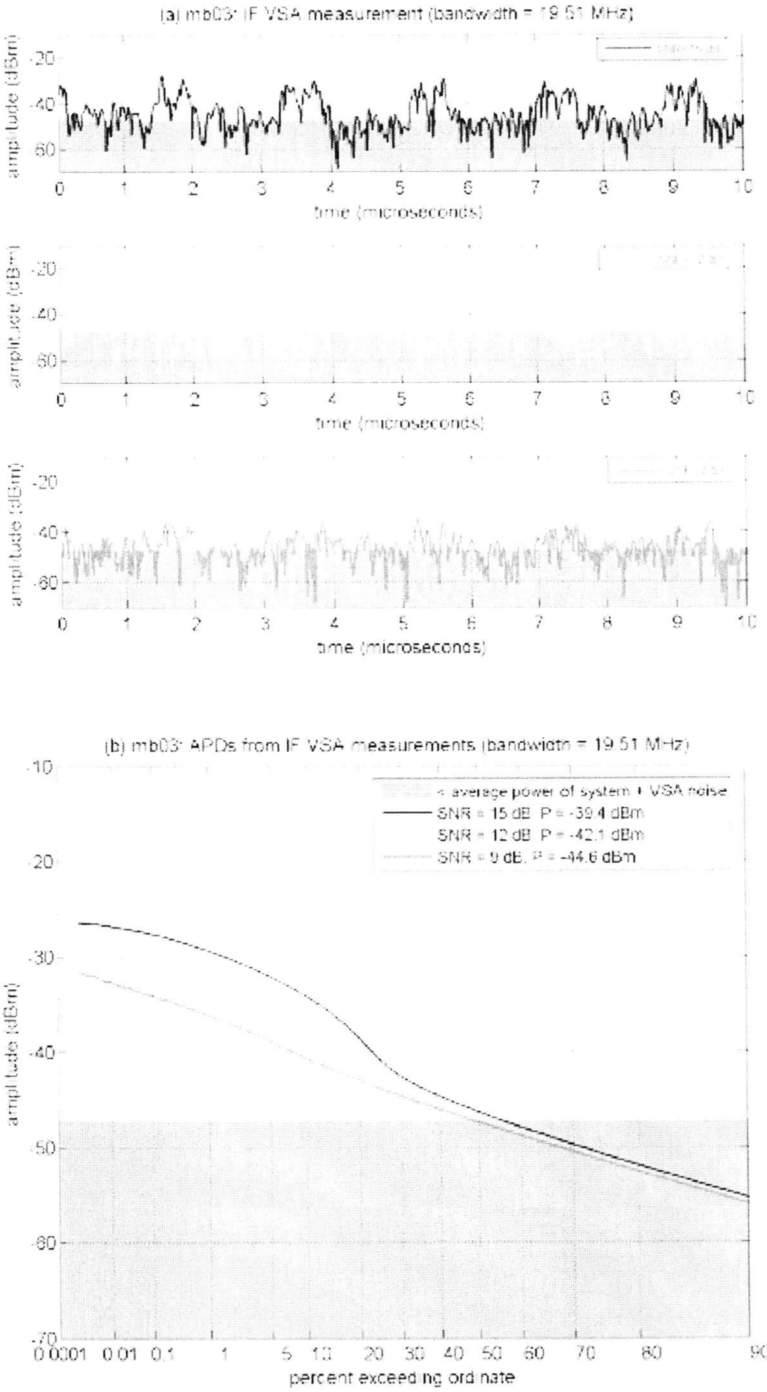

Figure B-50. IF amplitude analyses of MB-03 at INR_{TOV}.

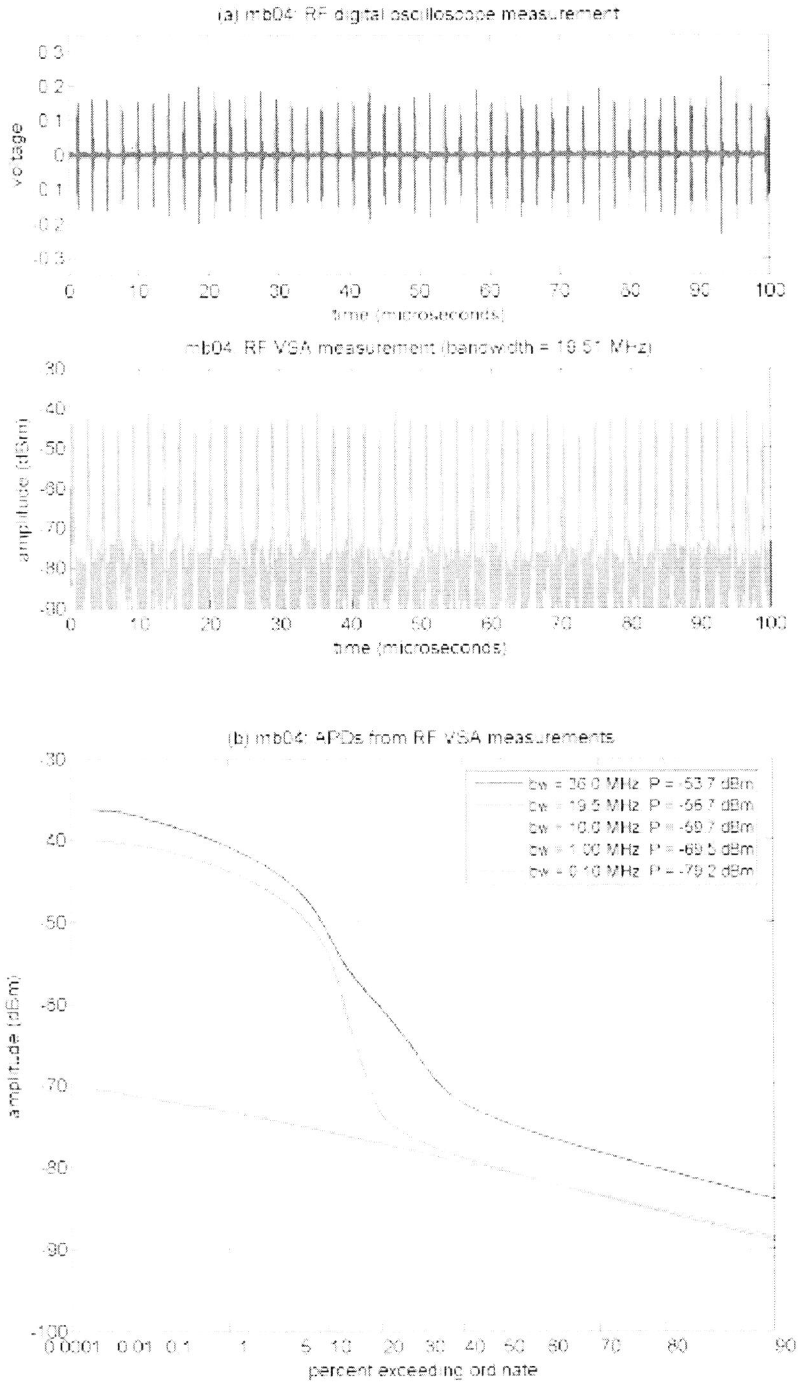

Figure B-51. RF amplitude analyses of MB-04 (b = 7, d = 1).

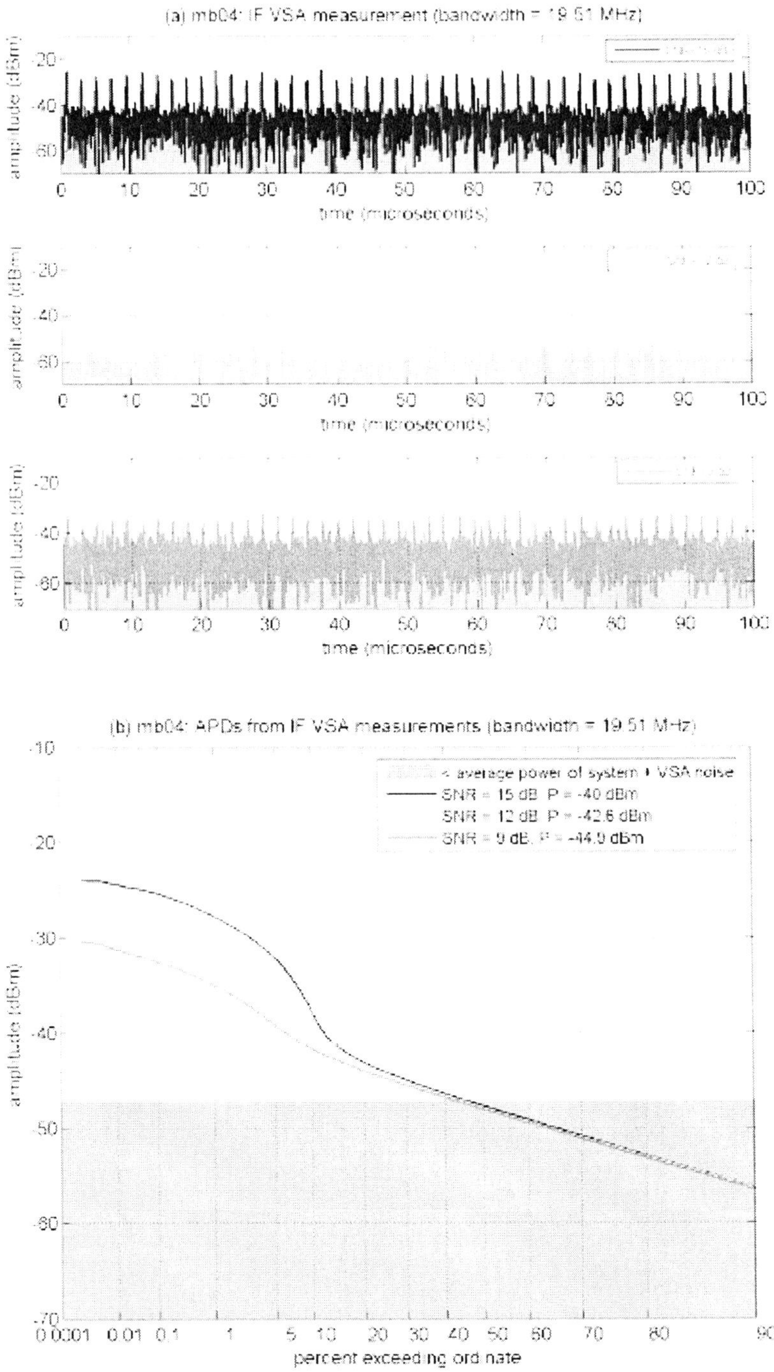

Figure B-52. IF amplitude analyses of MB-04 at INR_{TOV}.

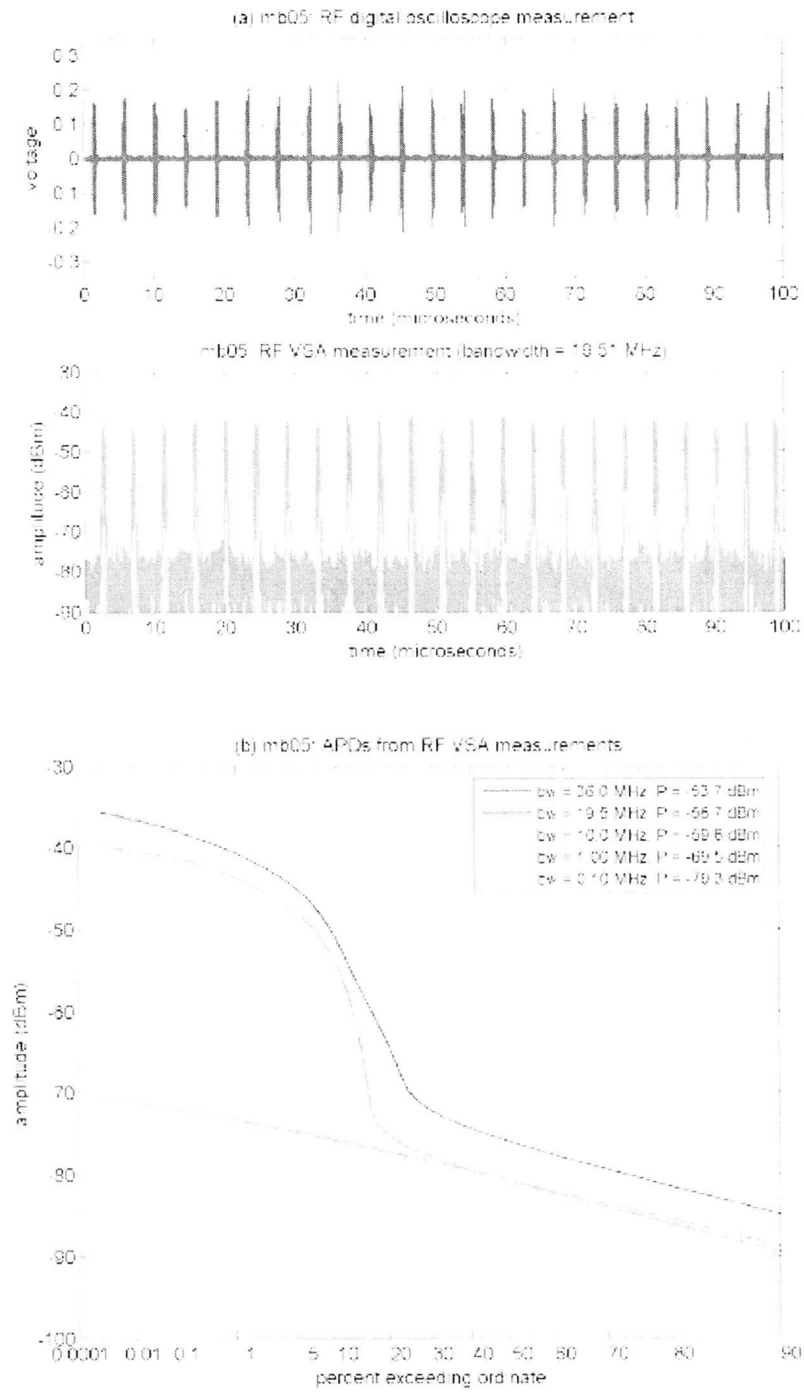

Figure B-53. RF amplitude analyses of MB-05 (b = 7, d = 2).

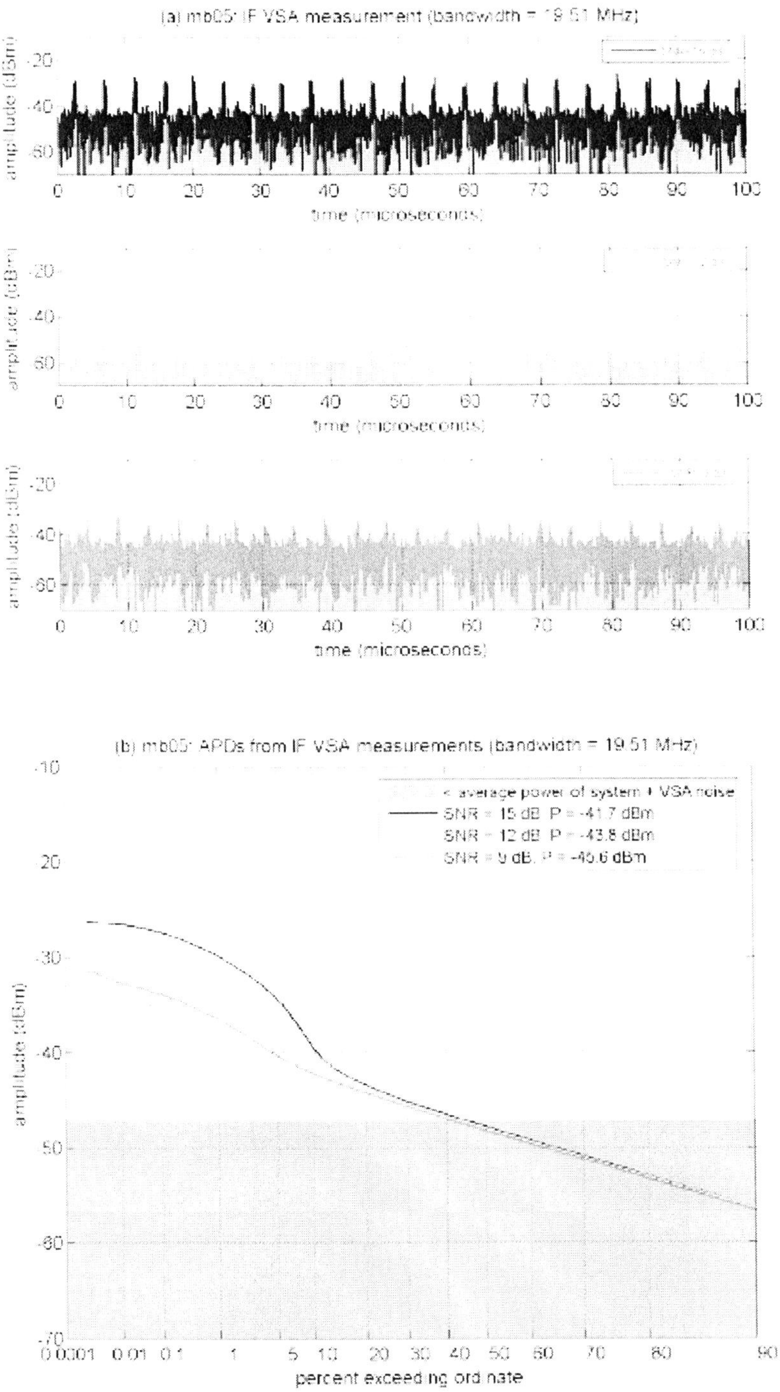

Figure B-54. IF amplitude analyses of MB-05 at INR_{TOV}.

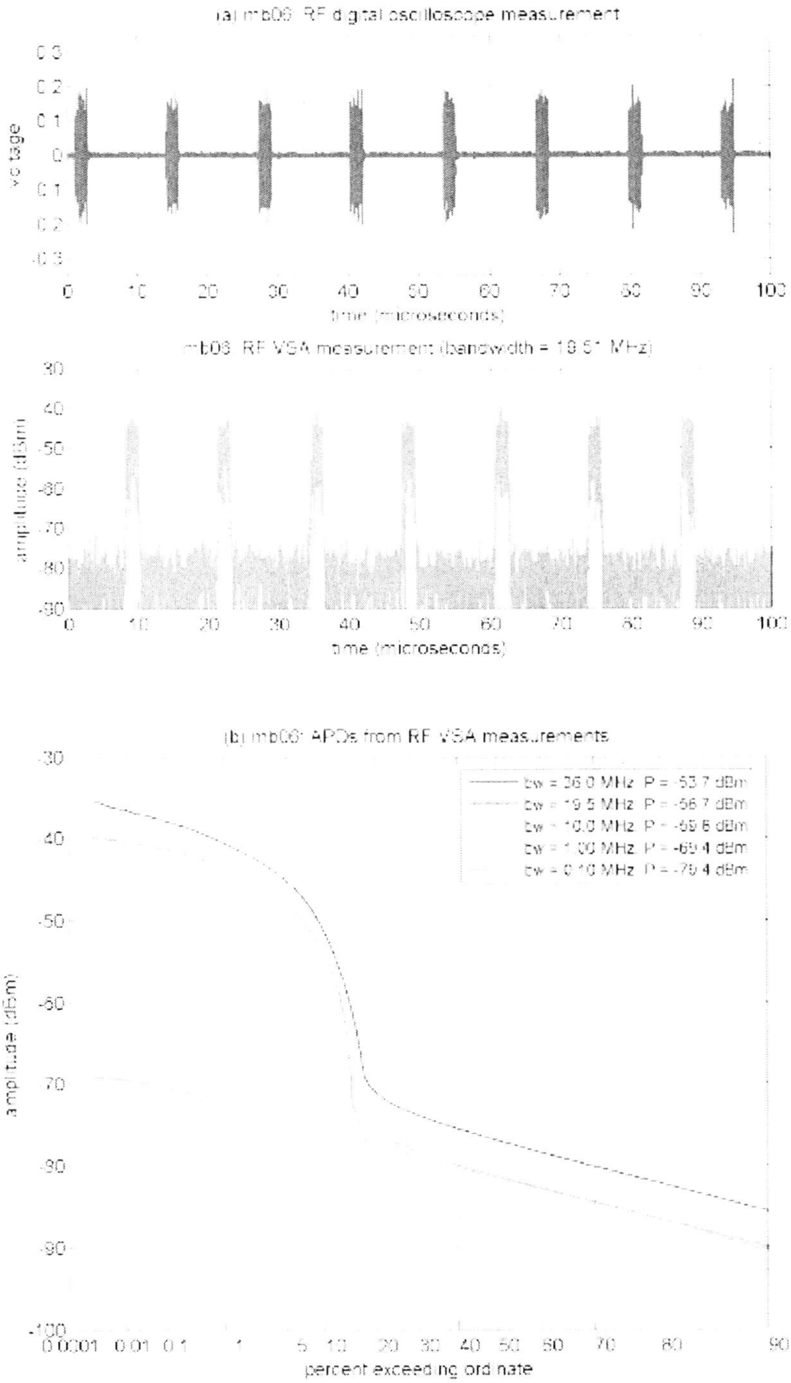

Figure B-55. RF amplitude analyses of MB-06 (b = 7, d = 6).

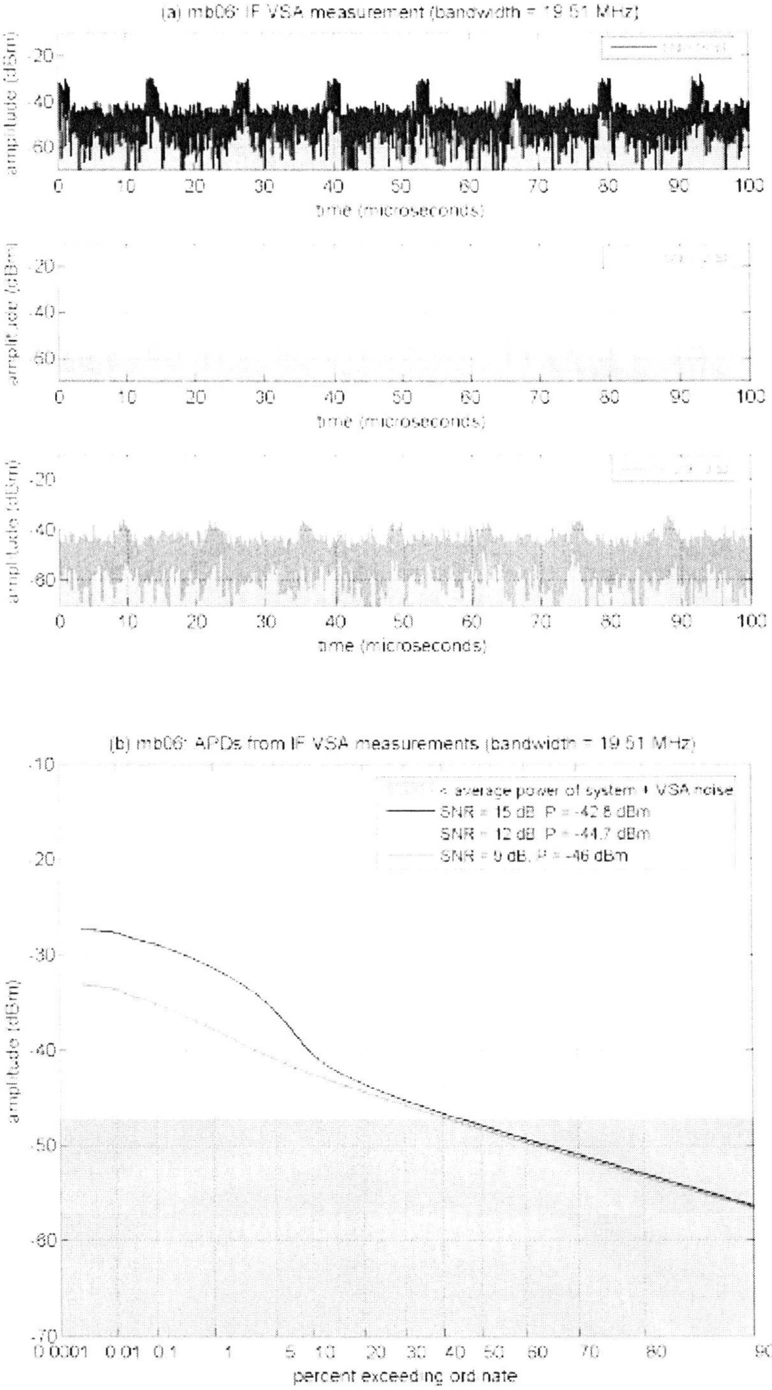

Figure B-56. IF amplitude analyses of MB-06 at INR_{TOV}.

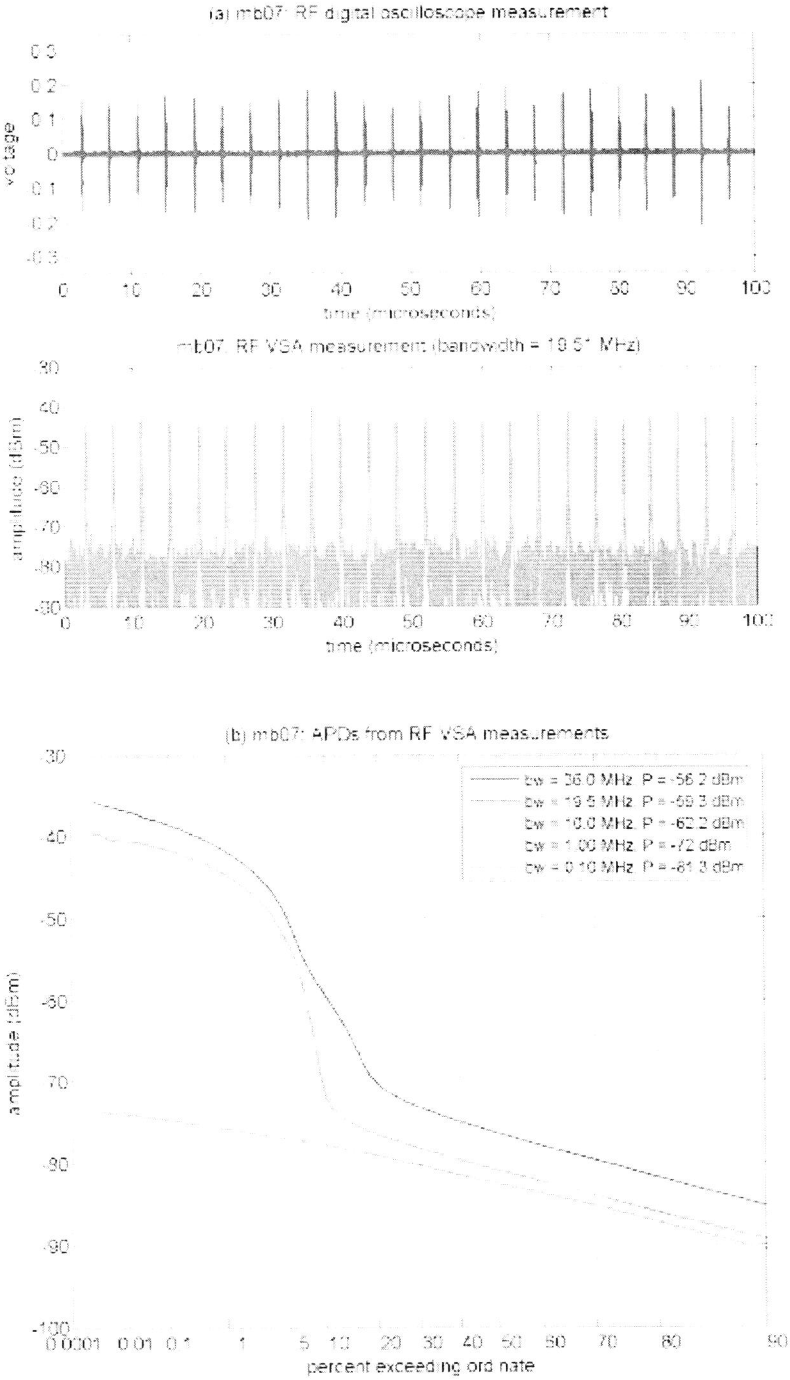

Figure B-57. RF amplitude analyses of MB-07 (b = 13, d = 1).

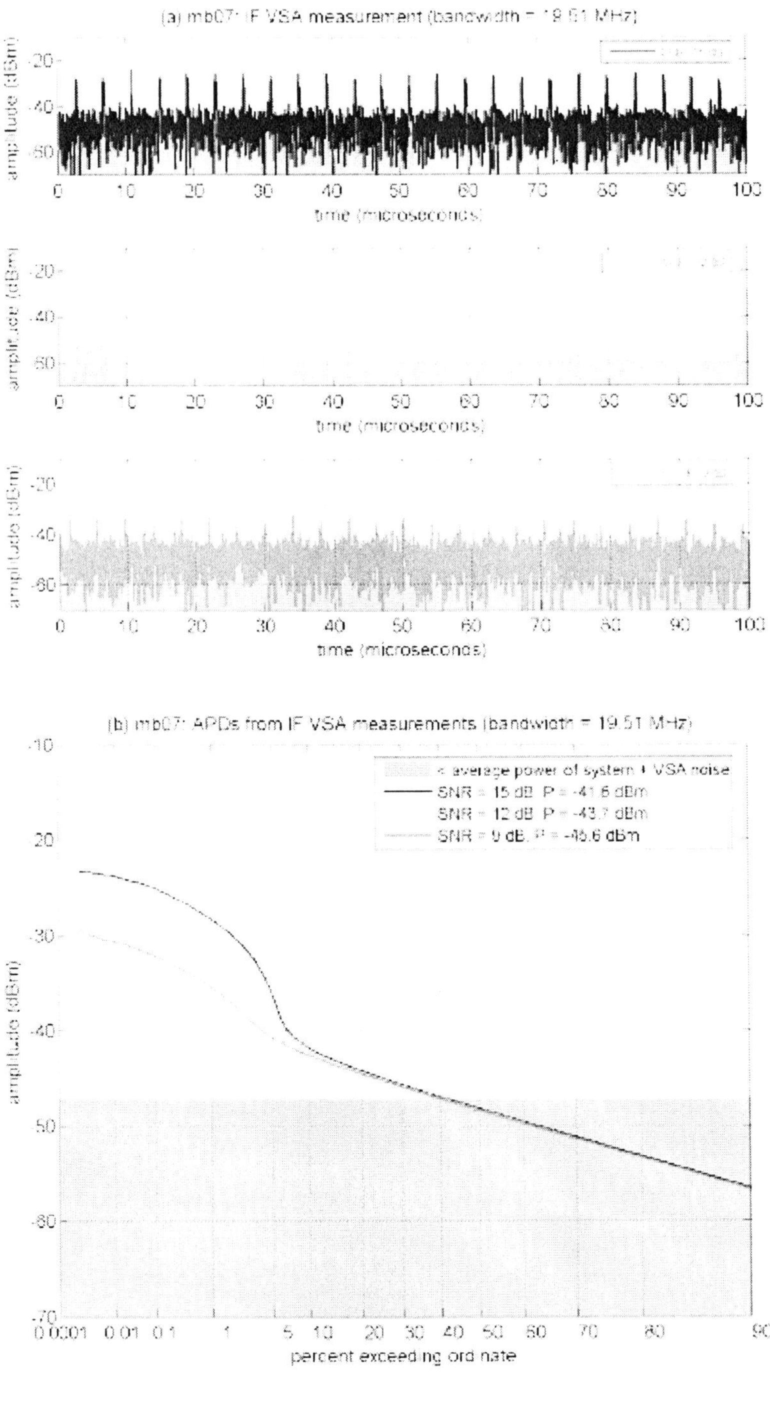

Figure B-58. IF amplitude analyses of MB-07 at INR_{TOV}.

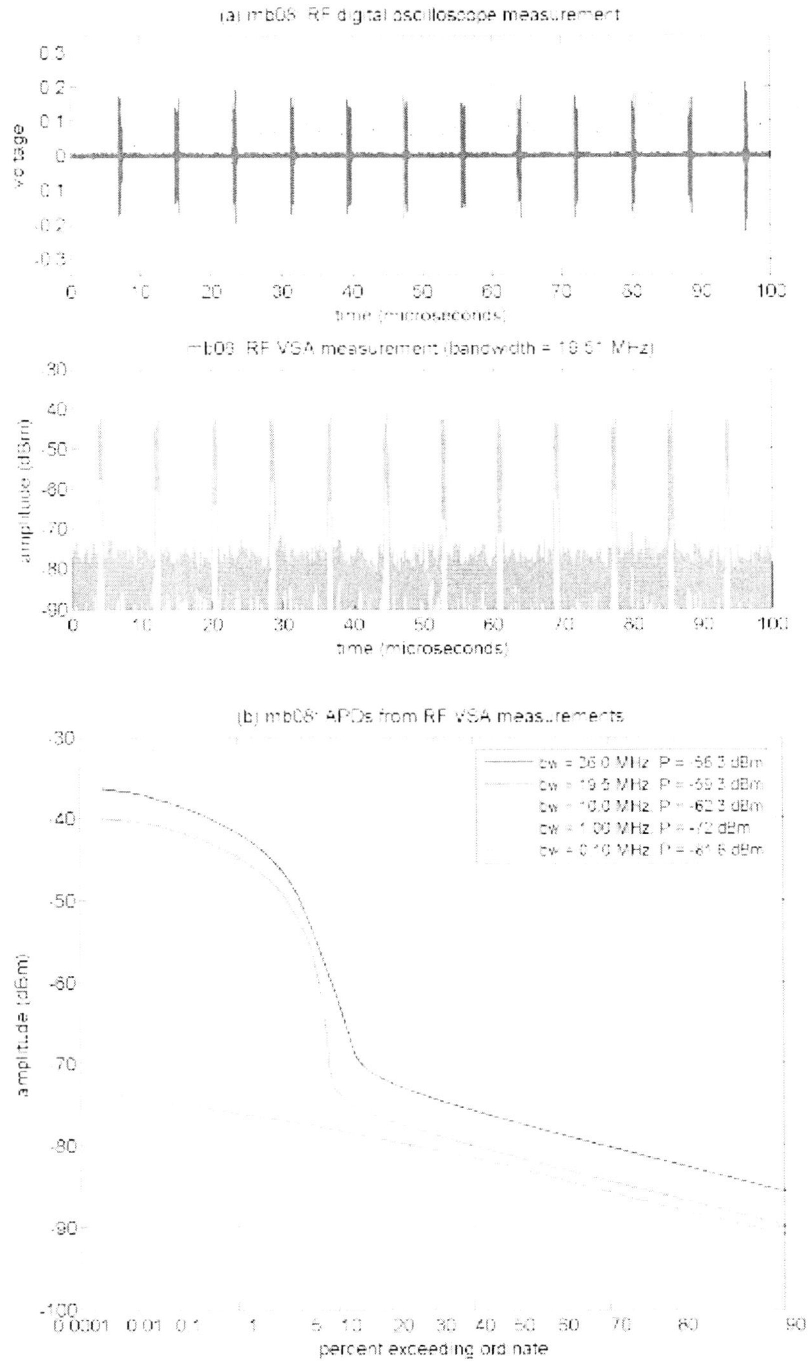

Figure B-59. RF amplitude analyses of MB-08 (b = 13, d = 2).

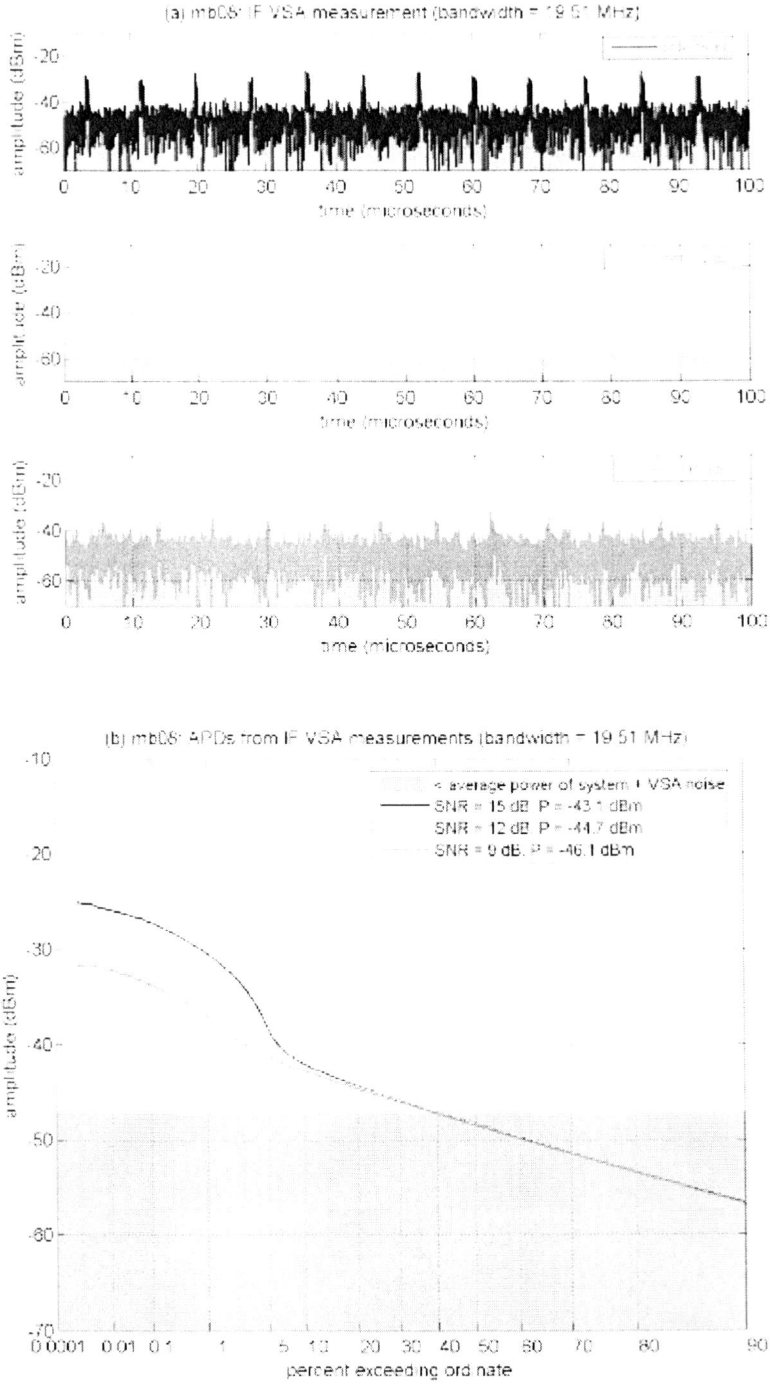

Figure B-60. IF amplitude analyses of MB-08 at INR_{TOV}.

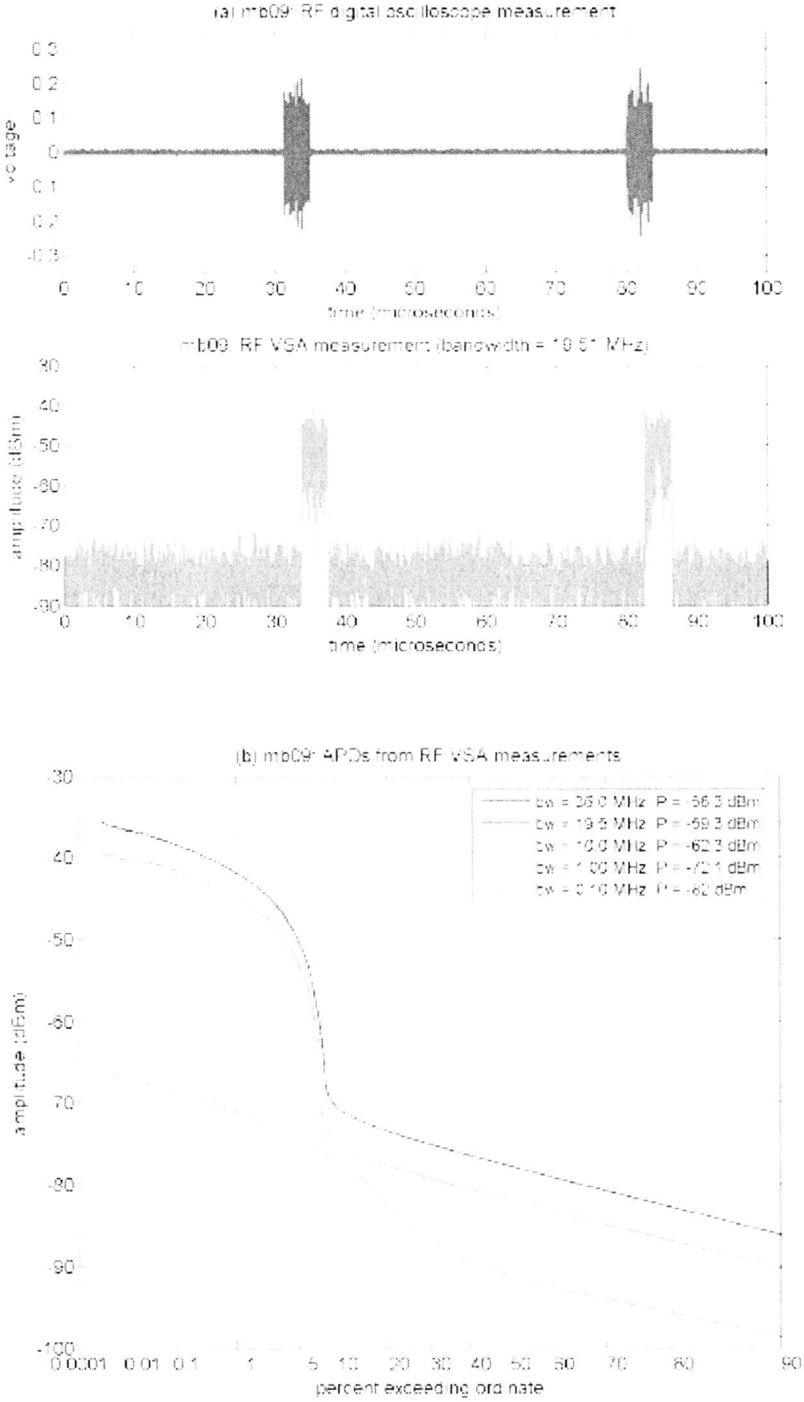

Figure B-61. RF amplitude analyses of MB-09 (b = 13, d = 12).

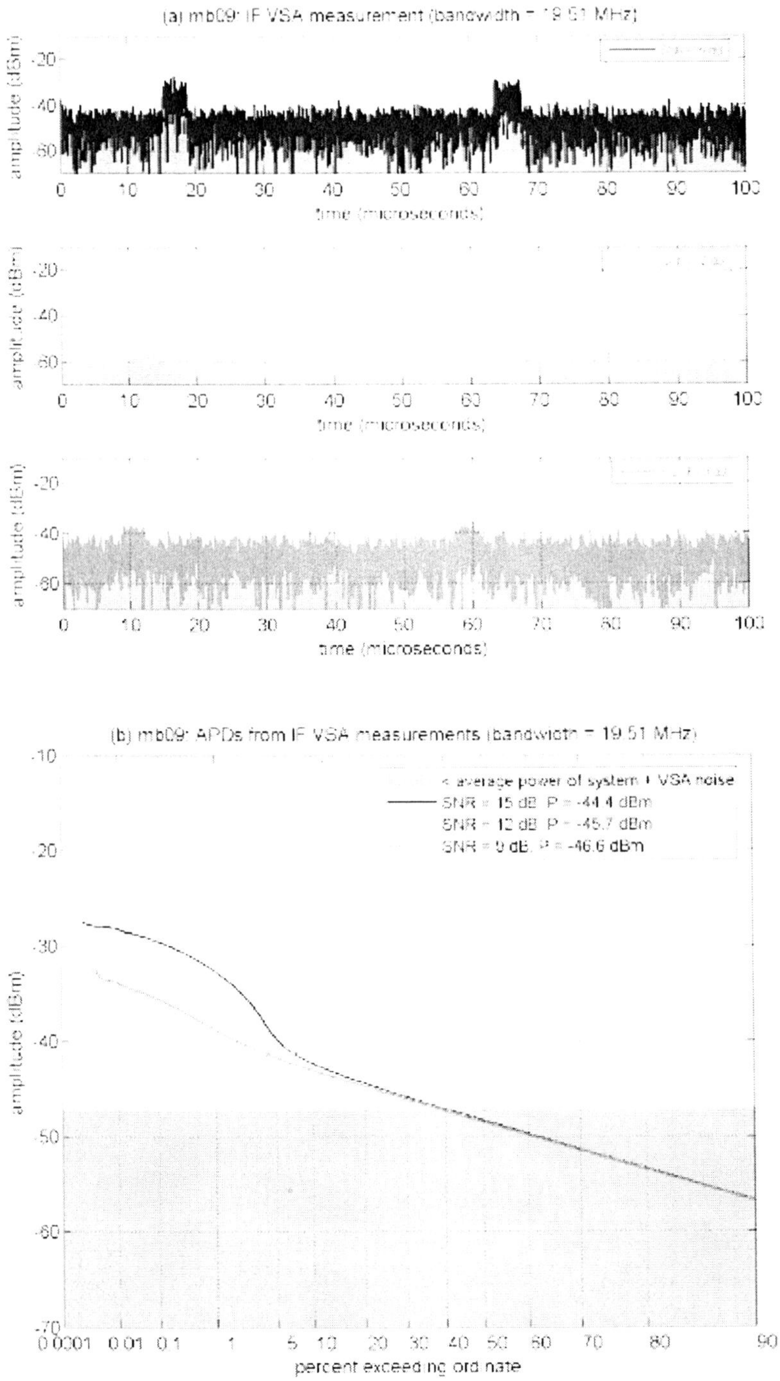

Figure B-62. IF amplitude analyses of MB-09 at INR_{TOV}.

Table B-4. Measured IF P/A (dB) of UWB Signals at INR_{TOV} and Band-Limited to B_{DTV} §

Type	Index	SNR = 9 dB	SNR = 12 dB	SNR = 15 dB
DP	01	23.4	?	?
	02	13.2	14.0	14.1
	03	8.9	7.9	7.2
	04	9.3	8.9	8.7
DS	01	9.6	9.4	9.4
	02	9.5	9.0	9.1
	03	9.1	8.8	8.6
	04	9.1	8.1	7.8
	05	8.9	8.4	7.7
	06	9.4	8.9	8.4
MB	01	9.6	9.3	9.0
	02	11.7	12.7	12.8
	03	11.9	12.4	12.5
	04	13.6	15.1	15.6
	05	12.8	14.5	15.2
	06	12.4	14.0	15.2
	07	15.0	16.6	17.8
	08	14.0	16.0	17.2
	09	12.7	14.6	16.2

§ Shaded cells highlight Rayleigh-amplitude statistics. Question marks indicate uncertainty in the measurement due to exceedingly high signal voltages overdriving the VSA.

B.4. Spectral Analyses

The spectral analyses provided in this subsection were derived from VSA measurements and intended to emulate the swept spectrum analyzer (SA) measurements described in table B-5. Complex samples measured in a 36-MHz VSA span were frequency shifted so that the frequency of interest was centered at 0 Hz. Next, the signal was band-limited to a resolution bandwidth (RBW) of 1 MHz or 0.1 MHz using a root-raised cosine filter with $\alpha = 0.35$. Mean, median, and peak statistics were computed for the resulting band-limited complex-baseband data.

Table B-5. Spectrum Analyzer Settings for Characterizing UWB Signals

Center Frequency (MHz)	Span (MHz)	Resolution Bandwidth (MHz)	Video Bandwidth (MHz)	Number of Frequency Steps	Sweep Time (seconds)	Detector
3950	600	1	50	3600	10.8	Rms, peak
3820	40	0.1		2400	72.0	

These RBWs are smaller than that of the victim digital television channel; however, results provide mean and peak estimates relevant to possible narrowband circuitry within the receiver. Frequencies of interest spanned the bandwidth of the satellite channel with a frequency step size set to one tenth of RBW to ensure that spectrum details were not missed.

Figures B-63 – B-81 provide spectral statistics for the UWB signals: Figures B-63 – B-66 correspond to DP-01 – DP-04, figures B-67 – B-72 correspond to DS-01 – DS-06, and figures B-73 – B-81 correspond to MB-01 – MB-09. SA measurements were made and show agreement with the VSA measurement results; SA measurements are not included in this chapter.

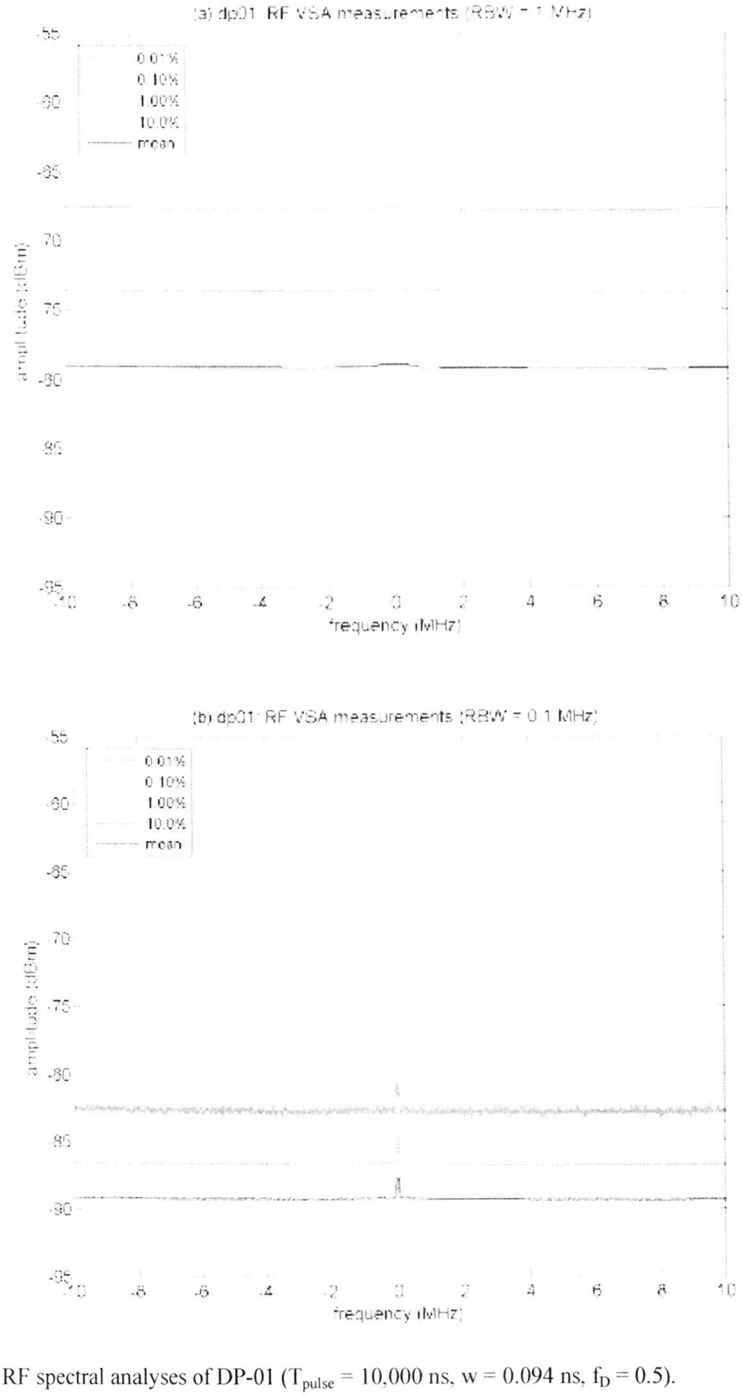

Figure B-63. RF spectral analyses of DP-01 (T_{pulse} = 10,000 ns, w = 0.094 ns, f_D = 0.5).

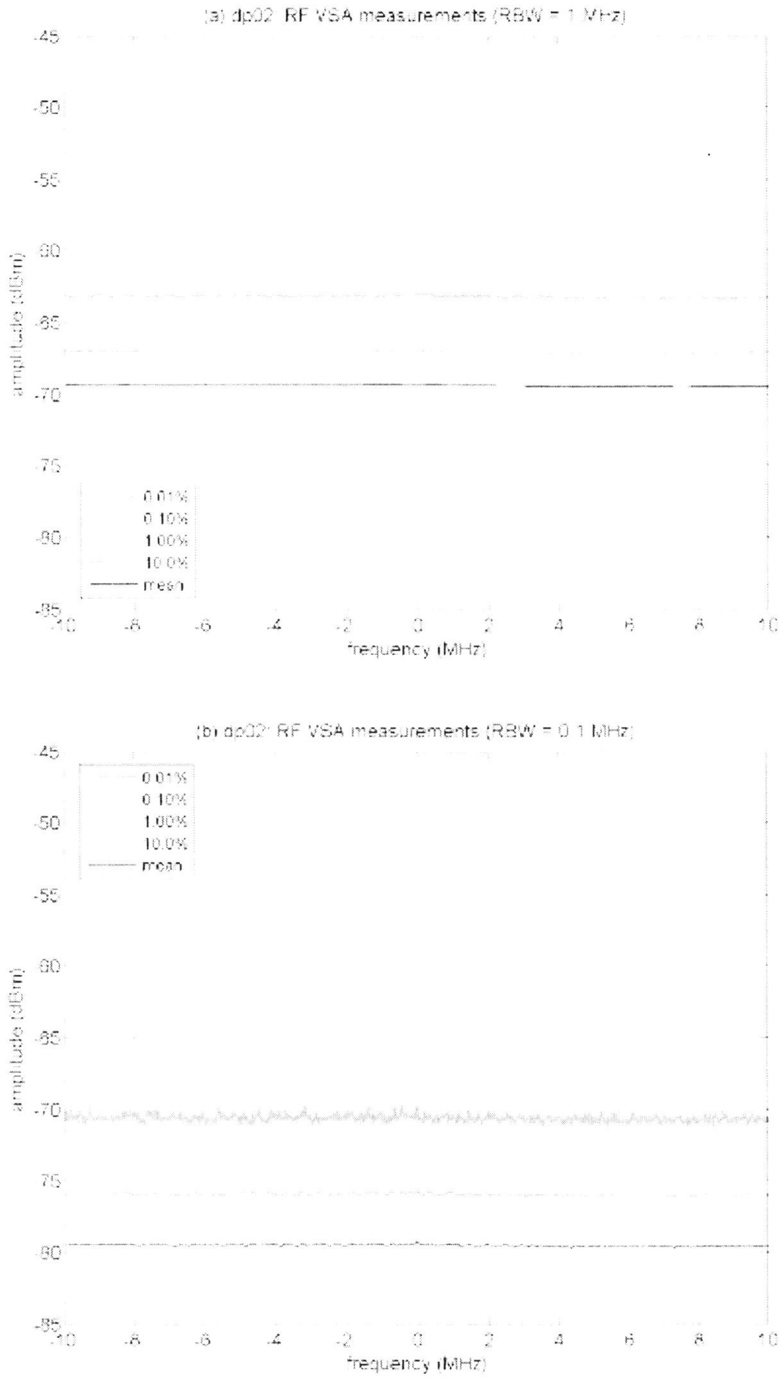

Figure B-64. RF spectral analyses of DP-02 (T_{pulse} = 1000 ns, w = 0.094 ns, f_D = 0.5).

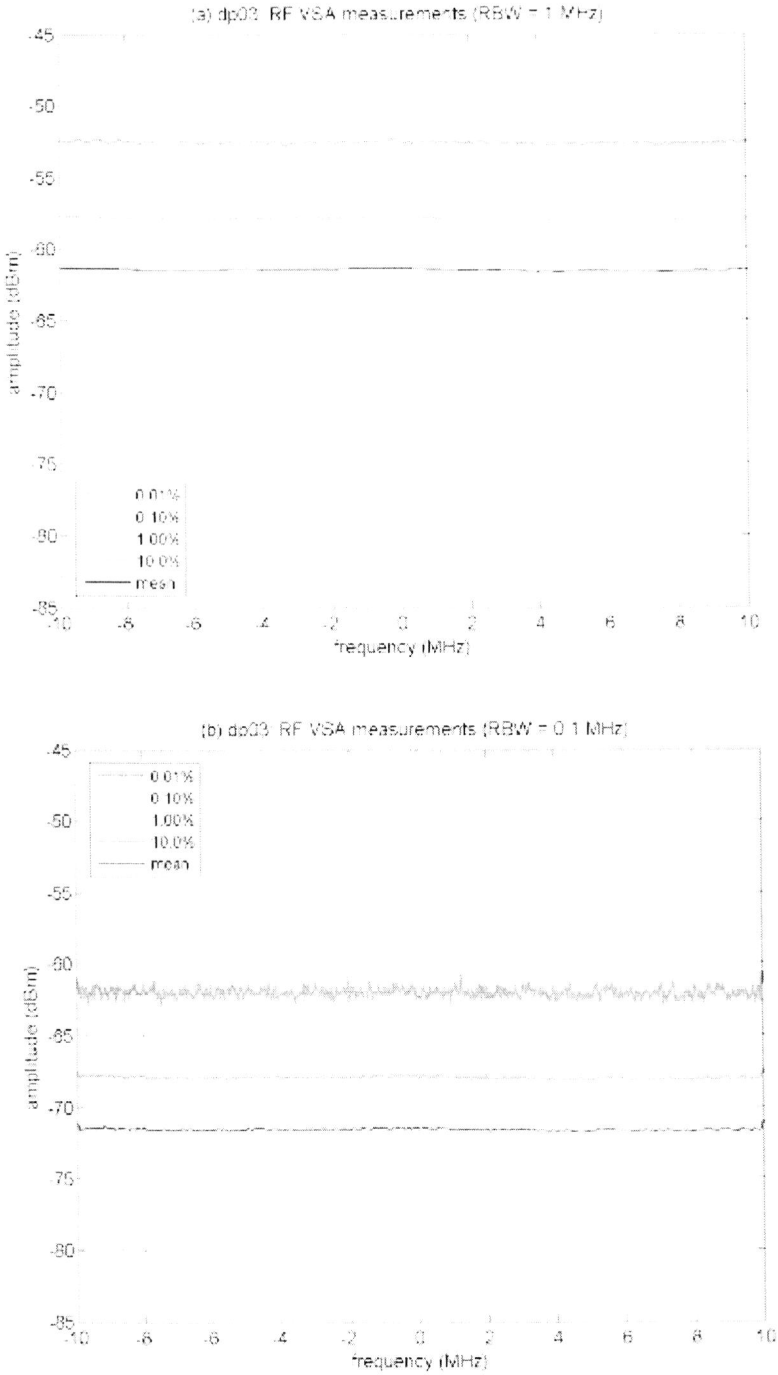

Figure B-65. RF spectral analyses of DP-03 (T_{pulse} = 100 ns, w = 0.094 ns, f_D = 0.5).

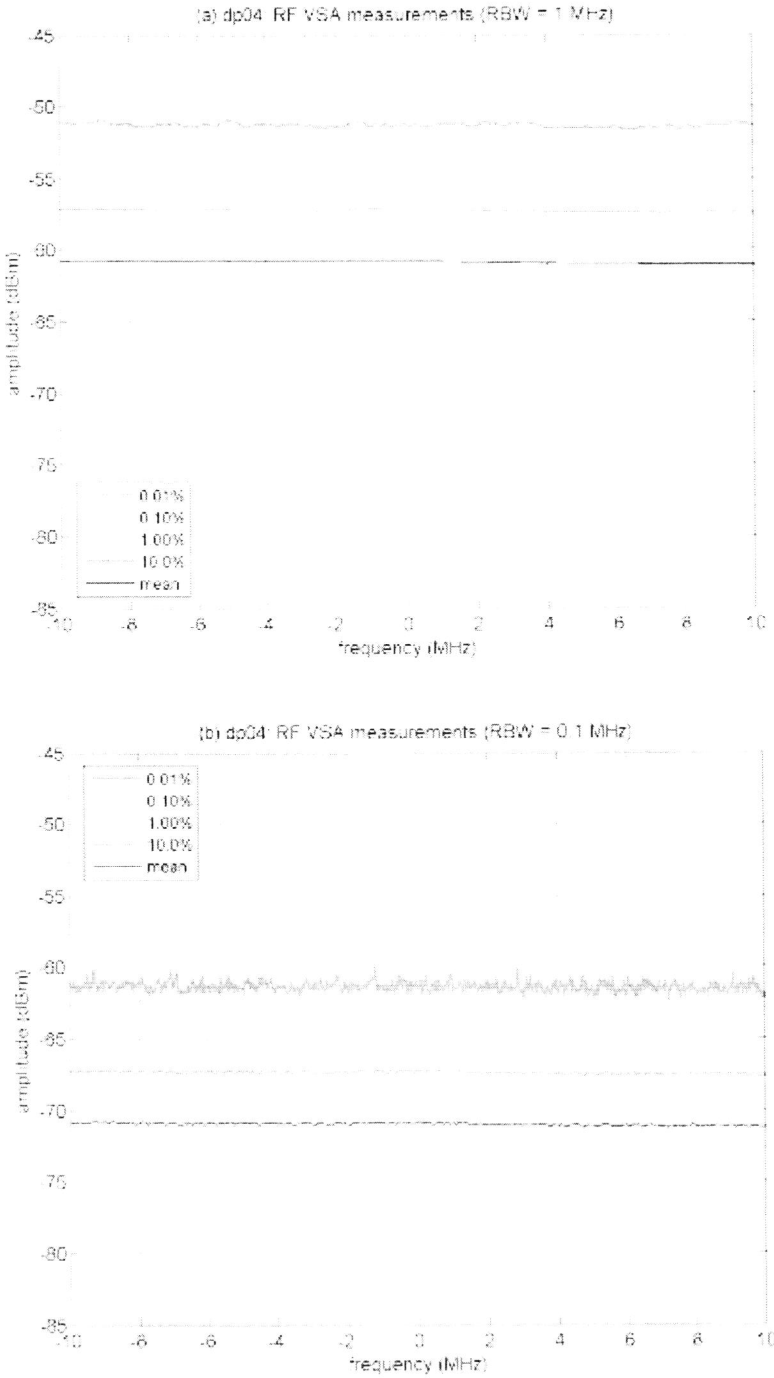

Figure B-66. RF spectral analyses of DP-04 (T_{pulse} = 10 ns, w = 0.094 ns, f_D = 0.5).

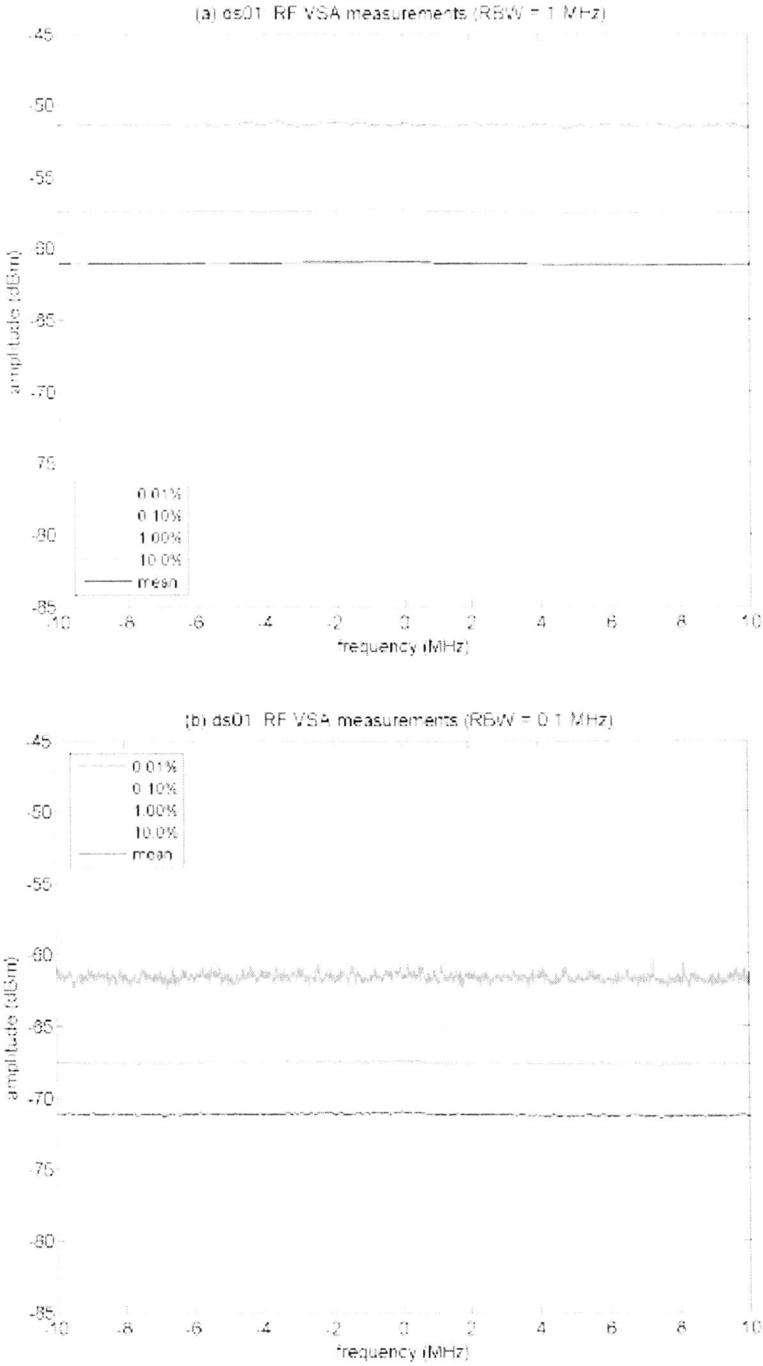

Figure B-67. RF spectral analyses of DS-01 (L = 1).

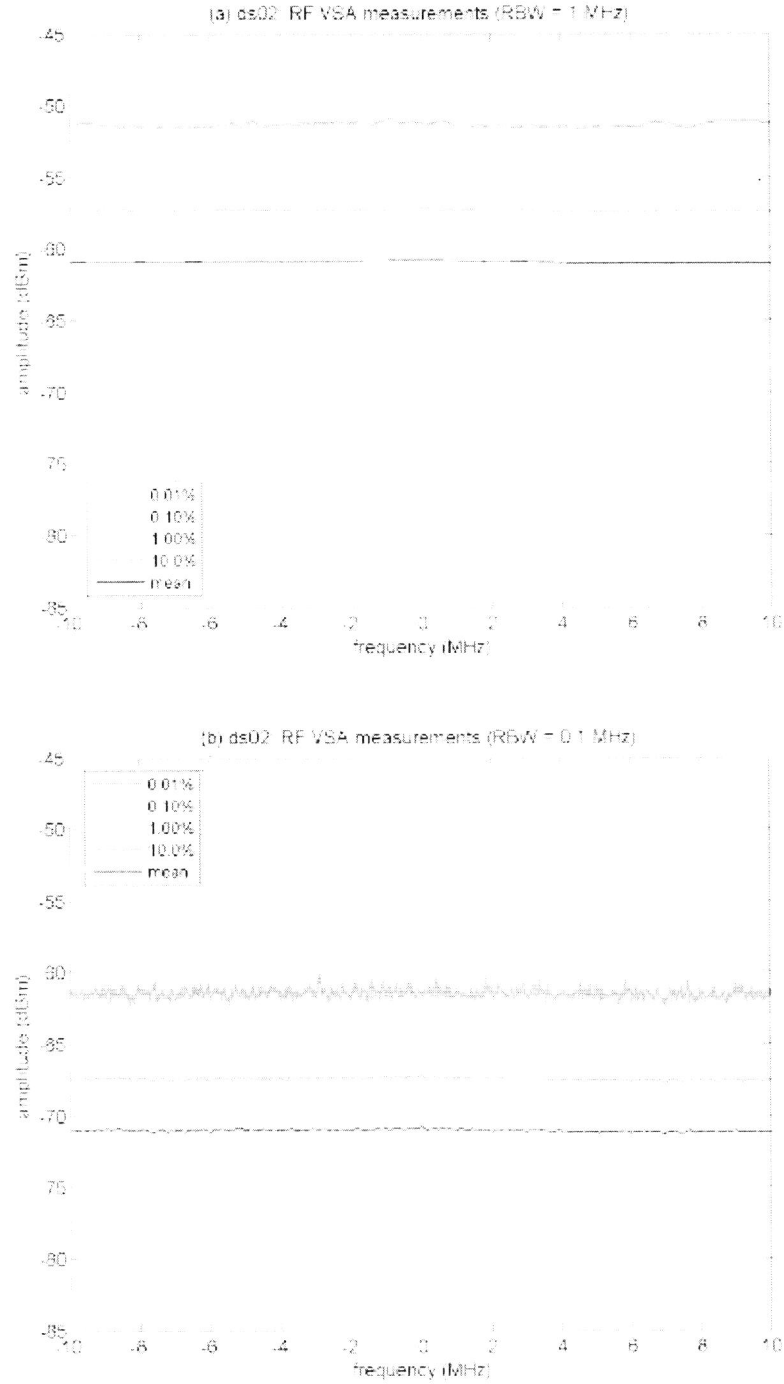

Figure B-68. RF spectral analyses of DS-02 (L = 3, sparse code).

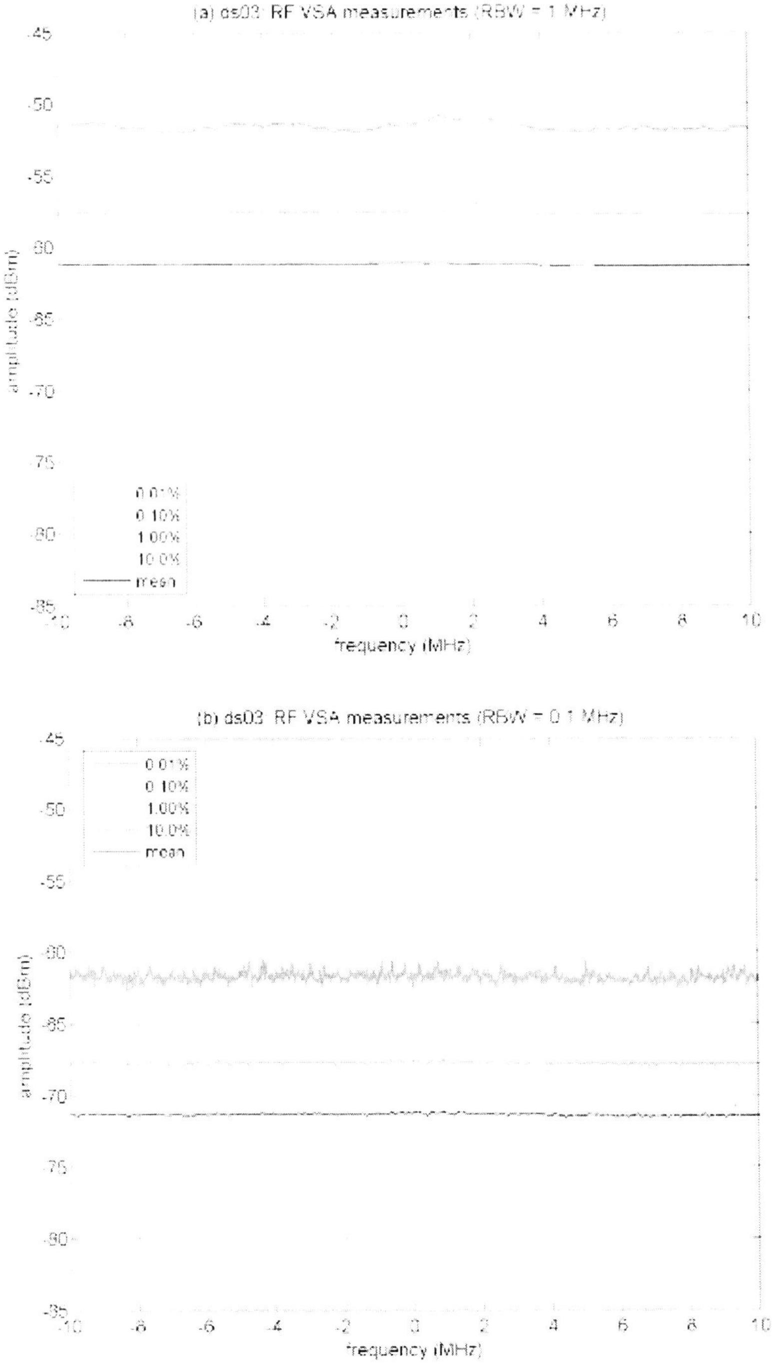

Figure B-69. RF spectral analyses of DS-03 (L = 6, sparse code).

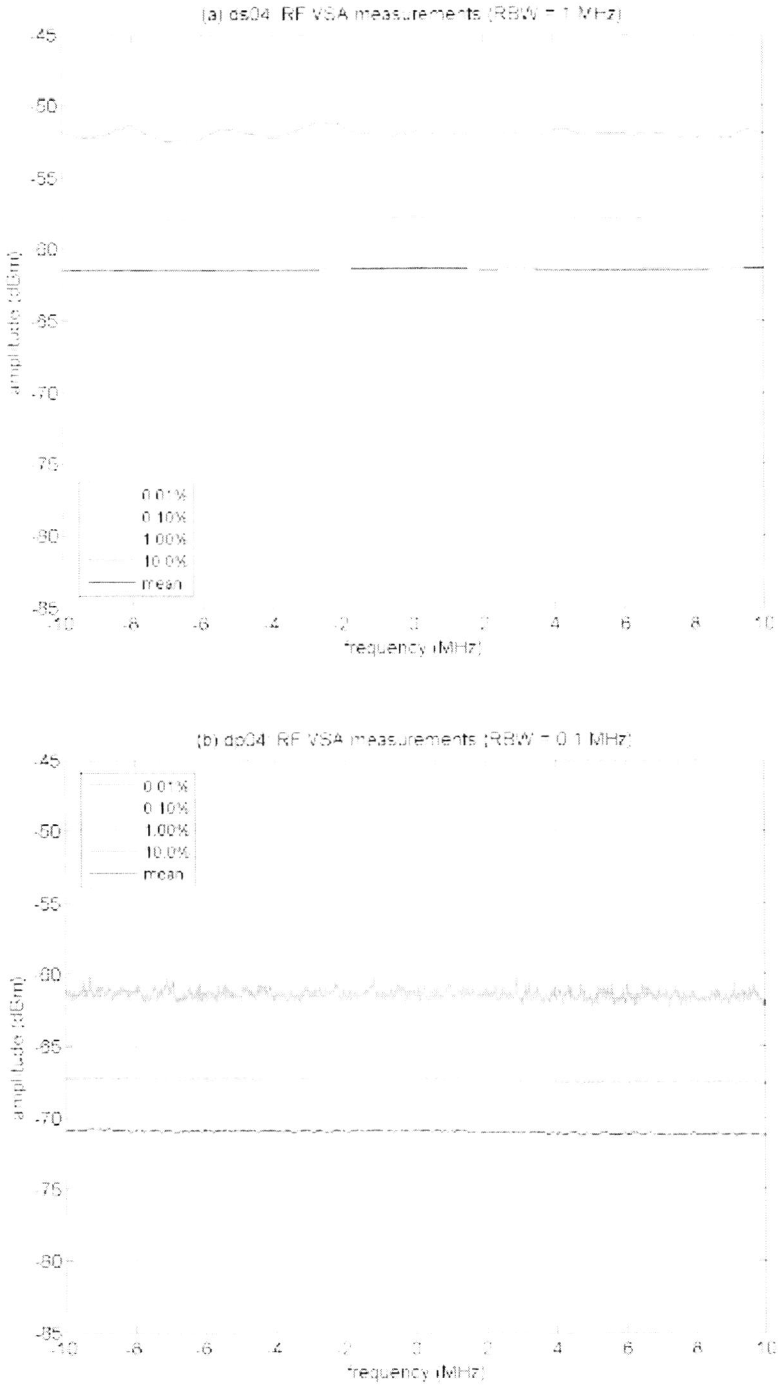

Figure B-70. RF spectral analyses of DS-04 (L = 12, DS code).

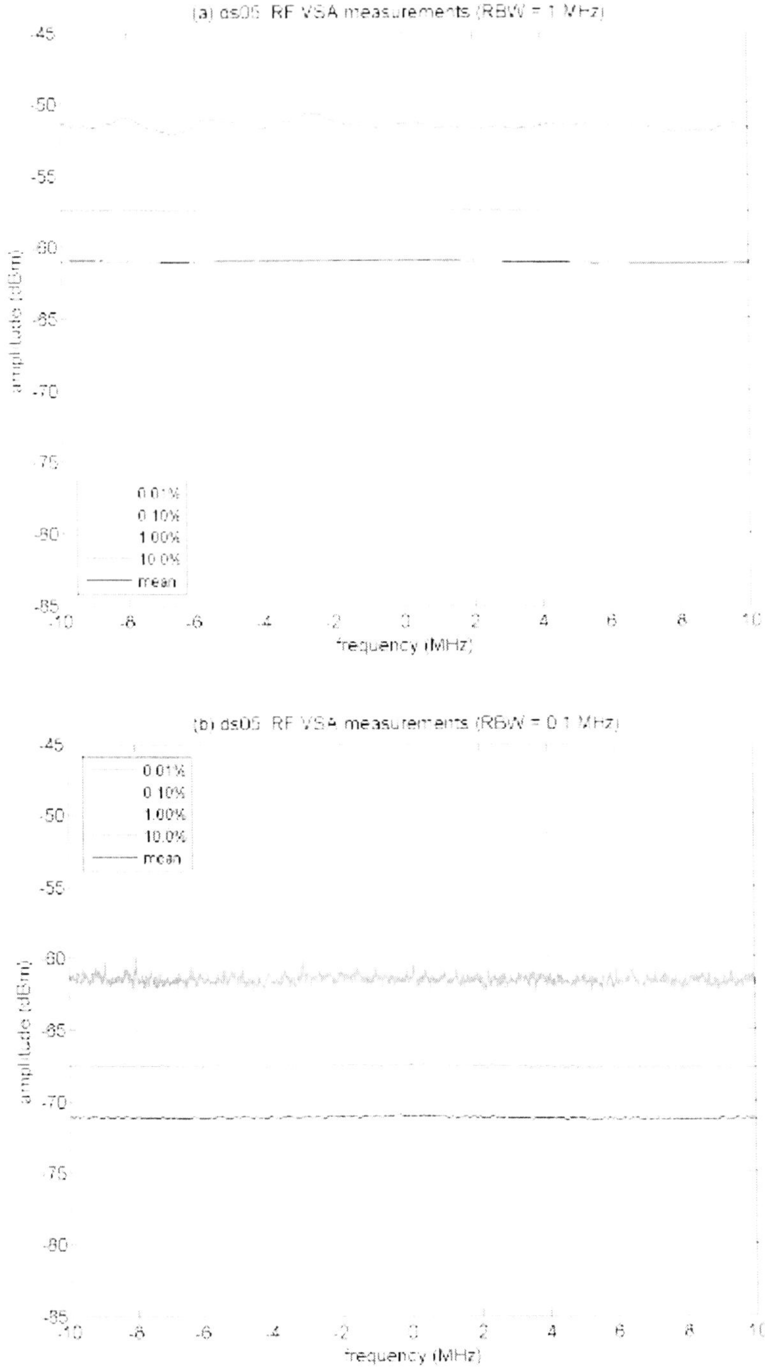

Figure B-71. RF spectral analyses of DS-05 (L = 12, sparse code).

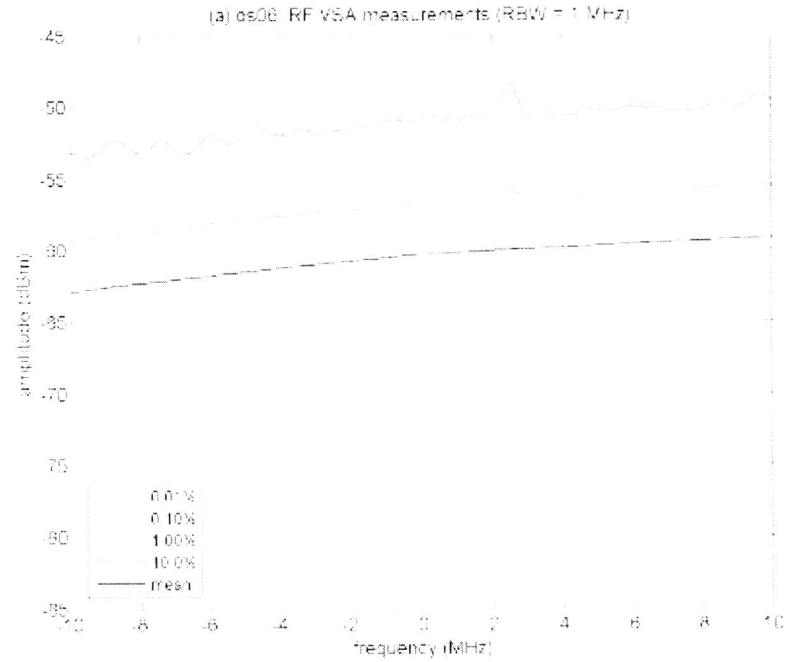

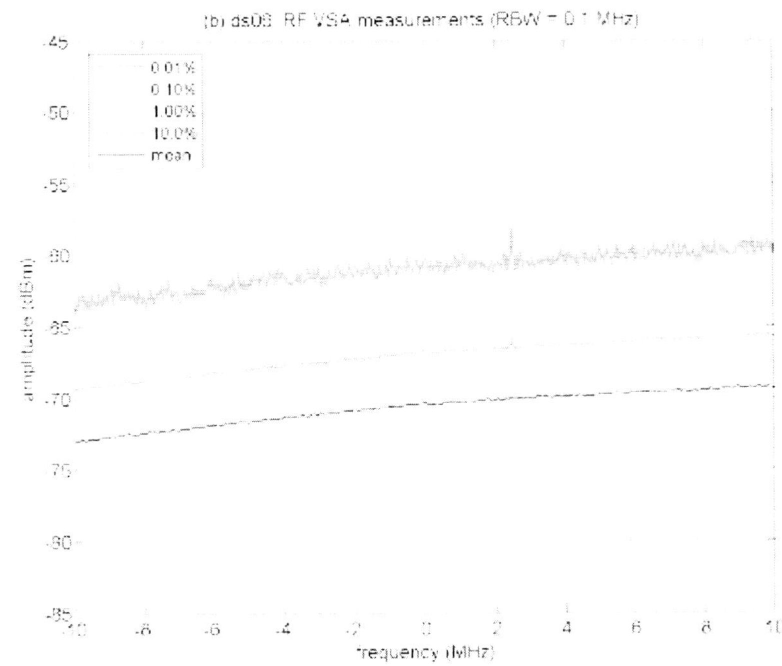

Figure B-72. RF spectral analyses of DS-06 (L = 24, DS code).

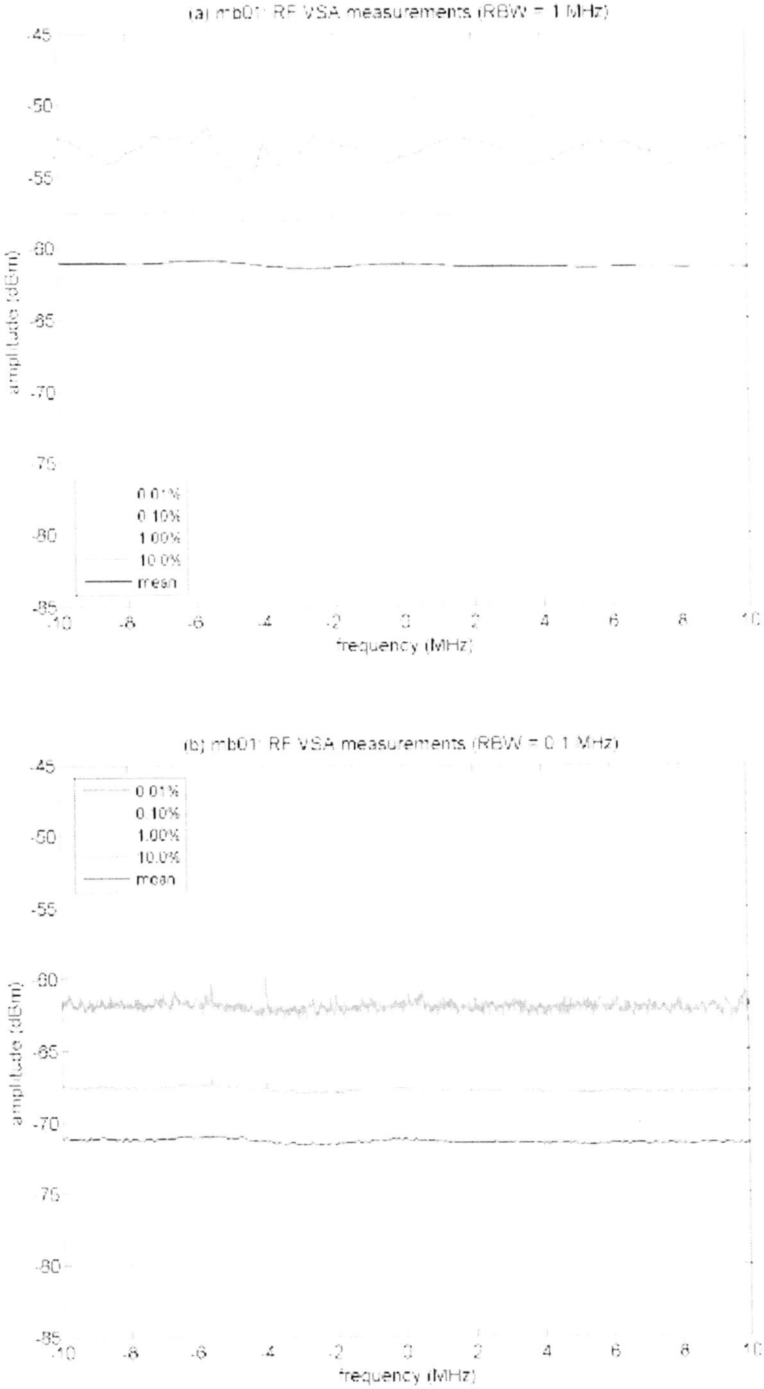

Figure B-73. RF spectral analyses of MB-01 (b = 1).

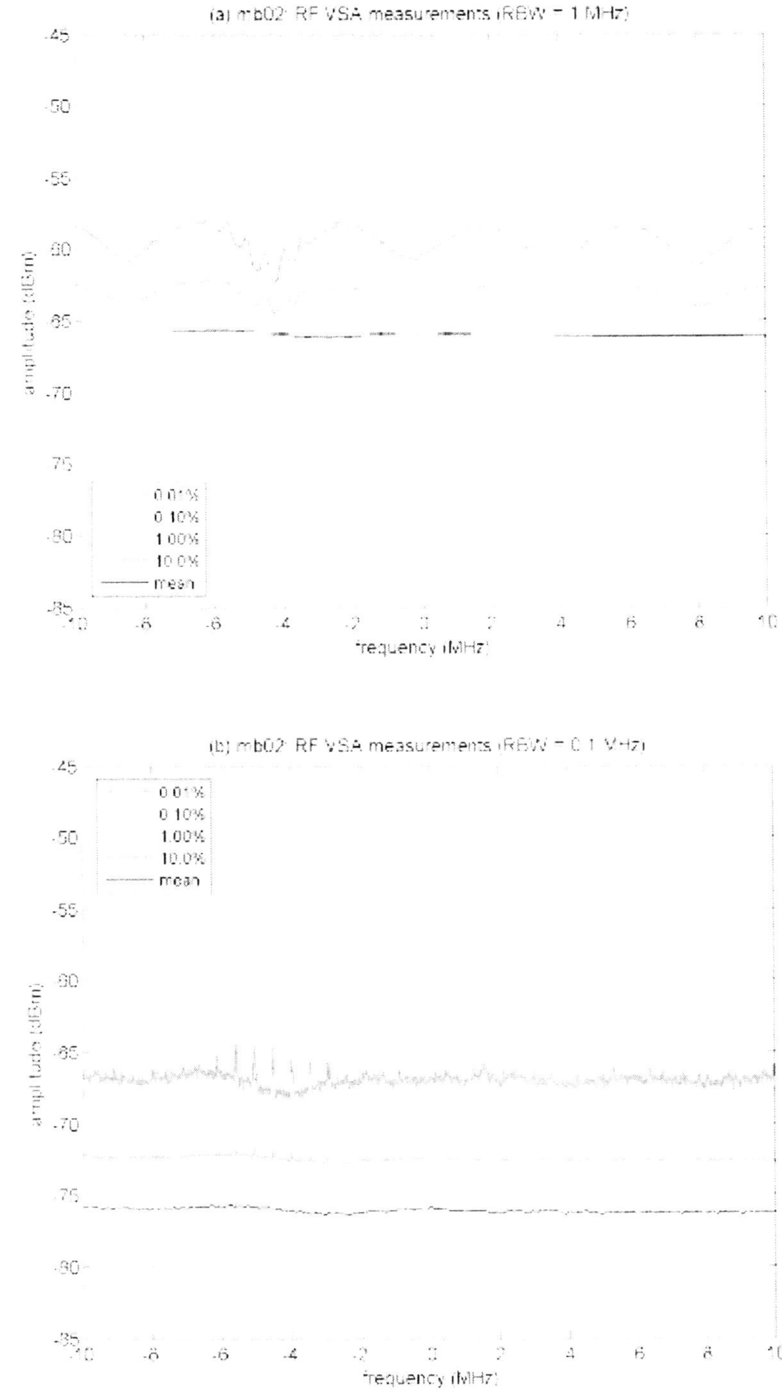

Figure B-74. RF spectral analyses of MB-02 (b = 3, d = 1).

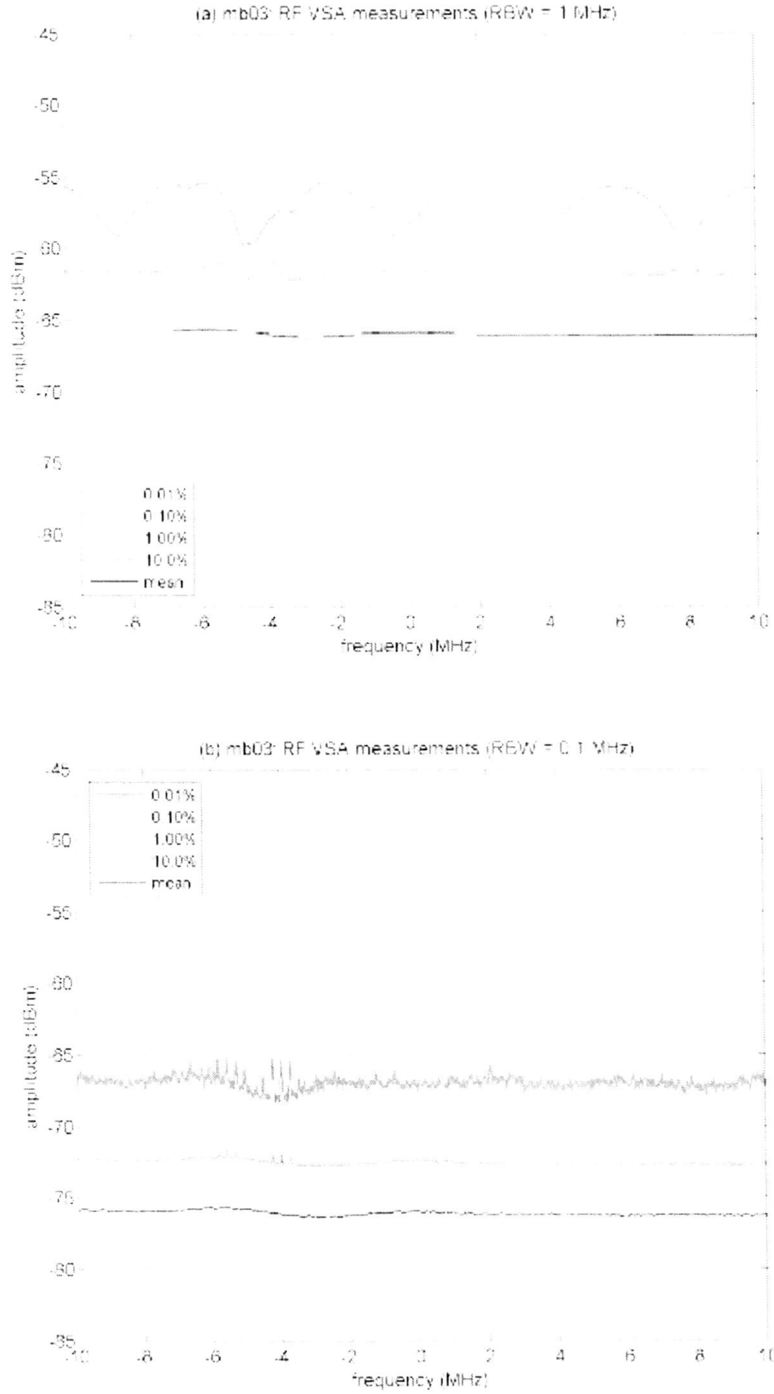

Figure B-75. RF spectral analyses of MB-03 (b = 3, d = 2).

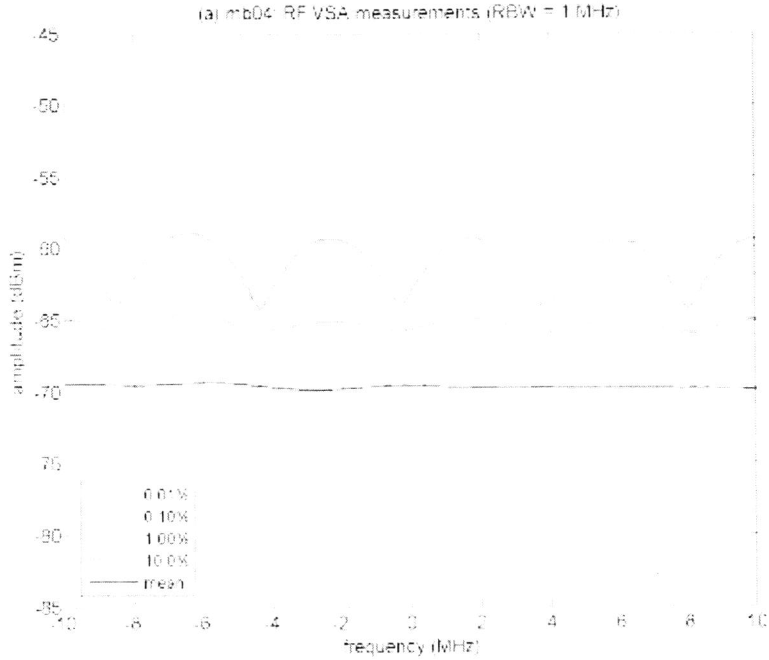

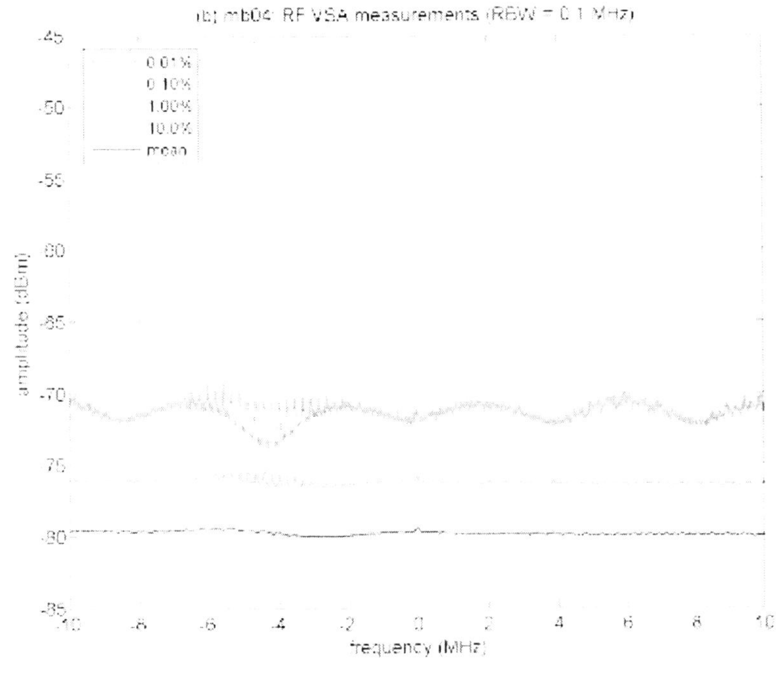

Figure B-76. RF spectral analyses of MB-04 (b = 7, d = 1).

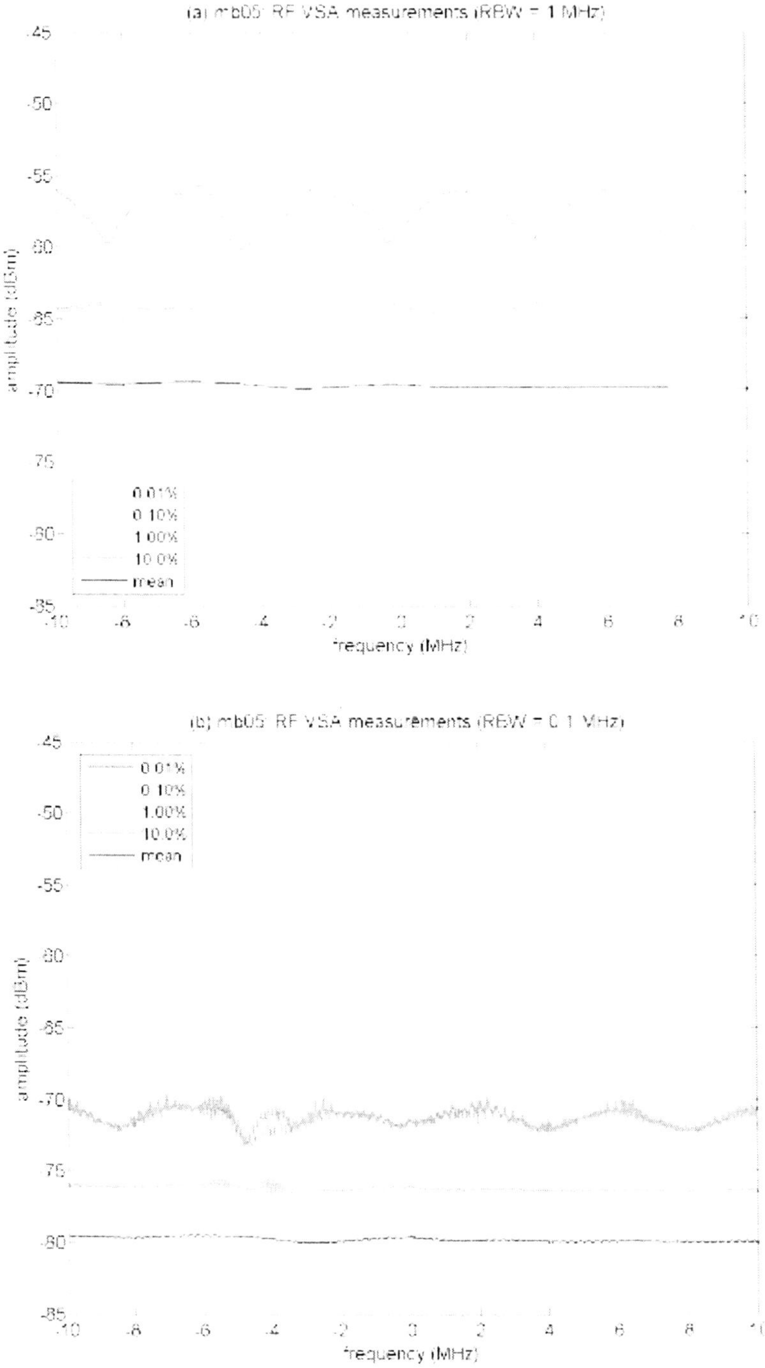

Figure B-77. RF spectral analyses of MB-05 (b = 7, d = 2).

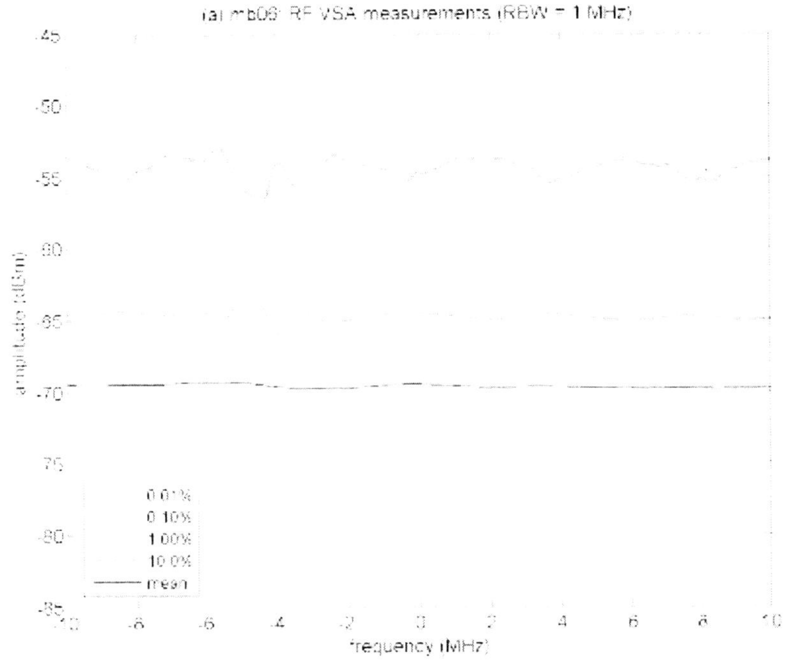

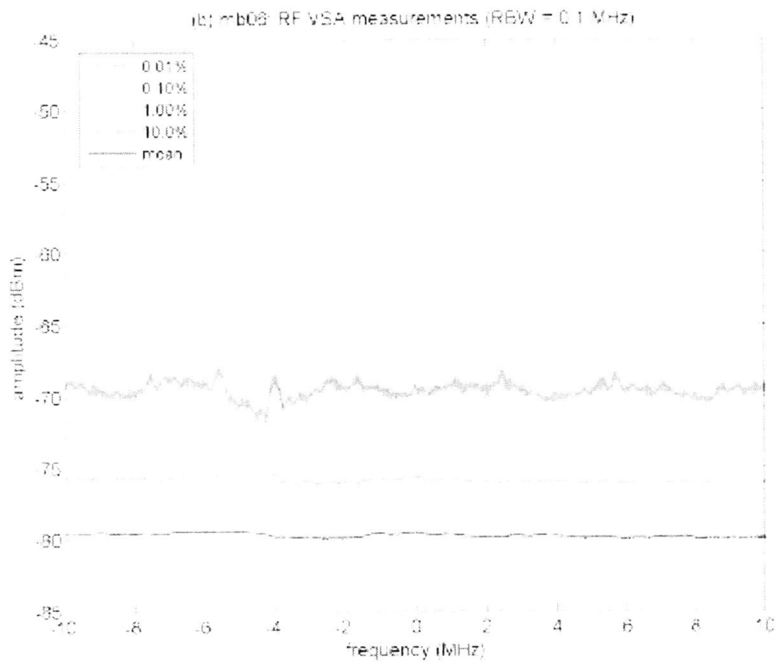

Figure B-78. RF spectral analyses of MB-06 (b = 7, d = 6).

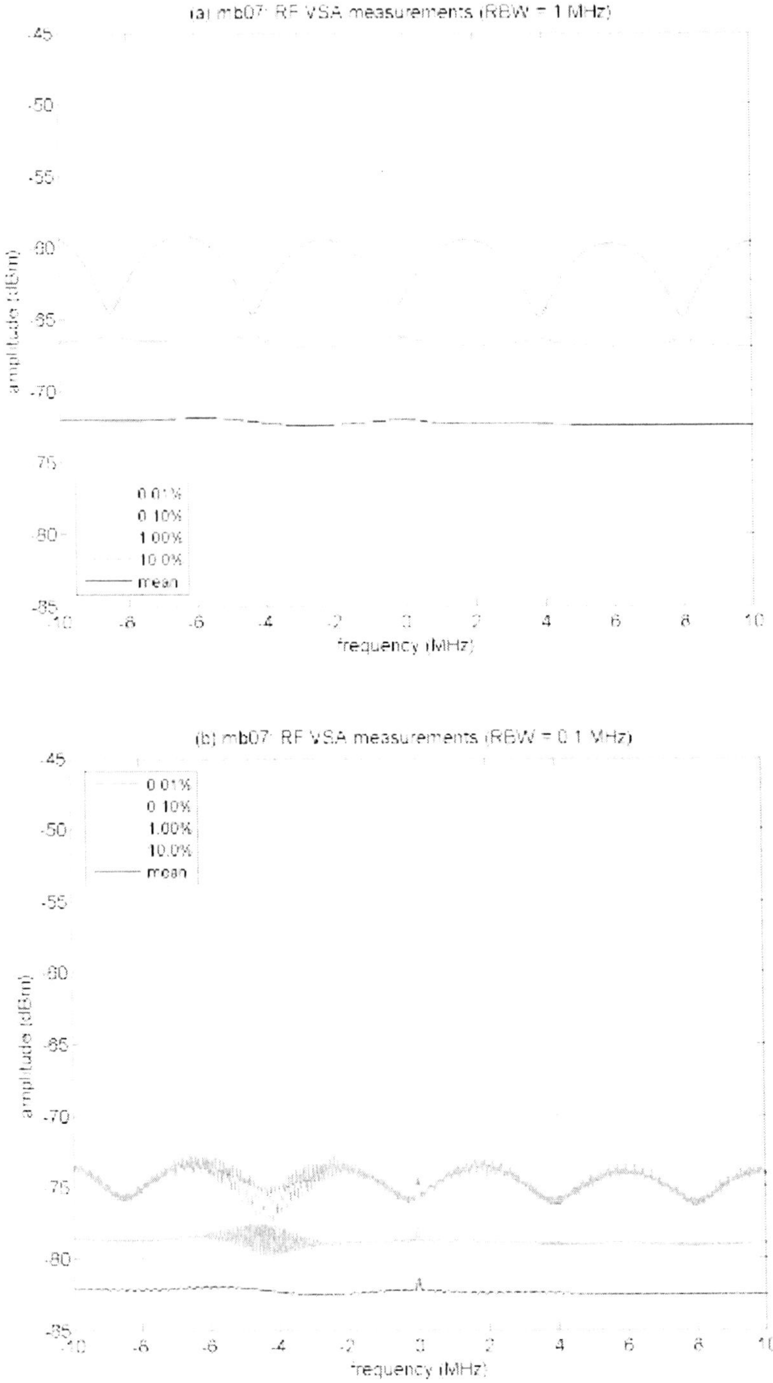

Figure B-79. RF spectral analyses of MB-07 (b = 13, d = 1).

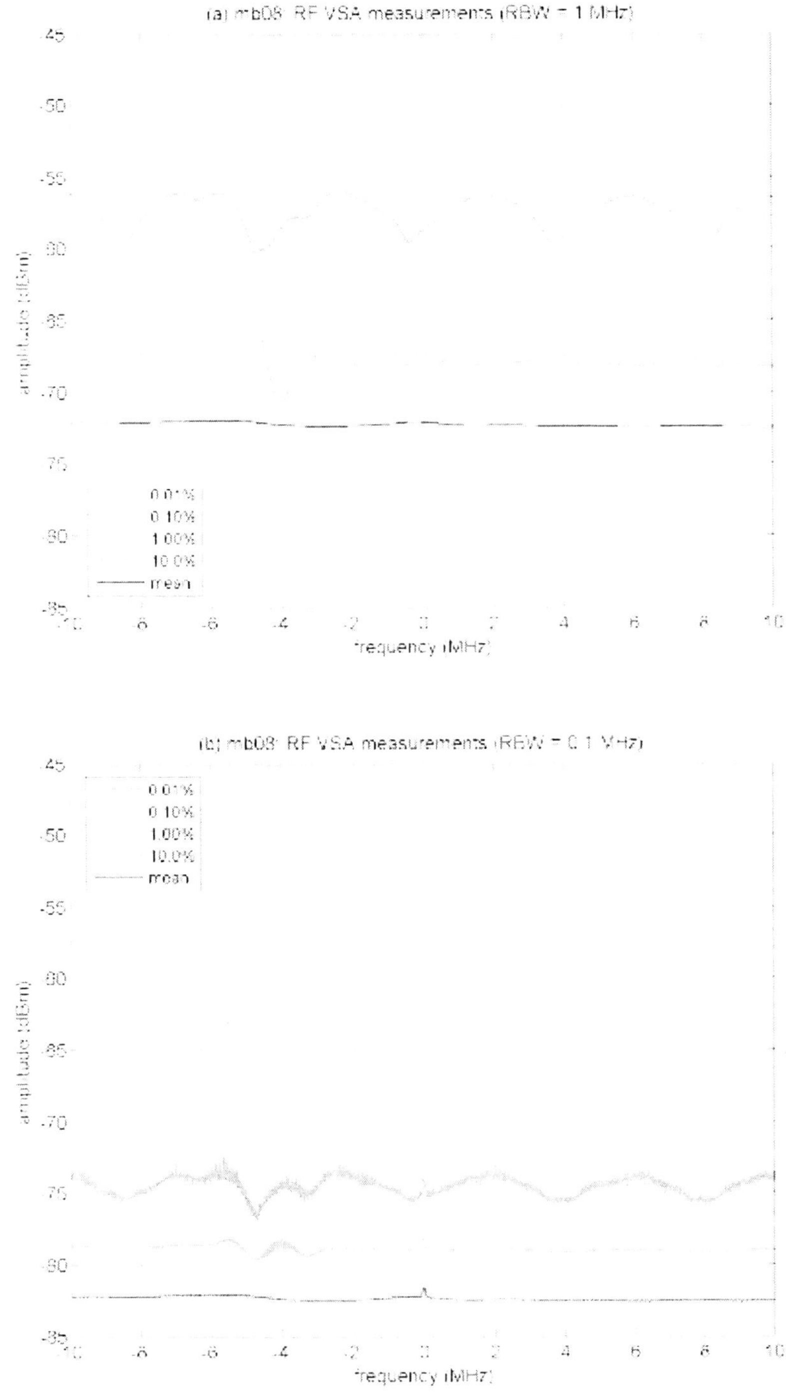

Figure B-80. RF spectral analyses of MB-08 (b = 13, d = 2).

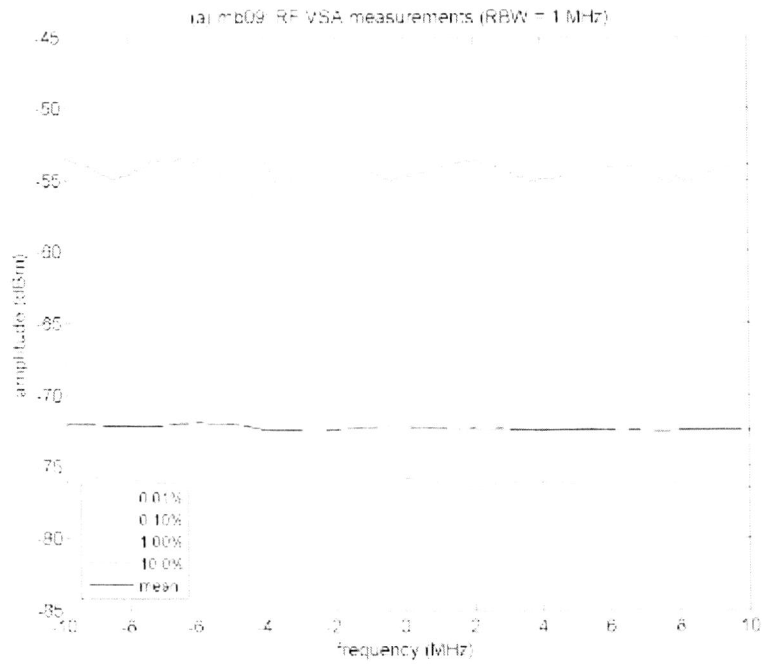

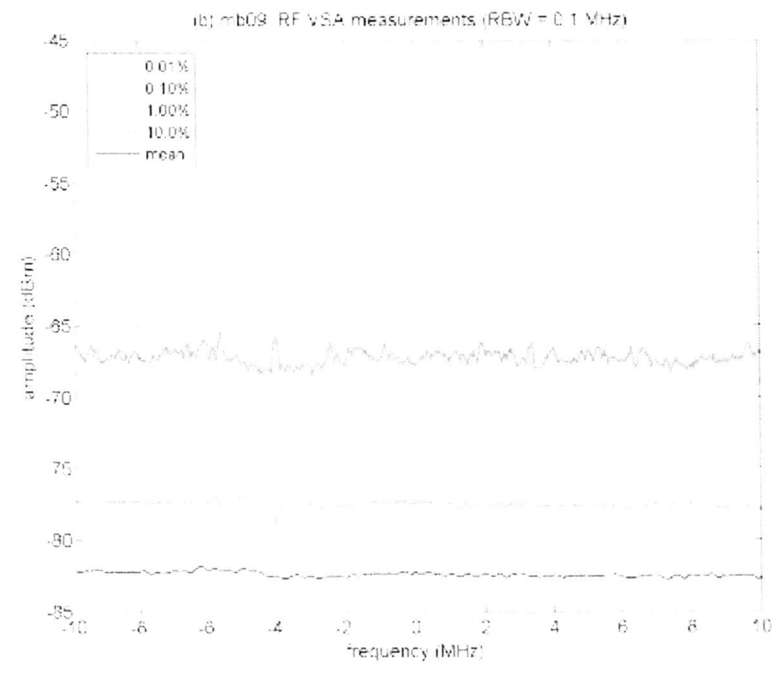

Figure B-81. RF spectral analyses of MB-09 (b = 13, d = 12).

The most notable trend was frequency-dependent variation in the MB peak amplitude statistics. This behavior is related to the location and modulation of the data and pilot tones of the MB signals, and is the primary difference between MB and GN(MB) signals. GN(MB) peak statistics are flat versus frequency as demonstrated theoretically in Section C.2 in Part 2 [7].

A notable example of this trend is MB-02 in a 0.1-MHz RBW. The pilot tone in the band of interest, located at -4.375 MHz, caused spikes in the peak spectra at intervals of ½ the reciprocal gating repetition rate, i.e., 0.533 MHz. Although this feature is evident in the 1-MHz RBW, the more prominent feature is the cyclic variation of all peak statistics. Interestingly, the nulls in the cyclic variation correspond to center frequencies of the data and pilot tones.

INDEX

A

B

C

D

E

F

G

H

I

J

K

L

S

T

U

V

W